Advanced Textbooks in Control and Signal Processing

Springer

London
Berlin
Heidelberg
New York
Barcelona
Hong Kong
Milan
Paris
Singapore
Tokyo

Series Editors

Professor Michael J. Grimble, Professor of Industrial Systems and Director
Professor Michael A. Johnson, Professor of Control Systems and Deputy Director

Industrial Control Centre, Department of Electronic and Electrical Engineering,
University of Strathclyde, Graham Hills Building, 50 George Street, Glasgow G1 1QE, U.K.

Other titles published in this series:

Genetic Algorithms: Concepts and Designs
K.F. Man, K.S. Tang and S. Kwong

Model Predictive Control
E. F. Camacho and C. Bordons

Introduction to Optimal Estimation
E.W. Kamen and J. Su

Discrete-Time Signal Processing
D. Williamson

Neural Networks for Modelling and Control of Dynamic Systems
M. Nørgaard, O. Ravn, N.K. Poulsen and L.K. Hansen

Modelling and Control of Robot Manipulators (2nd Edition)
L. Sciavicco and B. Siciliano

Fault Detection and Diagnosis in Industrial Systems
L.H. Chiang, E.L. Russell and R.D. Braatz

Soft Computing
L. Fortuna, G. Rizzotto, M. Lavorgna, G. Nunnari, M.G. Xibilia and R. Caponetto

Parallel Processing for Real-time Signal Processing and Control
M.O. Tokhi, M.A. Hossain and M.H. Shaheed
Publication due January 2003

T. Chonavel

Translated by Janet Ormrod

Statistical Signal Processing

Modelling and Estimation

Springer

Dr Thierry Chonavel, PhD
ENST de Bretagne,
Technopôle de Brest Iroise 29285, Brest Cedex, France

British Library Cataloguing in Publication Data
Chonavel, T.
  Statistical signal processing : modelling and estimation. -
  (Advanced textbooks in control and signal processing)
  1.Signal processing - Mathematical models 2.Signal
  processing - Statistical methods
  I.Title
  621.3'822
  ISBN 1852333855

Library of Congress Cataloging-in-Publication Data
Chonavel, T. (Thierry), 1963-
  Statistical signal processing : modelling and estimation / T. Chonavel.
    p. cm. -- (Advanced textbooks in control and signal processing)
  Includes bibliographical references and index.
  ISBN 1-85233-385-5 (alk. paper)
   1. Signal processing--Mathematics. 2. Statistics. I. Title. II. Series.
  TK5102.9 .C4835 2001
  621.382'2--dc21   2001020769

Additional material to this book can be downloaded from http://extras.springer.com.
ISSN 1439-2232
ISBN 1-85233-385-5 Springer-Verlag London Berlin Heidelberg
a member of BertelsmannSpringer Science+Business Media GmbH
http://www.springer.co.uk

Typesetting: Electronic text files prepared by author
Printed and bound by Athenæum Press Ltd, Gateshead, Tyne & Wear
69/3830-543210 Printed on acid-free paper SPIN 10783163

Series Editors' Foreword

The topics of control engineering and signal processing continue to flourish and develop. In common with general scientific investigation, new ideas, concepts and interpretations emerge quite spontaneously and these are then discussed, used, discarded or subsumed into the prevailing subject paradigm. Sometimes these innovative concepts coalesce into a new sub-discipline within the broad subject tapestry of control and signal processing. This preliminary battle between old and new usually takes place at conferences, through the Internet and in the journals of the discipline. After a little more maturity has been acquired by the new concepts then archival publication as a scientific or engineering monograph may occur.

A new concept in control and signal processing is known to have arrived when sufficient material has developed for the topic to be taught as a specialised tutorial workshop or as a course to undergraduates, graduates or industrial engineers. The *Advanced Textbooks in Control and Signal Processing* series is designed as a vehicle for the systematic presentation of course material for both popular and innovative topics in the discipline. It is hoped that prospective authors will welcome the opportunity to publish a structured presentation of either existing subject areas or some of the newer emerging control and signal processing technologies.

In communications, control engineering and related disciplines measured signals are almost always corrupted by noise or subject to limited random distortion. This means that if an experiment is repeated the same measured signals will not be obtained, and there is an uncertainty present in the signal. The main tools used to analyse and understand the mechanisms operating to produce this uncertainty are those of statistical signal processing. However, it is only over the last twenty years or so that a well-defined discipline of statistical signal processing has really emerged. The tools of this new discipline are built on foundations comprising probability theory, statistics, random processes and measure theory. Thierry Chonavel has written a book with strong roots in the basics of the discipline but which deals with fundamental problems in statistical signal processing. These important core problems of practical signal processing include Kalman and Wiener filtering, prediction, spectral identification, and non-parametric and parametric estimation. The approach to these standard problems is mathematical and rigorous: the reader is first led through chapters on random processes, power spectra and spectral representations before the core chapters of the book are reached. The closing chapters of the book generalise some of the methods and presents a broadening of the material, for example, higher-order statistical processes and adaptive estimation.

The mathematical approach of the text yields benefits in clarity and precision in definitions. Graduate students on Masters courses or studying for doctoral qualifications will find this text invaluable for communications, signal processing, control engineering courses and research. Engineers, and research workers who use statistical signal processing concepts are likely to find the book a good tutor for specific questions and an up-to-date reference source.

M.J. Grimble and M.A. Johnson
Industrial Control Centre
Glasgow, Scotland, U.K.
January 2002

Contents

List of Notation and Symbols

$\in, \subset, \cap, \cup$ element of, subset of, intersection, union

$\forall, \exists$ for all, there exists

$|x|, |M|$ modulus of a complex number, determinant of a matrix

v^T, M^T transpose of a vector, a matrix

x^*, v^*, M^* conjugate value of a complex scalar, a vector, a matrix

$P^*(z), \tilde{P}(z)$ if $P(z) = \sum_{k=1,p} p_k z^k$, $P^*(z) = \sum_{k=1,p} p_k^* z^k$ and $\tilde{P}(z) = z^p P^*(z^{-1})$

v^H, M^H transpose conjugate of a vector, a matrix

$[v]_i, [M]_{ij}$ i^{th} element of a vector, element (i, j) of a matrix

$< x, y >$ scalar products of x and y

$Tr(M)$ trace of a matrix

$sign(x)$ $sign(x) = +1, -1, 0$, if x is positive, negative, zero

$[x]$ round part of x

$\delta_{a,b}$ Kronecker symbol: $\delta_{a,b} = 1$ if $a = b$ and 0 otherwise

δ_a Dirac's measure carried by point a

$\nabla_x f$ gradient of function f with respect to the variable x

$diag(\lambda_1, \ldots, \lambda_n)$ diagonal matrix with diagonal terms $\lambda_1, \ldots, \lambda_n$

I, I_n identity matrix, identity matrix of size n

$\| X \|$ norm of X; the choice of the norm is defined by the context

$\mathcal{R}e[z], \mathcal{I}m[z]$ real part, imaginary part of z

$a = b \bmod[p]$ a is the remainder of b divided by p

inf, sup, min, max   infimum, supremum, minimum and maximum of a set

$\mathbb{N}, \mathbb{Z}, \mathbb{R}, \mathbb{C}$ sets of integers, relative integers, real numbers, complex numbers

$\mathbb{D}$ open unit disk

$[a, b]$ closed interval with end points a and b

$]a, b]$ $]a, b] = [a, b] - \{a\}$

$\mathcal{I}$ interval $]{-}1/2, 1/2]$

$\mathcal{C}^k$ set of functions with k continuous derivatives

$\mathcal{C}^\infty(K)$ set of functions with infinitely many derivatives, and compact support contained in K

$\mathcal{B}(\mathbb{R}^n)$ Borel σ-algebra of $\mathbb{R}^n$

$span\{(X_i)_{i \in I}\}$ vector space generated by finite linear combinations of the X_i

$\overline{span}\{(X_i)_{i \in I}\}$ complete closure of $span\{(X_i)_{i \in I}\}$

$\mathbb{I}_A$ index function of the set A

$\mathbb{E}[X]$ mathematical expectation of X

$\mathbb{E}[X|Y]$ expectation of X conditional to Y

$\mathrm{var}[X]$ variance of X

$\mathrm{cov}[X, Y]$ covariance of X and of Y

$X/H, X/Y$ orthogonal projection of X on the space H, on the space $vect\{Y\}$

$X \perp Y$ the variables X and Y are orthogonal, that is, $\mathbb{E}[(X - \mathbb{E}[X])(Y - \mathbb{E}[Y])^H] = 0$

$[h(z)]X_n$	value at time n of the output of the filter with transfer function $h(z)$ and process $(X_n)_{n \in \mathbb{Z}}$ as input
$f \star g$	convolution of functions f and g
$\oplus$	orthogonal sum of vector spaces
$\otimes$	product σ-algebra
$\lim_{x \to x_0} f(x)$	limit of $f(x)$ when x tends towards x_0
$\overset{m.s.}{=}, \overset{a.s.}{=}$	equality in the mean square sense, almost sure
$\overset{a.s.}{\to}, \overset{m.s.}{\to}, \overset{P}{\to}, \overset{L}{\to}$	almost sure convergence, mean square convergence, convergence in probability, in distribution
$\mathcal{U}_A$	uniform distribution on A
$\mathcal{N}(a, b)$	Gaussian distribution with mean a and variance b,
$\mathcal{IG}(a, b)$	inverse Gaussian distribution with parameters a and b
$\mathcal{P}(\lambda)$	Poisson distribution with parameter λ
$\mathcal{E}(\lambda)$	exponential distribution with parameter λ
$\chi^2(n)$	χ^2 distribution with n degrees of freedom
$X \sim \mathcal{N}(a, b)$	X is distributed as $\mathcal{N}(a, b)$
$X \sim f(x)$	the distribution of X is defined by the probability density function $f(x)$
$x \sim \mathcal{N}(a, b)$	x is a realisation of a random variable with distribution $\mathcal{N}(a, b)$
$x \sim f(x)$	x is a realisation of a random variable with probability density function $f(x)$
$\mathcal{F}, \mathcal{F}^{-1}$	Fourier transform, inverse Fourier transform

List of Abbreviations

AMI	Alternate Marked Inversion
AR	AutoRegressive
ARMA	AutoRegressive-Moving Average
BIBO	Bounded Input-Bounded Output
BLUE	Best Linear Unbiased Estimator
CRLB	Cramer-Rao Lower bound
EM	Expectation Maximisation
FFT	Fast Fourier Transform
LMS	Least Mean Square
LSP	Line Spectrum Pairs
MA	Moving Average
MAP	Maximum A Posteriori
MCMC	Monte Carlo Markov Chain
NRZ	Non-Return to Zero
ODE	Ordinary Differential Equation
PDF	Probability Density Function
PEF	Prediction Error Filter
PSD	Power Spectral Density
RLS	Recursive Least Square
RZ	Return to Zero
SAEM	Stochastic Approximation Expectation Maximisation

SEM   Stochastic Expectation Maximisation

SNR   Signal to Noise Ratio

SSB   Single Side Band

WSS   Wide Sense Stationary

1. Introduction

1.1 Foreword

This book presents an introduction to statistical signal processing. It mainly deals with the modelling and spectral estimation of wide sense stationary processes, and their filtering. Some digressions from this theme are aimed at pointing out the existence of techniques that generalise the methods only involving second order statistics, that is, those that only rely on the mean and autocovariance functions of processes. Indeed, it is often necessary to use tools that more completely account for statistical properties of signals in order to tackle new fields of application in statistical signal processing.

This book is particularly designed for graduate students. In particular, it corresponds to courses taught to students both in telecommunications and applied statistics. It can also be used by engineers, researchers, and professors interested in statistical signal processing. This work is intended to present the theoretical bases of this discipline in a synthetic framework.

The mathematical prerequisites needed to tackle this course are essentially the basics in probability theory and statistics, and some notions of random processes and measure theory. Furthermore, it is assumed that the reader has some knowledge of deterministic signal processing.

1.2 Motivation for a Book in Signal Processing

More often than not, the results of deterministic signal processing theory are insufficient for tackling certain problems. Indeed, in many applications, the signals under consideration show some randomness, in the sense that reproducing the experimental procedure does not lead to the same observation. This is due to the presence of perturbations that the user cannot entirely master. These perturbations can be of differing types: additive noise or distortion phenomena acting on the received signals, errors associated with the measuring devices, and so on. Therefore, the observation involves some uncertainty. Probability theory and stochastic processes theory offer a satisfactory framework in which to model this lack of knowledge. Moreover, in practical applications of signal processing, the term *signal* is often used indifferently to designate a random process or one of its sample paths.

Thus, one can imagine that generally some processing needs to be undertaken, for example an average over several successive experiments, in order to obtain an estimation of the parameters of interest acceptable to the user.

Our interest in the stochastic modelling of signals lies in the possibility of using the results related to the theory of statistics in order to carry out this kind of estimation. In particular, this theory enables us to envisage the relevance of the estimation method chosen. In this way, describing the data by means of a stochastic model, it appears that the aim of the averaging mentioned above is to approach the parameters of interest by virtue of the law of large numbers, when these parameters are given by the mathematical expectation of the averaged quantities.

In some cases, only one single experiment is available, like for example when attempting to extract statistical information from the reading of a seismic recording obtained during an earthquake. In contexts like these, it is very useful when the ergodic property is confirmed, since this property enables us to reasonably estimate mathematical expectations by time averages computed from the only trajectory at hand for the observed process.

Studying deterministic signals highlights the importance of the Fourier transform. The Fourier transform allows us to evaluate the distribution of the energy, or the power, of deterministic signals as a function of frequency. Furthermore, many signal transforms used in technical devices and for modelling natural phenomena are filtering operations. These filterings that are convolutional transforms of signals appear as simple multiplication operations when working in the frequency domain. We shall see that a very large, commonly-used class of stochastic processes, namely the class of wide sense stationary processes, can be represented in the frequency domain. This representation enables the problem of filtering to be treated simply. It also makes it possible to evaluate, through the notion of the spectrum of a random process, the way in which the power of a process is, on average, distributed as a function of frequency.

In practice, in order to be able to use the above-mentioned results when considering non-stationary processes, it is often assumed that the stationary property is satisfied if one only works on relatively short signal observation times. Note, however, that recent developments in time-frequency methods [54, 51] and in particular wavelet transform techniques [52, 53, 55], enable the study of non-stationary processes to be broached with greater precision. However, this topic is beyond the scope of this book.

1.3 A Few Classical Problems in Statistical Signal Processing

In order to better understand the importance of studying the spectrum of wide sense stationary processes and the filtering of these processes, we now present a few very classical problems that fall within the scope of this book.

1) Let us consider an electronic oscillator with power supplied by means of a noise diode, and whose behaviour we wish to study. The signal observed at the output of this oscillator only approximately resembles a sinusoid and is not exactly periodic. In the spectral domain we may envisage the way in which, on average, the power of the signal supplied by the oscillator is distributed as a function of frequency. In particular, this enables the central frequency of the oscillator to be evaluated and performance indices to be determined, such as the spectrum bandwidth of the oscillator. This mean distribution of power as a function of frequency, called the *spectrum* (Chapter 4), and the way in which it may be estimated (Chapters 12 and 13) are not wholly intuitive when, like here, it is desirable to model the signal observed as a random process. We shall also see that the convergence properties of classical spectral estimators like the periodogram (Chapter 12) may prove to be disappointing and that some precautions have to be taken when performing spectrum estimation.

2) It is often difficult to use a physical model to characterise the interactions that lead to the formation of a signal and to the shape of its spectrum. Using parametric spectral models such as rational spectral densities (Chapter 10) leads to descriptions that are often simpler to estimate (Chapter 13) but also more compact than modelling based on the analysis of the underlying physical phenomena that generated the signal. For instance, in order to realise speech synthesis systems, we want to generate sounds that possess spectral properties similar to those of speech, by means of rational transfer function filters (Chapter 10). By using such filters with sequences of uncorrelated random variables or periodic sequences of impulses at the input, according to the type of sound we wish to generate, we can produce a comprehensible spoken message.

3) During a radio emission, a signal transmitted on the radio channel is corrupted by additive noise at the receiver end. If we consider that the spectrum of the transmitted signal only occupies a limited frequency bandwidth and that the corrupting additive noise occupies a much wider bandwidth, we can try placing a filter at the receiver input in order to limit the received noise power. If the filter bandwidth is too large, the received noise power might be high. On the other hand, if the filter only selects a very narrow bandwidth of frequencies, then the signal of interest might be highly distorted by the action of the filter. Wiener filtering (Chapter 9) represents a satisfactory compromise in this kind of situation.

4) In order to perform digital encoding or digital transmission of a speech signal, it is usually sampled and then quantised. In order to reduce the amount of data to be transmitted, before quantisation, we generally begin by uncorrelating the signal by means of a filter (the prediction error filter, Chapter 8). Then the signal at the output of this filter, called the prediction error signal, is quantised and sampled. This signal has a variance smaller than that of the initial signal, so it is possible, for a fixed level of degradation introduced

by the quantisation, to diminish the stored or transmitted amount of data. The prediction error filter may be calculated using Levinson's fast algorithm (Chapter 11), for intervals of time over which the speech signal is assumed to be stationary. In addition, the coefficients of the prediction error filter may be represented by a set of equivalent parameters, called reflection coefficients, whose values are in the interval [-1,1] (Chapter 11). This property facilitates the quantisation of the parameters of the prediction error filter, and allows the stability of the speech signal reconstruction filter to be controlled easily.

5) A transmission channel may be modelled as a filter whose frequency response modulus is supplied by the knowledge of the spectrum of the received signal. Unfortunately, the phase of the frequency response of this filter may not be accessible from this information alone. In this case, higher order statistics (Chapter 14) or Bayesian methods (Chapter 15) enable ambiguity about the phase of the frequency response to be avoided.

6) In the previous problem, it may be that the transmission channel is characterised by a transfer function that evolves over time, and this is typically the case with radio-mobile communications. Adaptive filtering techniques like Kalman filtering (Chapter 9), or stochastic approximation methods (Chapter 16), must then be envisaged in order to follow these variations, and to recover the transmitted message.

1.4 Why This Book?

Signal processing occupies a place whose frontiers with fundamental sciences, like mathematics or physics, and engineering sciences are fairly fuzzy. In the engineering sciences, signal processing plays a part, especially in electronics, computer science, automatic control, or mathematics applied to fields ranging from finance to biology. The result is that several approaches are possible for the presentation of a book about stochastic signal processing, depending on the point of view of the author. However, this difference in approach becomes less marked as signal processing becomes a clearly-identified discipline.

What is special about this book? Let us indicate here that one essential objective of this book is to formalise and bring together elements of statistical signal processing that are sometimes encountered in more mathematically-oriented works.

Starting with wide sense stationary processes, exposing spectrum modelling and estimation and filtering is a conventional way to introduce statistical signal processing. The problem lies in choosing whether one concentrates essentially on the study of the mathematical bases underlying the many techniques that can be used in practice, or whether one wishes to make a structured and more or less exhaustive inventory of these techniques.

In this book we have endeavoured to give some theoretical bases of statistical signal processing. Consequently, we have had to limit the number of particular practical techniques and the number of examples presented. It was our priority to clearly lay down the bases of the discipline within the framework of an introductory overview.

Therefore, you will find the proof of many results given, or at least corresponding references. Some results about spectral identification or parametric spectral estimation have been described with approaches slightly different from what is often found in the literature on statistical signal processing. This has enabled us to introduce certain tools of important practical interest, and in our opinion in a coherent, natural manner.

However, as this book is intended in particular for engineers, we restrict ourselves to using only mathematical tools that are standard for engineers. For the same reason, it is clear that we cannot restrict ourselves to solely expounding general principles, and we have to envisage the more usual techniques of solving the associated practical problems. The approximately equivalent development we have given to aspects of modelling and to those relating to estimation is evidence of this preoccupation. Moreover, although the space devoted to the presentation of practical tools might seem small, this is partly linked to the fact that the preliminary effort for modelling random signals often enables the solution to the corresponding practical problems to be presented concisely. One could, for example, refer to items such as *algorithm*, *filter*, or *method* in the index to get an idea of the signal processing tools that are presented in the book. There may not be many of them, but we hope to have chosen the most significant ones and to have introduced them into the book coherently.

On the basis of the knowledge acquired in the context of spectrum modelling and estimation, and of wide sense stationary process filtering, we have slightly changed our viewpoint in the last three chapters. To be more precise, we have attempted to broaden the processing possibilities by envisaging the possible contribution of higher order statistics methods, Bayesian and Monte Carlo methods, and stochastic optimisation methods. Unlike the previous chapters, and unlike works more specifically dedicated to these approaches, we have above all endeavoured here to give a concise presentation of their principles, before mentioning how to use them for estimating filter transfer functions, a major problem in signal processing.

Rather than multiplying numerical examples, we have opted to propose a set of very simple programs, written in MATLAB language, which should allow readers to visualise some of the results presented. The simple and concise syntax of this language should make it easier to identify the various program parameters and should be an incentive to experimentation on the basis of the codes supplied. These programs are to be found on the CD-ROM supplied with the book. The files on the CD-ROM are in HTML format. It also contains probability course notes that are slightly too long to be presented as a mere appendix.

1.5 Book Contents

In Chapter 2, after some basics about stochastic processes, we shall look more particularly at wide sense stationary stochastic processes. In Chapter 3 we define their power spectrum, in Chapter 4 we study the notion of spectral representation, and in Chapter 5 we examine their filtering.

Next, in Chapter 6, we examine some types of processes that are of particular interest for signal processing. We shall recall some properties of Gaussian processes, used in particular to model many noise phenomena. We also mention Poisson processes, useful for studying certain queuing problems. Then we present the notions of white noise, followed by cyclostationary processes and circular processes, encountered in particular in telecommunications.

The transforms a signal can undergo are, of course, not limited to linear transforms. But in the case of non-linear transforms, there are no general results concerning the second order properties of the transformed processes. In Chapter 7, some non-linear or time-dependent transforms that are especially important in electronics and in signal communications are presented in order to illustrate this problem.

Taking into account the evolution of signal processing towards digital methods, we often merely consider discrete time processes in the rest of our presentation. However, it should be noted that for all the notions mentioned and the results presented, similar formulations exist for the case of continuous time processes.

The linear prediction of wide sense stationary processes plays an important role in filtering and in modelling. In Chapter 8, we examine in particular under what conditions a process may be seen as the output of a causal filter with a white noise input as well as the problem of finite order linear prediction.

The results of the theory of linear prediction then enable us in Chapter 9 to simply study two particular filtering techniques: Wiener filtering and Kalman filtering. From the observation of a trajectory of a process Y that depends on a process X, these techniques consider the problem of evaluating the corresponding trajectory of X by minimising an error variance criterion. We also present the notion of a matched filter, which allows a noisy deterministic signal to be detected.

Next, in Chapter 10, we investigate processes whose spectrum is given by a rational function. Such processes may be seen as the output of a rational transfer function filter with white noise at the input. While remaining relatively simple, this modelling allows us to satisfactorily approximate a large number of phenomena, which justifies the particular emphasis that we give them here.

In practice, evaluating the spectrum of a process is undertaken from the knowledge, or the estimation, of a limited set of its Fourier coefficients. Therefore, with a view to possible applications to spectral estimation, Chapter 11 considers the problem of the spectral identification of a process from knowledge of its first Fourier coefficients. For processes with a rational spectrum, this problem of identification is relatively simple. More generally, solving the problem does not lead to a single solution, and we study the set of spectra that match the autocovariance coefficient constraints.

Then, in Chapter 12, we examine the non-parametric spectral estimation of stationary processes. We begin by recalling some notions of non-parametric statistical estimation, before tackling the elementary properties of conventional estimators of mean and autocovariance functions for second order stationary processes. We then study the non-parametric periodogram spectral estimator.

After some basics about parametric estimation, we show in Chapter 13 how the parametric estimation of rational spectra using a minimum mean square approach may be undertaken simply. We then consider the case of Gaussian processes. In this case, the asymptotic expression of the likelihood criterion has a simple expression in terms of the spectrum of the process. The asymptotic approximation of the likelihood thus obtained can be used to perform the maximum likelihood estimator of the spectral parameters.

Certain properties of wide sense stationary processes may be extended to processes that are stationary at higher orders. For such processes, Chapter 14 presents the notion of a cumulant spectrum that generalises the conventional definition of a spectrum as the Fourier transform of the autocovariance function. We also indicate the possible use of higher order statistics in the context of spectral estimation and rational transfer function estimation.

In Chapter 15, Bayesian estimation techniques are considered. These enable some *a priori* information to be taken into account concerning the parameters to estimate. Unfortunately, computing the Bayesian estimators often comes up against problems of integration and of optimisation that have no analytical solution. Monte Carlo methods offer practical solutions to these problems. Presenting these methods gives us the opportunity to recall some important properties of Markov processes that appear in many areas of signal processing. We show how Monte Carlo methods may be implemented in the context of state space model filtering, and for estimating rational transfer functions.

To estimate the parameters associated with non-stationary phenomena, we are often led to optimise a criterion that is evolving over time. Stochastic optimisation algorithms are used in this kind of situation. We devote Chapter 16 to these, looking more especially at LMS and RLS algorithms often encountered in signal processing. We also indicate some results relative to the Ordinary Differential Equation (ODE) method, which represents a general approach for studying the behaviour of adaptive algorithms.

1.6 Acknowledgment

To end this introduction, I should like to acknowledge my fellow professors, lecturers, researchers, technicians and engineers, whose wealth and variety of scientific competences have shaped my vision of statistical signal processing. I should particularly like to thank Philippe Loubaton, Professor at the University of Marne La Vallée, whose courses in wide sense stationary processes are an essential reference for the elaboration of this book, and Alain Hillion, Scientific Director at ENST Bretagne, whose teaching in mathematics served for writing the basics in probabilities presented on the CD-ROM.

I also wish to thank Jean-Marc Tetu, systems engineer at ENST Bretagne, for his assistance on the computing side.

2. Random Processes

Purpose In this chapter, we recall some notions relating to random processes, looking more particularly at second order properties of processes.

2.1 Basic Definitions

Unless otherwise stated, we assume that the random variables brought into play are zero mean and complex-valued. Let $(\Omega, \mathcal{A}, dP)$ be a probability space. We recall that the vector space $L^2(\Omega, \mathcal{A}, dP)$ of finite variance random variables defined on $(\Omega, \mathcal{A}, dP)$ is a Hilbert space: it is a complete normed vector space, equipped with the scalar product defined by $< X, Y > = \mathbb{E}[XY^*]$. We shall denote the corresponding norm by

$$\| X \| = \sqrt{< X, X >}. \tag{2.1}$$

The proof that $L^2(\Omega, \mathcal{A}, dP)$ is a complete space is given in Appendix B.

The random variables considered in what follows will generally be taken in $L^2(\Omega, \mathcal{A}, dP)$. Henceforth, many problems like those of calculating a conditional expectation or a linear regression may be seen as problems of geometry.

2.1.1 Definition

We consider:

- a probability space $(\Omega, \mathcal{A}, dP)$;
- a set of indices T (usually $\mathbb{N}$, $\mathbb{Z}$, $\mathbb{R}$,...) called *time space*;
- a measurable space $(E, \mathcal{E})$ called *state space*. We shall generally consider that $(E, \mathcal{E}) = (\mathbb{R}, \mathcal{B}(\mathbb{R}))$ or $(E, \mathcal{E}) = (\mathbb{C}, \mathcal{B}(\mathbb{C}))$, $\mathcal{B}(E)$ designating the σ-algebra of Borel sets of E;
- a function $(X_t)_{t \in T}$, such that for every $t \in T$, X_t is a measurable application (or, in other words, a random variable) from Ω to E, with respective σ-algebra $\mathcal{A}$ and $\mathcal{E}$. $(X_t)_{t \in T}$ is called a *random function*.

The above elements define a random process. When there is no ambiguity about the different sets and about the probability measure P, we simply designate a random process by the random function $X = (X_t)_{t \in T}$.

The rest of Section 2.1 may be omitted at first reading.

2.1.2 Probability Distribution of a Random Process

Let $X = (X_t)_{t \in T}$ denote a random function mapping Ω onto E^T: $X(\omega) = (X_t(\omega))_{t \in T}$. X is measurable for the σ-algebras $\mathcal{A}$ of Ω and $\mathcal{E}^{\otimes T}$ of E^T, where $\mathcal{E}^{\otimes T}$ is defined as follows:

Definition 2.1 *For a family $(E_t, \mathcal{E}_t)_{t \in T}$ of measurable spaces, we define the product σ-algebra $\otimes_{t \in T} \mathcal{E}_t$ on the product space $\prod_{t \in T} E_t$ as the σ-algebra generated by the sets of $\prod_{t \in T} \mathcal{E}_t$ of the form $\prod_{t \in T} \Gamma_t$, where $\Gamma_t = E_t$ for all $t \in T$ except for a finite number of them for which $\Gamma_t \in \mathcal{E}_t$. If $E_t = E$ and $\mathcal{E}_t = \mathcal{E}$ for all $t \in T$, we note $\otimes_{t \in T} \mathcal{E} = \mathcal{E}^{\otimes T}$, and $(E^T, \mathcal{E}^{\otimes T}) = (E, \mathcal{E})^T$.*

Then, the σ-algebra $\mathcal{E}^{\otimes T}$ is generated by the events $\prod_{t \in T} \Gamma_t$, where all the sets Γ_t except a finite number of them coincide with E. In addition, let us recall that a measure is entirely defined by the values that it takes for the elements of a generating family of the σ-algebra[1] on which it is defined. Thus, the probability distribution P_X of X, defined by

$$\forall B \in \mathcal{E}^{\otimes T}, P_X(B) = P(X^{-1}(B)), \tag{2.2}$$

is entirely characterised by knowledge of the set of probabilities

$$\left\{ P\left((X_{t_1}, \ldots, X_{t_n}) \in \prod_{k=1,n} \Gamma_{t_k}\right); n \in \mathbb{N}, t_k \in T, \Gamma_{t_k} \in \mathcal{E}, \text{ for } k = 1, n \right\}, \tag{2.3}$$

where

$$\left((X_{t_1}, \ldots, X_{t_n}) \in \prod_{k=1,n} \Gamma_{t_k}\right) = \{\omega \in \Omega; X_{t_k}(\omega) \in \Gamma_{t_k}, \text{ for } k = 1, n\}. \tag{2.4}$$

This is a nice result, which shows that the distribution of a random process $X = (X_t)_{t \in T}$ is entirely characterised by the set of the distributions of all the random vectors of finite size made up of the random variables X_t.

2.1.3 Kolmogorov's Consistency Theorem

We denote by $\Pi(T)$ the set of all finite subsets of T, and we consider a family $(P_I)_{I \in \Pi(T)}$ of probability measures defined on the corresponding measurable spaces $(E, \mathcal{E})^I$. We might wonder whether this family characterises the distribution of some random process, that is, whether there exists a probability space $(\Omega, \mathcal{A}, P)$ such that we can define a random function $(X_t)_{t \in T}$ for which

---

[1] Let us recall that for a measurable space $(E, \mathcal{E})$ a family of subsets of E is a generating family of $\mathcal{E}$, if $\mathcal{E}$ is the smallest σ-algebra of E that contains the elements of this family.

$$\forall I \in \Pi(T), I = (t_1, \ldots, t_n), \ \forall \Gamma_1, \ldots, \Gamma_n \in \mathcal{E}^n,$$

$$P_I(\Gamma_1, \ldots, \Gamma_n) = P((X_{t_1}, \ldots, X_{t_n}) \in \Gamma_1 \times \ldots \times \Gamma_n). \tag{2.5}$$

To answer this question, we define the notion of a coherent family of probability distributions:

Definition 2.2 *Let* $(P_I)_{I \in \Pi(T)}$ *denote a family of probability distributions. It is said to satisfy the symmetry property if for all I in $\Pi(T)$ and any permutation $\sigma(I)$ of I,*

$$dP_{\sigma(I)}((x_i)_{i \in \sigma(I)}) = dP_I((x_i)_{i \in I}). \tag{2.6}$$

We define the compatibility property for $(P_I)_{I \in \Pi(T)}$ *as follows: if for any I and J in $\Pi(T)$, with $J \subset I$,*

$$\int_{(x_i)_{i \in I-J}} dP_I((x_i)_{i \in I}) = dP_J((x_i)_{i \in J}). \tag{2.7}$$

We can now recall the important following theorem:

Theorem 2.1 (Kolmogorov) *Let T be a set of indices and let $(P_I)_{I \in \Pi(T)}$ be a family of distributions defined on the corresponding measurable spaces $(E, \mathcal{E})^I$. Then, we may define a process $(X_t)_{t \in T}$ such that $(P_I)_{I \in \Pi(T)}$ characterises the distribution of this process if and only if $(P_I)_{I \in \Pi(T)}$ satisfies the above symmetry and compatibility properties. The distribution of $X = (X_t)_{t \in T}$ is thus defined uniquely.*

Proof See, for example, [1], Chapter 7.

2.2 Second Order Processes

We consider complex-valued processes, indexed by a set T $(T = \mathbb{N}, \mathbb{Z}, \mathbb{R}, \ldots)$. We assume that they are second order processes, that is,

$$\forall t \in T, \ \| X_t \|^2 = \mathbb{E}[|X_t|^2] < \infty, \tag{2.8}$$

and that $(E, \mathcal{E}) = (\mathbb{R}, \mathcal{B}(\mathbb{R}))$. We then define the mean function of the process X by

$$m_X(t) = \mathbb{E}[X_t] = \int_{\mathbb{R}} x \, dP_{X_t}(x), \tag{2.9}$$

where P_{X_t} represents the probability measure of X_t. The autocovariance function of X is given by

$$R_X(t_1, t_2) = \mathbb{E}[(X_{t_1} - m_X(t_1))(X_{t_1} - m_X(t_2))^*]$$

$$= \mathbb{E}[X_{t_1} X_{t_2}^*] - m_X(t_1) m_X^*(t_2), \tag{2.10}$$

and its autocorrelation function is defined by

$$\rho_X(t_1, t_2) = \frac{R_X(t_1, t_2)}{\sqrt{R_X(t_1, t_1)}\sqrt{R_X(t_2, t_2)}}. \tag{2.11}$$

Remarks

1) Often, the autocovariance and autocorrelation functions are simply referred to as *covariance* and *correlation* functions. The prefix 'auto' marks the distinction between the autocovariance function of a process X and the cross-covariance function of two processes X and Y, defined by

$$R_{XY}(t_1, t_2) = \mathbb{E}[X_{t_1} Y_{t_2}^*] - m_X(t_1) m_Y^*(t_2). \tag{2.12}$$

2) The definition of the autocorrelation function that we have adopted here may somewhat differ from one author to another. The above definition can be found, for instance, in [18] while in [6] the autocorrelation function is defined by $R_X(t_1, t_2) = \mathbb{E}[X_{t_1} X_{t_2}^*]$. With the latter definition, it is clear that the autocovariance and autocorrelation functions are similar when considering zero mean processes.

We note that inequality (2.8) implies the existence of the mean and autocovariance functions of the process under consideration (see Exercise 2.2).

A function $r(t_i, t_j)$ is said to be of the positive type if

$$\forall M \in \mathbb{N}, \forall t_1, ..., t_M \in T, \forall a_1, ..., a_M \in \mathbb{R}, \sum_{i,j=1}^{M} a_i a_j^* r(t_i, t_j) \geq 0. \tag{2.13}$$

It is clear that the autocovariance function of a process is of the positive type. Indeed,

$$\sum_{i,j=1}^{M} a_i a_j^* R_X(t_i, t_j) = \| \sum_i a_i X_{t_i} \|^2 \geq 0. \tag{2.14}$$

This property has important consequences, as will be seen in the next chapter.

For two processes X and Y, we define the cross-covariance function of X and Y by

$$R_{XY}(t_1, t_2) = \mathbb{E}[(X(t_1) - m_X(t_1))(Y(t_2) - m_Y(t_2))^*]$$

$$= \mathbb{E}[X(t_1) Y(t_2)^*] - m_X(t_1) m_Y^*(t_2). \tag{2.15}$$

This function is not generally of the positive type.

2.3 Classical Operations in $L^2(\Omega, \mathcal{A}, dP)$

In the following, $(X_t)_{t \in \mathbb{R}}$ is a second order process. We are now going to recall some basic notions and operations in $L^2(\Omega, \mathcal{A}, dP)$ that are commonly used in statistical signal processing.

2.3.1 Mean Square Convergence

Definition 2.3 *We say that X_t converges in the mean square sense towards a random variable X when t tends towards t_0, with $t_0 \in \mathbb{R} \cup \{-\infty, +\infty\}$, and we note $\lim_{t \to t_0} X_t \overset{m.s.}{=} X$, if*

$$\lim_{t \to t_0} \| X_t - X \|^2 = 0. \tag{2.16}$$

Often, it is easy to verify the mean square convergence using the following theorem:

Theorem 2.2 (Loeve) *The mean square limit at point t_0 exists if and only if $\mathbb{E}[X_t X_{t'}^*]$ has a finite limit when t and t' tend towards t_0, independently one from the other.*

Proof We assume that $\lim_{t \to t_0} X_t \overset{m.s.}{=} X$.

$$|\mathbb{E}[X_t X_{t'}^* - X X^*]|$$

$$= |\mathbb{E}[X_t (X_{t'} - X)^* + (X_t - X) X^*]|$$

$$\leq \| X_t \| \| X_{t'} - X \| + \| X \| \| X - X_t \| \tag{2.17}$$

$$\leq (\| X_t - X \| + \| X \|) \| X_{t'} - X \| + \| X \| \| X - X_t \|,$$

and the right-hand terms tend towards 0 when t and t' tend towards t_0. Therefore

$$\lim_{t,t' \to t_0} \mathbb{E}[X_t X_{t'}^*] = \| X \|^2 . \tag{2.18}$$

Now we assume that $\lim_{t,t' \to t_0} \mathbb{E}[X_t X_{t'}^*] = \alpha$, with $\alpha < \infty$.

$$\| X_t - X_{t'} \|^2 = \| X_t \|^2 + \| X_{t'} \|^2 - \mathbb{E}[X_t X_{t'}^*] - \mathbb{E}[X_{t'} X_t^*]. \tag{2.19}$$

As each of the right-hand terms tends towards α when $t, t' \to t_0$, $\lim_{t,t' \to t_0} \| X_t - X_{t'} \|^2 = 0$. Therefore, for any sequence $(t_n)_{n \in \mathbb{N}}$ that tends towards t_0, $(X_{t_n})_{n \in \mathbb{N}}$ is a Cauchy sequence and therefore converges in $L^2(\Omega, \mathcal{A}, dP)$. Let X be the limit of such a sequence. It is easy to verify that this limit is independent of the sequence chosen and that $\lim_{t \to t_0} X_t \overset{m.s.}{=} X$, which establishes the converse of the theorem. $\square$

We note that the convergence of $\mathbb{E}[X_t \overline{X}_{t'}^*]$ towards $\| X \|^2$ when $\lim_{t \to t_0} X_t \overset{m.s.}{=} X$, established in the proof, is in fact a straightforward consequence of the continuity of the scalar product.

2.3.2 Mean Square Continuity

Definition 2.4 *We say that the process X is continuous in the mean square sense, or mean square continuous, at point t if*

$$\forall \varepsilon > 0, \ \exists \delta > 0, \ \forall t', \ |t - t'| < \delta \Rightarrow \| X_t - X_{t'} \| < \varepsilon. \tag{2.20}$$

It appears that the mean square continuity at point t in fact means that $\lim_{t' \to t} X_{t'} \overset{m.s.}{=} X_t$. We may characterise the mean square continuity, thanks to the following result:

Theorem 2.3 *For a zero mean process X, $\lim_{t' \to t} X_{t'} \overset{m.s.}{=} X_t$ if and only if $R_X(t_1, t_2)$ is continuous at point $(t_1, t_2) = (t, t)$.*

Proof

$$|R_X(t_1, t_2) - R_X(t, t)| = \mathbb{E}[X_{t_1} X_{t_2}^* - X_t X_t^*]$$

$$= \mathbb{E}[(X_{t_1} - X_t)(X_{t_2}^* - X_t^*)] - \mathbb{E}[X_t (X_t - X_{t_2})^*]$$

$$+ \mathbb{E}[(X_{t_1} - X_t)X_t^*]. \tag{2.21}$$

Hence, if X is mean square continuous at point t, the three right-hand terms tend towards 0 when $(t_1, t_2) \to (t, t)$ (from Cauchy-Schwarz's inequality), and $R_X(t_1, t_2)$ is continuous to the point $(t_1, t_2) = (t, t)$.

The converse is obtained immediately by considering the equality

$$\| X_t - X_{t'} \|^2 = R_X(t, t) - R_X(t, t') - R_X(t', t) + R_X(t', t'). \ \square \tag{2.22}$$

2.3.3 Mean Square Derivative

Definition 2.5 *We say that the process X is mean square derivable at point t if $h^{-1}(X_{t+h} - X_t)$ converges in the mean square sense when $h \to 0$.*

Of course, the existence of the mean square derivative implies mean square continuity. We denote by X_t' the derivative of X at point t. We have the following result:

Theorem 2.4 *The process X is mean square derivable at point t if and only if $\frac{\partial^2}{\partial t_1 \partial t_2} R_X(t_1, t_2)$ exists and is finite at $(t_1, t_2) = (t, t)$. Moreover,*

$$R_{X'}(t_1, t_2) = \frac{\partial^2 R_X(t_1, t_2)}{\partial t_1 \partial t_2},$$

$$R_{X'X}(t_1, t_2) = \frac{\partial R_X(t_1, t_2)}{\partial t_1}, \tag{2.23}$$

$$R_{XX'}(t_1, t_2) = \frac{\partial R_X(t_1, t_2)}{\partial t_2}.$$

Proof We use Loeve's theorem. The mean square derivability of X is equivalent to the convergence of

$$\mathbb{E}[(\frac{X_{t+h} - X_t}{h})(\frac{X_{t+h'} - X_t}{h'})^*] \tag{2.24}$$

when h and h' tend towards 0.

$$\mathbb{E}[(\frac{X_{t+h} - X_t}{h})(\frac{X_{t+h'} - X_t}{h'})^*]$$

$$= \frac{1}{h'}[\frac{R_X(t+h, t+h') - R_X(t, t+h')}{h} - \frac{R_X(t+h, t) - R_X(t, t)}{h}], \tag{2.25}$$

and the limit, if it exists, is equal to the value of $\frac{\partial^2}{\partial t_1 \partial t_2} R_X(t_1, t_2)$ at the point $(t_1, t_2) = (t, t)$. The proof of the rest of the theorem is straightforward. $\square$

2.3.4 Mean Square Integration

Construction of integrals $\int_I \mathbf{g}(t, \tau) \mathbf{X}_\tau d\tau$ Let $\mathcal{P}_n = \{A_k^n; k = 1, N_n\}$ be a sequence of partitions of an interval I, where the A_k^n are intervals of I. We denote by Δ_k^n the length of A_k^n, and we assume that $\lim_{n \to \infty}(\max_{k=1, N_n} \Delta_k^n) = 0$. We define the mean square integral, denoted by $Y_t = \int_I g(t, \tau) X_\tau d\tau$ as the mean square limit, if it exists, of the sequence of random variables

$$Y_t^n = \sum_{k=1, N_n} g(t, \tau_k^n) X_{\tau_k^n} \Delta_k^n, \tag{2.26}$$

where τ_k^n is any point of A_k^n, the limit having to be independent of the choice of the partition. We notice the similarity of this construction with that of Riemann's integral. From Loeve's theorem, it is clear that if X is zero mean the integral $Y_t = \int_I g(t, \tau) X_\tau d\tau$ is defined in the mean square sense if and only if the integral

$$\int_{I \times I} g(t, \tau_1) g(t, \tau_2)^* R_X(\tau_1, \tau_2) d\tau_1 d\tau_2 \tag{2.27}$$

is defined in the Riemann sense. Moreover, if the integral Y_t is defined for all values of t in an interval, we obtain a process on this interval whose autocovariance function is given by

$$R_Y(t_1, t_2) = \int_{I \times I} g(t_1, \tau_1) g(t_2, \tau_2)^* R_X(\tau_1, \tau_2) d\tau_1 d\tau_2. \tag{2.28}$$

Mean square integration with variable bound We consider the random variable

$$Y_t = \int_a^t \phi(\tau) X_\tau d\tau, \tag{2.29}$$

where $\phi(t)$ is continuous, and X_t is mean square continuous.

Theorem 2.5 *If $\phi(t)$ is continuous, and X_t is mean square continuous, Y_t $= \int_a^t \phi(\tau)X_\tau d\tau$ is mean square derivable, with derivative equal to $\phi(t)X(t)$.*

Proof From Fubini's theorem (see Appendix A) and Cauchy-Schwarz's inequality,

$$\| \frac{Y_{t+h} - Y_t}{h} - \phi(t)X_t \|^2$$

$$= \mathbb{E}[\frac{1}{h^2} \int_{\tau,\tau'\in[t,t+h]} (\phi(\tau)X_\tau - \phi(t)X_t)(\phi(\tau')X_{\tau'} - \phi(t)X_t)^* d\tau d\tau']$$

$$= \frac{1}{h^2} \int_{\tau,\tau'\in[t,t+h]} \mathbb{E}[(\phi(\tau)X_\tau - \phi(t)X_t)(\phi(\tau')X_{\tau'} - \phi(t)X_t)^*] d\tau d\tau'$$

$$\leq \frac{1}{h^2} \left(\int_{\tau\in[t,t+h]} \| \phi(\tau)X_\tau - \phi(t)X_t \| d\tau \right)^2 .$$

$$(2.30)$$

Moreover, we note that

$$\| \phi(\tau)X_\tau - \phi(t)X_t \| \leq |\phi(\tau)| \| X_\tau - X_t \| + |\phi(\tau) - \phi(t)| \| X_t \| .$$

$$(2.31)$$

Hence, from the continuity of ϕ and the mean square continuity of X, $\phi(t)X_t$ is mean square continuous. Consequently, the right-hand term of (2.31) tends towards 0 when h tends towards 0, thus yielding the result stated. $\square$

Partial integration formula Let $\phi(t)$ denote a function of class C^1 and let X_t be a mean square continuous and mean square derivable process, with continuous derivative. Then, it is easy to check that

$$\int_{t_0}^t \phi'(\tau)X_\tau d\tau = [\phi(t)X_t - \phi(t_0)X_{t_0}] - \int_{t_0}^t \phi(\tau)X_\tau' d\tau, \qquad (2.32)$$

and

$$(\phi(t)X_t)' = \phi'(t)X_t + \phi(t)X_t'. \qquad (2.33)$$

Remark (operations on sample paths) We recall that the sample paths of a process X are the functions $t \to X_t(\omega)$, where $\omega \in \Omega$ is fixed. Sample paths are also called *trajectories* or *realisations*. We can also define the continuity, derivative or integration for the sample paths of a process, when these sample paths are almost surely continuous, derivable or integrable.

In what follows, we shall consider only the definitions of these notions in the mean square sense, except for the definition of strict ergodicity presented in the following section.

2.4 Stationarity and Ergodicity

2.4.1 Stationary Processes

Definition 2.6 *A process $X = (X_t)_{t \in T}$ is said to be strict sense stationary if $\forall n \in \mathbb{N}$ and $\forall t_1, \ldots, t_n \in T$, the distribution of $\mathbf{X}_t = [X_{t+t_1}, \ldots, X_{t+t_n}]^T$ does not depend on t. A process is said to be Wide Sense Stationary (WSS), or second order stationary, if the mean and the autocovariance of $\mathbf{X}_t$ do not depend on t:*

$$\forall t, \tau \in T, \qquad m_X(t + \tau) \qquad\quad = m_X(t),$$

$$\forall t_1, t_2, \tau \in T, \ R_X(t_1 + \tau, t_2 + \tau) = R_X(t_1, t_2). \tag{2.34}$$

Note that a process stationary in the strict sense is also stationary in the wide sense. In fact, statistical properties defined in the strict sense are stronger than properties in the wide sense. The equivalence between these two types of properties corresponds to the case where X is a Gaussian process.

We shall say that a process is stationary up to the order m if, for any vector $\mathbf{X}_t = [X_{t+t_1}, \ldots, X_{t+t_n}]^T$, all the moments up to the order m do not depend on t.

For a Wide Sense Stationary (WSS) process X, $R_X(t_1, t_2) = R_X(t_1 - t_2, 0)$, and we shall then simply denote by $R_X(t_1 - t_2)$ the autocovariance function of X. Moreover, it is clear that for a WSS process the autocovariance and autocorrelation functions are equal up to one factor. Indeed, in this case, $\rho_X(t) = R_X^{-1}(0) R_X(t)$.

2.4.2 Ergodic Processes

Ergodicity is an important property. For a stationary process X it enables us to satisfactorily estimate expressions of the form $\mathbb{E}[\phi(X_{t_1}, \ldots, X_{t_n})]$, where ϕ is a measurable function, from observing a single sample path of the process.

Definition 2.7 *A process $X = (X_t)_{t \in \mathbb{R}}$ is strict sense ergodic if $\forall n \in \mathbb{N}$ and $\forall t_1, \ldots, t_n \in T$, the limits of sample path integrals of the form*

$$(2\tau)^{-1} \int_{-\tau}^{\tau} \phi(X_{t+t_1}, \ldots, X_{t+t_n}) dt, \tag{2.35}$$

if they exist when τ tends towards $+\infty$, are almost surely independent of the sample path of X under consideration.

This definition assumes, of course, that almost all the sample paths of X are measurable functions of the variable t.

For a strict sense ergodic stationary process with continuous trajectories, it then appears that

$$\lim_{\tau \to \infty} \frac{1}{2\tau} \int_{-\tau}^{\tau} \phi(X_{t+t_1}, \ldots, X_{t+t_n}) dt \overset{\text{a.s.}}{=} \mathbb{E}[\phi(X_{t_1}, \ldots, X_{t_n})]. \qquad (2.36)$$

This property, known as Birkhoff-Von Neumann's theorem, can be formulated in the case of a discrete stationary process by relations of the type

$$\lim_{N \to \infty} \frac{1}{2N+1} \sum_{l=-N,N} \phi(X_{l+t_1}, \ldots, X_{l+t_n}) \overset{\text{a.s.}}{=} \mathbb{E}[\phi(X_{t_1}, \ldots, X_{t_n})]. \qquad (2.37)$$

Second order ergodicity, weaker than strict sense ergodicity, is characterised by the fact that for functions ϕ of the form

$$\phi(X_{t+t_1}, \ldots, X_{t+t_n}) = \sum_k \alpha_k X_{t+t_k} + \sum_{m,n} \beta_{m,n} X_{t+t_m} X_{t+t_n}^*, \qquad (2.38)$$

time averages $(2\tau)^{-1} \int_{-\tau}^{\tau} \phi(X_{t+t_1}, \ldots, X_{t+t_n}) dt$, where the integral is defined here in the mean square sense, or means $(2N+1)^{-1} \sum_{l=-N}^{N} \phi(X_{l+t_1}, \ldots, X_{l+t_n})$ in the discrete case, converge in $L^2(\Omega, \mathcal{A}, dP)$ towards constant values. We note that this mean square convergence condition is weaker than the almost sure convergence condition of strict sense ergodicity.

To reveal the second order ergodicity of a process stationary up to the fourth order, we can use Slutsky's theorem, stated here for a discrete time process (an equivalent result can be derived in the continuous case).

Theorem 2.6 (Slutsky). *A process* $X = (X_n)_{n \in \mathbb{Z}}$, *stationary up to the fourth order, is second order ergodic if and only if*

$$\lim_{N \to \infty} \frac{1}{N} \sum_{n=1,N} R_X(n) = 0, \qquad (2.39)$$

$$\forall l \in \mathbb{N}, \ \lim_{N \to \infty} \frac{1}{N} \sum_{n=1,N} \mathbb{E}[X_{n+l}^c X_n^{c*} X_l^{c*} X_0^c] - R_X^2(l) = 0, \qquad (2.40)$$

and $X_n^c = X_n - m_X$.

Proof For a WSS process, second order ergodicity is expressed by relations

$$\lim_{N \to \infty} \frac{1}{N} \sum_{n=1,N} X_n^c \overset{\text{m.s.}}{=} 0, \qquad (2.41)$$

and

$$\forall l, \ \lim_{N \to \infty} \frac{1}{N} \sum_{n=1,N} X_{n+l}^c X_n^{c*} \overset{\text{m.s.}}{=} R_X(l). \qquad (2.42)$$

We begin by showing that (2.39) is equivalent to (2.41). We note that from the Cauchy-Schwartz inequality,

$$\left|\frac{1}{N}\sum_{n=1,N} R_X(n)\right| = \left|\mathbb{E}\left[\left(\frac{1}{N}\sum_{n=1,N} X_n^c\right)X_0^{c*}\right]\right|$$

$$\leq R_X(0)^{1/2}\left\|\frac{1}{N}\sum_{n=1,N} X_n^c\right\|. \tag{2.43}$$

Hence, (2.41) implies (2.39). Conversely, we assume that (2.39) is satisfied. First, we remark that

$$\forall \varepsilon > 0, \exists N_0, \forall N, N_1, N \geq N_1 \geq N_0 \Rightarrow \left|\frac{1}{N}\sum_{n=N_1,N} R_X(n)\right| < \varepsilon, \tag{2.44}$$

following (2.39) and the relation

$$\left|\frac{1}{N}\sum_{n=N_1,N} R_X(n)\right|$$

$$= \left|\frac{1}{N}\sum_{n=1,N} R_X(n) - \frac{1}{N}\sum_{n=1,N_1-1} R_X(n)\right| \tag{2.45}$$

$$< \left|\frac{1}{N}\sum_{n=1,N} R_X(n)\right| + \left|\frac{1}{N_1-1}\sum_{n=1,N_1-1} R_X(n)\right|.$$

Consequently, for fixed ε, we can find N_0 such that for $N \geq N_1 \geq N_0$,

$$\left\|\frac{1}{N}\sum_{n=1}^{N} X_n^c\right\|^2$$

$$= \frac{1}{N^2}\sum_{n=-N,N} (N - |n|)R_X(n)$$

$$= \frac{1}{N^2}\left(NR_X(0) + 2\mathcal{R}e\left[\sum_{n=1}^{N}(N-n)R_X(n)\right]\right)$$

$$\leq \frac{1}{N^2}\left(NR_X(0) + 2\left|\sum_{n=1}^{N_1-1}(N-n)R_X(n)\right| + 2\left|\sum_{n=N_1}^{N}(N-n)R_X(n)\right|\right)$$

$$\leq \frac{1}{N^2}\left(NR_X(0) + 2\left|\sum_{n=1}^{N_1-1}(N-n)R_X(n)\right| + 2(N-N_1)N\varepsilon\right), \tag{2.46}$$

because $\sum_{n=N_1,N}(N-n)R_X(n) = \sum_{k=1,N-N_1}\left(\sum_{n=N_1,N-k} R_X(n)\right)$ and from (2.44). It is clear that when $N \to \infty$ the right-hand term of the last inequality tends towards 2ε. As ε may be chosen with an arbitrarily small value, it appears that relation (2.41) is satisfied.

It remains to be shown that (2.40) is equivalent to (2.42). To do this, we simply proceed as in the first part of the proof, considering the process $Y_n = X_{n+l}^c X_n^{c*}$ whose autocovariance coefficients are given by

$$R_Y(n) = \mathbb{E}[X_{n+l}^c X_n^{c*} X_l^{c*} X_0^c] - R_X^2(l). \quad\Box \tag{2.47}$$

Remark When the stationary property of X is satisfied only up to the second order, we can again apply the theorem, but replacing relation (2.40) by

$$\forall l, \lim_{N\to\infty} \frac{1}{N^2} \sum_{(n_1,n_2)=1:N} \mathbb{E}[X_{n_1+l}^c X_{n_1}^{c*} X_{n_2+l}^{c*} X_{n_2}^c] - R_X^2(l) = 0. \tag{2.48}$$

Exercises

2.1 Show that the sum and the product of two autocovariance functions are autocovariance functions.

2.2 (Gaussian process) Let $X = (X_t)_{t\in\mathbb{R}}$ denote a set of random variables such that for all n in $\mathbb{N}$ and all $t_1,\dots,t_n$ in $\mathbb{R}$, the distribution of $\mathbf{X} = [X_{t_1},\dots,X_{t_n}]^T$ has a probability density function given by

$$f_{\mathbf{X}}(\mathbf{x}) = (2\pi)^{-N/2}|\Gamma_{\mathbf{X}}|^{-1/2}\exp[-\frac{1}{2}(\mathbf{x}-m_{\mathbf{X}})^T\Gamma_{\mathbf{X}}^{-1}(\mathbf{x}-m_{\mathbf{X}})], \tag{2.49}$$

where we assume here that, for any vector $\mathbf{X}$, $|\Gamma_{\mathbf{X}}| \neq 0$.
   a) Show that $m_{\mathbf{X}} = \mathbb{E}[\mathbf{X}]$ and $\Gamma_{\mathbf{X}} = cov(\mathbf{X})$.
   b) Using Kolmogorov's theorem (Theorem 2.1) show that X is a random process. Such a process is called a Gaussian process.

2.3 Check that if inequality (2.8) holds, then the mean function and the autocovariance function of X exist.

2.4 Check formulas (2.32) and (2.33).

2.5 We consider a WSS process $X = (X_t)_{t\in\mathbb{R}}$. Show that $R_X(0) \geq |R_X(t)|$ for all t in $\mathbb{R}$, and that if $R_X(T) = R_X(0)$, then X is mean square periodic of period T, that is $\| X_{t+T} - X_t \| = 0$.

2.6 We consider X and Y, two WSS processes such that X_{t_1} and Y_{t_2} are independent for all t_1 and t_2 in $\mathbb{R}$. Show that $Z = XY$ is a WSS process.

2.7 Show that if an autocovariance function $(R(t))_{t\in\mathbb{R}}$ is continuous at $t = 0$, then it is continuous for all t in $\mathbb{R}$.

2.8 (White noise) Show that sequences $(X_n)_{n\in\mathbb{Z}}$ of independent random variables with the same distribution define strict sense stationary processes and that sequences of uncorrelated random variables with the same mean and the same variance define WSS processes.

2.9 (Harmonic process) We consider $X = (X_n)_{n \in \mathbb{Z}}$, where $X_n = \sum_{k=1,p} \xi_k e^{2i\pi n f_k}$. The ξ_k are random variables and the f_k are distinct frequencies. Show that $X = (X_n)_{n \in \mathbb{Z}}$ is WSS stationary if and only if the ξ_k are zero mean uncorrelated random variables.

2.10 Show that sequences $(X_n)_{n \in \mathbb{Z}}$ of independent random variables with the same distribution and $\mathbb{E}[|X_1|] < \infty$ define strict sense ergodic processes. Show that Gaussian processes such that $\lim_{n \to \infty} R_X(n) = 0$ define strict sense ergodic processes.

2.11 Check that a WSS process $X = (X_n)_{n \in \mathbb{Z}}$ defined by $X_n = \sum_{k=1,p} \xi_k e^{2i\pi n f_k}$ is not wide sense ergodic.

2.12 We consider $Y_t = X_t - X_{t-\tau}$, where $X = (X_t)_{t \in \mathbb{R}}$ is a WSS process. Check that Y is WSS and calculate $R_Y(t)$ in terms of $R_X(t)$.

2.13 Check that the inverse of a covariance matrix, when it exists, is a covariance matrix.

2.14 (Toeplitz matrix) Let $X = (X_n)_{n \in \mathbb{Z}}$ denote a WSS process and denote by $T_{X,N}$ the autocovariance matrix of $\mathbf{X}_N = [X_0, \ldots, X_N]^T$.

a) Check that $T_{X,N}$ is a Toeplitz matrix, that is $[T_{X,N}]_{a+1,b+1} = [T_{X,N}]_{a,b}$ for any a and b in $\{1, \ldots, N\}$.

b) We assume that $T_{X,N}$ is a positive definite matrix. Show that $T_{X,N+1}$ is a positive singular matrix if and only if $R_X(N+1)$ belongs to a circle. Calculate the centre and the radius of this circle.

2.15 (Successive transmissions of a symbol) We consider an infinite sequence of random variables, $X = (X_n)_{n \in \mathbb{N}}$, such that $P(X_0 = 1) = p_0 = 1 - P(X_0 = -1)$ and $P(X_n = X_{n-1}) = p = 1 - P(X_n = -X_{n-1})$. This process models the transmission of a digital symbol X_0 iteratively through successive transmission channels. p represents the probability of making an error when transmitting the symbol from a transmitter to the following receiver.

a) Is X stationary?

b) What is the probability of recovering the transmitted symbol without error at the nth receiver, that is $P(X_n = X_0)$?

c) What is the asymptotic distribution of X_n when n tends to ∞? Explain this result.

3. Power Spectrum of WSS Processes

Purpose We wish to evaluate the mean distribution of the power of a Wide Sense Stationary (WSS) process as a function of frequency.

3.1 Spectra with a Density: Power Spectral Density (PSD)

For a zero mean, WSS, process X whose autocovariance function is integrable, that is $\int_{\mathbb{R}} |R_X(t)| dt < \infty$, we define the Power Spectral Density (PSD) $S_X(f)$ of X as the Fourier transform of its autocovariance function:

$$S_X(f) = \int_{\mathbb{R}} R_X(t) e^{-2i\pi f t} dt. \tag{3.1}$$

In the discrete case, if $\sum_{n \in \mathbb{Z}} |R_X(n)| < \infty$ this definition becomes

$$S_X(f) = \sum_{n \in \mathbb{Z}} R_X(n) e^{-2i\pi n f}. \tag{3.2}$$

From Fubini's theorem, it is therefore clear that

$$R_X(n) = \int_{\mathcal{I}} e^{2i\pi n f} S_X(f) df, \tag{3.3}$$

with $\mathcal{I} =]-1/2, 1/2]$. Indeed,

$$\int_{\mathcal{I}} e^{2i\pi n f} S_X(f) df = \sum_{m \in \mathbb{Z}} R_X(m) \int_{\mathcal{I}} e^{2i\pi (n-m) f} df$$

$$= R_X(n). \tag{3.4}$$

The link between this definition and the mean distribution of the power of the process as a function of frequency is not immediately obvious. In order to give an intuitive justification for this, we can, for example, notice that in the discrete case if $\sum_{n=0,\infty} |R_X(n)| < \infty$, then

$$S_X(f) = \lim_{n \to \infty} \| \frac{1}{\sqrt{n}} \sum_{k=1}^{n} X_k e^{-2i\pi k f} \|^2. \tag{3.5}$$

Indeed,

$$\| \frac{1}{\sqrt{n}} \sum_{k=1}^{n} X_k e^{-2i\pi k f} \|^2 = \sum_{k=-n,n} (1 - \frac{|k|}{n}) R_X(k) e^{-2i\pi k f} \tag{3.6}$$

and tends towards $S_X(f)$ when n tends towards $+\infty$, from Lebesgue's dominated convergence theorem. $S_X(f)$ therefore appears as the variance of the Fourier transform of the process at frequency f, when the size n of the transformed vector tends towards $+\infty$.

The function $\hat{S}_{X,n}(f) = |n^{-1/2} \sum_{k=1}^{n} X_k e^{-2i\pi k f}|^2$ is called the order n periodogram of X, and will be studied in the context of spectral estimation of stationary processes.

It is clear that for many processes, like $X_t = \xi e^{2i\pi t f_0}$, where $\xi \in L^2(\Omega, \mathcal{A}, dP)$, the sequence of the autocovariances is not integrable (here $R_X(t) = \| \xi \|^2 e^{2i\pi t f_0}$), and then the integral (3.1) is not defined. When processing deterministic signals we generally introduce distribution theory in order to be able to define the Fourier transform of functions that, as functions, do not have a Fourier transform. Although we can do the same thing here, we shall merely introduce the notion of spectral measure that is a tool sufficient for addressing many problems in statistical signal processing.

3.2 Spectral Measure

Let us first recall that an autocovariance function is of the positive type (see relation (2.13)). We thus define the notion of spectral measure by means of the following result:

Theorem 3.1 (Bochner) *For a function $R(t)$ defined on $\mathbb{R}$,*

$((R(t))_{t \in \mathbb{R}}$ is of the positive type)

$$\Leftrightarrow \exists! \mu, \text{ positive measure, } \forall t \in \mathbb{R}, R(t) = \int_{\mathbb{R}} e^{2i\pi f t} d\mu(f), \tag{3.7}$$

and for a sequence $(R(n))_{n \in \mathbb{Z}}$ indexed by $\mathbb{Z}$,

$(R(n))_{n \in \mathbb{Z}}$ is of the positive type

$$\Leftrightarrow \exists! \mu, \text{ positive measure, } \forall n \in \mathbb{Z}, R(n) = \int_{\mathcal{I}} e^{2i\pi n f} d\mu(f). \tag{3.8}$$

Proof The proof of this result for the discrete case will be presented within the study of spectra compatible with a fixed set of autocovariances (see Appendix N; see also Exercise 3.4). Here we shall merely establish that μ is unique. We consider, for example, the continuous case (the discrete case may be treated in the same way). Let us assume that there exist two bounded measures such that

$$\forall t, \quad \int_{\mathbb{R}} e^{2i\pi ft} d\mu_1(f) = \int_{\mathbb{R}} e^{2i\pi ft} d\mu_2(f). \tag{3.9}$$

For any bounded measure μ, the family of functions $\chi_t(f) = e^{2i\pi ft}$ is a generating family of $L^2(\mathbb{R}, \mathcal{B}(\mathbb{R}), d\mu)$, that is, any element of $L^2(\mathbb{R}, \mathcal{B}(\mathbb{R}), d\mu)$ can be expressed as the limit of a sequence of finite linear combinations of functions χ_t. Therefore, from (3.9), and as for any element Δ of $\mathcal{B}(\mathbb{R})$ $\mathbb{I}_\Delta(f)$ that belongs to $L^2(\mathbb{R}, \mathcal{B}(\mathbb{R}), d\mu)$, it results that

$$\forall \Delta \in \mathcal{B}(\mathbb{R}), \quad \int_{\mathbb{R}} \mathbb{I}_\Delta(f) d\mu_1(f) = \int_{\mathbb{R}} \mathbb{I}_\Delta(f) d\mu_2(f), \tag{3.10}$$

that is, $\forall \Delta \in \mathcal{B}(\mathbb{R})$, $\mu_1(\Delta) = \mu_2(\Delta)$, and hence $\mu_1 = \mu_2$. □

For a WSS process X and $R(t) = R_X(t)$, the corresponding measure $\mu = \mu_X$ obtained in Bochner's theorem defines the spectral measure, or spectrum, of X. Moreover, if μ_X is absolutely continuous with respect to Lebesgue's measure, $d\mu_X(f) = S_X(f)df$, where $S_X(f)$ is the PSD of X.

Remark For WSS processes with mean $m_X \neq 0$, the Fourier transform of the function $R_X(t) + m_X$, or of the sequence $(R_X(n) + m_X)_{n\in\mathbb{Z}}$ in the discrete case, should be considered to define the spectrum, yielding an additional contribution $m_X\delta_0$ at frequency 0.

Examples The spectral measure of a discrete process X defined by a sequence of independent random variables with the same distributions is constant and equal to the variance σ_X^2 of X_n. Indeed, it is clear that the autocovariance coefficients of X are of the form $R_X(n) = \sigma_X^2\delta_{0,n}$, and that these coefficients coincide with the Fourier coefficients of the measure with constant PSD equal to σ_X^2 on $\mathcal{I}$.

For a harmonic process defined by $X_t = \sum_{k=1,p} \xi_k e^{2i\pi f_k t}$, it is clear that the second order stationarity property of X is only satisfied if the random variables ξ_k are uncorrelated. If in addition these variables are zero mean,

$$R_X(t) = \sum_{k=1,p} \|\xi_k\|^2 e^{2i\pi f_k t}$$

$$= \int_{\mathbb{R}} e^{2i\pi ft} \Big(\sum_{k=1,p} \|\xi_k\|^2 \delta_{f_k}\Big). \tag{3.11}$$

Consequently, the spectral measure of X is $d\mu_X(f) = \sum_{k=1,p} \|\xi_k\|^2 \delta_{f_k}$. It is therefore a discrete measure carried by the points $f_1, \ldots, f_p$.

Cross-spectral measure The stationary processes X and Y are said to be jointly stationary if $\mathbb{E}[X_{t+\tau} Y_\tau]$ is independent of τ. In this case, we define the cross-spectral measure μ_{XY} of (X, Y) by

$$R_{XY}(t) = \mathbb{E}[X_{t+\tau}Y_\tau^*] - \mathbb{E}[X_t]\mathbb{E}[Y_t^*]$$

$$= \int_{\mathbb{R}} e^{2i\pi ft} d\mu_{XY}(f). \tag{3.12}$$

Generally, the measure μ_{XY} is not positive.

Thus, if X and Y are jointly stationary with absolutely continuous spectra, the PSD of $aX + bY$, where $a, b \in \mathbb{C}$, is of the form

$$S_{aX+bY}(f) = |a|^2 S_X(f) + |b|^2 S_Y(f) + ab^* S_{XY}(f) + a^* b S_{YX}(f). \tag{3.13}$$

We also note that since $R_{XY}(t) = R_{YX}^*(-t)$, $S_{XY}(f) = S_{YX}^*(f)$.

Exercises

3.1 Let $R(t) = e^{-\alpha|t|}$ for $t \in \mathbb{R}$, with $\alpha > 0$.
   a) Check that R is a function of the positive type.
   b) Study the corresponding spectrum when α tends to 0.

3.2 (Periodogram spectral estimator) Let $X = (X_t)_{t \in \mathbb{R}}$ denote a zero mean WSS process with a PSD denoted by $S_X(f)$. We assume that X is mean square integrable and we define

$$\hat{S}_{X,T}(f) = \frac{1}{2T} \left| \int_{[-T,T]} X_t e^{-2i\pi ft} dt \right|^2. \tag{3.14}$$

a) Show that

$$\mathbb{E}[\hat{S}_{X,T}(f)] = \int_{[-2T,2T]} \left(1 - \frac{|t|}{2T}\right) R_X(t) e^{-2i\pi ft} dt. \tag{3.15}$$

b) If $\int_{\mathbb{R}} |R_X(t)| dt < \infty$, show that $\hat{S}_{X,T}(f)$ converges in mean to $S_X(f)$, that is

$$\lim_{T \to \infty} \mathbb{E}\left[|\hat{S}_{X,T}(f) - S_X(f)|\right] = 0. \tag{3.16}$$

$\hat{S}_{X,T}(f)$ is known as the periodogram estimator of $S_X(f)$.
   c) Check that the convergence is uniform on $\mathcal{I}$.

3.3 (Spectrum inversion formula) We consider a WSS process, $X = (X_n)_{n \in \mathbb{Z}}$, and $(R_X(n))_{n \in \mathbb{Z}}$ its autocovariance sequence.
   a) Let $-1/2 < f_1 < f_2 < 1/2$ and note

$$H(f) = \mathbb{I}_{]f_1,f_2[}(f) + (1/2)\mathbb{I}_{\{f_1\}}(f) + (1/2)\mathbb{I}_{\{f_2\}}(f),$$

$$h_n = \int_{\mathcal{I}} e^{-2i\pi nf} H(f) df. \tag{3.17}$$

Show that $\sum_{|n|\le N} h_n e^{-2i\pi nf}$ converges to $H(f)$ at any point f in $\mathcal{I}$.
(Hint: use Dirichlet's criterion, see Appendix F.)

b) Show that $\sum_{|n|\le N} h_n R_X(n)$ converges to $\int_{\mathcal{I}} H(f) d\mu_X(f)$.

c) Let $F_{\mu_X}(f)$ denote the distribution function associated with $d\mu_X(f)$: $F_{\mu_X}(f) = \mu_X(]-1/2, f])$. Show that for all f_1, f_2 in $]-1/2, 1/2[$,

$$\tfrac{1}{2}[F_{\mu_X}(f_2) + F_{\mu_X}(f_2^-)] - \tfrac{1}{2}[F_{\mu_X}(f_1) + F_{\mu_X}(f_1^-)]$$

$$= (f_2 - f_1)R_X(0) + \lim_{N\to\infty} \sum_{\substack{|n|\le N \\ n\ne 0}} R_X(n) \left(\frac{e^{-2i\pi nf_2} - e^{-2i\pi nf_1}}{-2in\pi} \right).$$

$$(3.18)$$

3.4 (Bochner's theorem) We consider a sequence $(R(n))_{n\in\mathbb{Z}}$ of the positive type.

a) Letting $N > 0$, check that

$$g_N(f) = \sum_{|n|\le N} (1 - \frac{|n|}{N})R(n)e^{-2i\pi nf} \qquad (3.19)$$

is a positive function on $\mathcal{I}$.

b) Using Helly's selection theorem (see Appendix A), show that there exists a positive measure μ such that $R(n) = \int_{\mathcal{I}} e^{2i\pi nf} d\mu(f)$ for all n in $\mathbb{Z}$.

3.5 Let $V = (V_n)_{n\in\mathbb{Z}}$ be a sequence of independent random variables with the same distribution. We assume that $\mathbb{E}[V_n] = 0$ and $\| V_n \|^2 = \sigma^2 < \infty$.

a) We define $X_n = \sum_{l=1,q} b_l V_{n-l}$, for all n in $\mathbb{Z}$. Calculate the autocovariance function and the spectrum of the process $X = (X_n)_{n\in\mathbb{Z}}$.

b) We define $Y = (Y_n)_{n\in\mathbb{Z}}$ by the difference equation $Y_n + aY_{n-1} = V_n$, with $|a| < 1$. Express Y_n as a function of $(V_{n-k})_{k\ge 0}$. Calculate the autocovariance function and the spectrum of Y.

c) Same question as in b) but with a a random variable independent of $(V_n)_{n\in\mathbb{Z}}$ and such that $P(|a| < 1) = 1$.

3.6 Let X and Y denote two jointly stationary WSS processes.

a) Show that

$$|R_{XY}(t)| \le \frac{1}{2}[R_X(0) + R_Y(0)]. \qquad (3.20)$$

b) We assume that X and Y have absolutely continuous spectra. Prove that $d\mu_{XY}(f)$ is an absolutely continuous measure and that for any Borel set A

$$|\int_A S_{XY}(f)df| \le \sqrt{\int_A S_X^2(f)df} \sqrt{\int_A S_Y^2(f)df}. \qquad (3.21)$$

What is the value of $S_{XY}(f)$ outside the support of $S_X(f)$ or of $S_Y(f)$?

3.7 Let $X = (X_t)_{t \in \mathbb{R}}$ denote a WSS process and $Y_t = X_{t-\tau}$, where τ is a constant.

a) Express the cross-covariance R_{XY} and the cross-spectrum $d\mu_{XY}$ in terms of R_X and $d\mu_X$ respectively.

b) How can τ be recovered from knowledge of R_{XY} ?

3.8 We consider $X_t = A\cos(2\pi f_0 t + \varphi)$, where A is a zero mean random variable and $\varphi \sim \mathcal{U}_{[0,2\pi]}$. A and φ are independent. Show that $X = (X_t)_{t \in \mathbb{R}}$ is a WSS and wide sense ergodic process and calculate its autocovariance and its spectrum.

3.9 We consider $X = (X_n)_{n \in \mathbb{Z}}$, with $X_n = \xi e^{2i\pi n\nu}$. ξ and ν are independent random variables and we assume that the probability measure of ν is carried by $\mathcal{I}$.

a) Show that X is a WSS process.

b) Letting $\phi_\nu(t) = \mathbb{E}[e^{i\nu t}]$ denote the characteristic function of ν, calculate $R_X(n)$ and $S_X(f)$.

3.10 (Karuhnen-Loeve expansion) First let us recall the following result (see, for instance, [32])

Theorem 3.2 *We consider an autocovariance function $R(t_1, t_2)$, belonging to $L^2(\mathbb{R}^2, \mathcal{B}(\mathbb{R}^2), dt_1 \times dt_2)$. Then there exists an orthonormal sequence of functions $(f_n)_{n \in \mathbb{N}}$ of $L^2(\mathbb{R}, \mathcal{B}(\mathbb{R}), dt)$ $(\int_{\mathbb{R}} f_n(t) f_m^*(t) dt = \delta_{m,n})$ and a sequence of real numbers $(\lambda_n)_{n \in \mathbb{N}}$, such that*

$$R(t_1, t_2) = \sum_{n \in \mathbb{N}} \lambda_n f_n(t_1) f_n^*(t_2). \tag{3.22}$$

a) Let $X = (X_t)_{t \in \mathbb{R}}$ denote a zero mean second order process. Show that X can be written as

$$X_t \overset{m.q.}{=} \sum_{n \in \mathbb{N}} \xi_n f_n(t), \tag{3.23}$$

where the f_n are orthonormal functions and the ξ_n are orthogonal zero mean random variables: $\mathbb{E}[\xi_n \xi_m^*] = \lambda_n \delta_{m,n}$.

b) Now we assume that X is a WSS process. Note that in this case

$$R_X(t) = \sum_{n \in \mathbb{N}} \lambda_n f_n(t + \tau) f_n^*(\tau) \tag{3.24}$$

for any choice of τ. If X is mean square periodic with period 1, that is if $\| X_{t+1} - X_t \| = 0$ for all $t \in \mathbb{R}$, show that $R_X(t)$ is of the form $R_X(t) = \sum_{n \in \mathbb{Z}} \rho_n e^{2i\pi nt}$. What is the form of the Karuhnen-Loeve expansion in this case?

c) Let $R_X(t) = \sum_{n \in \mathbb{N}} \lambda_n f_n(t + c) f_n^*(c)$ denote the Karhunen-Loeve expansion of a WSS process X, the PSD of which is $S_X(f) = B^{-1} \mathbb{I}_{[-B,B]}(f)$. Show that

$$\int_{\mathbb{R}} \frac{\sin(2\pi n B(t-u))}{2\pi n B(t-u)} f_n(u)\,du = \lambda_n f_n(t). \tag{3.25}$$

Functions $(f_n)_{n \in \mathbb{N}}$ that are defined by equations (3.25) are known as *prolate spheroidal wave functions* [45].

3.11 (Polya's condition) Show that if the function $R(t)$ satisfies the conditions $R(-t) = R(t)$, $R''(t) > 0$ for $t > 0$, and $\lim_{t \to \infty} R(t) = 0$, then $R(t)$ has a positive Fourier transform.

4. Spectral Representation of WSS Processes

Purpose We want to obtain a generalisation of the expression of purely harmonic processes $X_t = \sum_{k=1,p} \xi_k e^{2i\pi f_k t}$ in the case of WSS processes that would be of the form $X_t = \int_{\mathbb{R}} e^{2i\pi f t} d\hat{X}(f)$, where the nature of $d\hat{X}(f)$ should be specified. This representation will enable the filtering of stationary processes to be studied in a simple way in the next chapter.

4.1 Stochastic Measures and Stochastic Integrals

With a view to obtaining a frequency representation of wide sense stationary processes, we begin by defining stochastic measures.

4.1.1 Definition

A function $\hat{Z}$ defined on the Borel σ-algebra $\mathcal{B}(\mathbb{R})$ and with values in $L^2(\Omega, \mathcal{A}, dP)$ is called a stochastic measure if it verifies the following properties:

$$\forall \Delta \in \mathcal{B}(\mathbb{R}), \quad \mathbb{E}[\hat{Z}(\Delta)] = 0,$$

$$\forall (\Delta_n)_{n \in \mathbb{N}} \in \mathcal{B}(\mathbb{R})^{\mathbb{N}}, \text{ if } \Delta_n \cap \Delta_m = \emptyset \text{ for } m \neq n,$$

$$\hat{Z}(\cup_{n \in \mathbb{N}} \Delta_n) = \sum_{n \in \mathbb{N}} \hat{Z}(\Delta_n), \tag{4.1}$$

$$\forall \Delta_1, \Delta_2 \in \mathcal{B}(\mathbb{R}), \text{ if } \Delta_1 \cap \Delta_2 = \emptyset, \quad \mathbb{E}[\hat{Z}(\Delta_1)\hat{Z}(\Delta_2)^*] = 0.$$

4.1.2 Measure $\mu_{\hat{Z}}$ Associated with $\hat{Z}$

We define on $\mathcal{B}(\mathbb{R})$ the positive measure $\mu_{\hat{Z}}$ associated with the stochastic measure $\hat{Z}$ by

$$\mu_{\hat{Z}}(\Delta) = \| \hat{Z}(\Delta) \|^2 . \tag{4.2}$$

The σ-additivity property of $\mu_{\hat{Z}}$ stems directly from the fact that if $\Delta_n \cap \Delta_m = \emptyset$ for $m \neq n$,

$$\mu_{\hat{Z}}(\bigcup_n \Delta_n) = \mathbb{E}[|\hat{Z}(\bigcup_n \Delta_n)|^2]$$

$$= \mathbb{E}[\textstyle\sum_{m,n} \hat{Z}(\Delta_m)\hat{Z}^*(\Delta_n)]$$

$$= \mathbb{E}[\textstyle\sum_n |\hat{Z}(\Delta_n)|^2]$$

$$= \textstyle\sum_n \mu_{\hat{Z}}(\Delta_n).$$

(4.3)

As for the positivity of $\mu_{\hat{Z}}$, it is immediate.

We often use the differential notations

$$d\hat{Z}(f) = \hat{Z}([f, f + df[)$$

$$\text{and } d\mu_{\hat{Z}}(f) = \| d\hat{Z}(f) \|^2 .$$

(4.4)

4.1.3 Principle of the Method

In order to better understand the approach followed in the rest of this chapter to construct the spectral representation of processes, also known as *Cramer's representation*, we summarise it here. We have defined above the notion of stochastic measure. In order to be able to obtain the spectral representation of a process X, that is, a representation of the form

$$X_t = \int_{\mathbb{R}} e^{2i\pi ft} d\hat{X}(f),$$

(4.5)

we shall first define the integral expressions of the form $\int_{\mathbb{R}} \phi(f) d\hat{Z}(f)$, where ϕ is a function that will have to belong to a certain class to be defined. Then, for a given process X, we shall have to find the stochastic measure $\hat{X}$ that gives representation (4.5). To do this, we shall first show that the application T_X associating X_t with $e^{2i\pi ft}$ is extended to an isomorphism $\tilde{T}_X$ of Hilbert spaces. We shall then see that the representation of interest is obtained by defining the stochastic measure $\hat{X}$ by $\hat{X}(\Delta) = \tilde{T}_X(\mathbb{I}_\Delta)$, $\forall \Delta \in \mathcal{B}(\mathbb{R})$, and by extension we shall obtain

$$\tilde{T}_X(\phi) = \int_{\mathbb{R}} \phi(f) d\hat{X}(f),$$

(4.6)

for a large class of functions ϕ, leading to relation (4.5) when $\phi(f) = e^{2i\pi ft}$.

4.1.4 Construction of Stochastic Integrals $\int_{\mathbb{R}} \phi(f) d\hat{Z}(f)$

We shall begin by defining the stochastic integral for measurable step functions carried by a compact set, before extending the definition to a wider class of functions.

A measurable step function on the compact set K is a function ϕ for which there exists a partition $(\Delta_k)_{k=1,p}$ of K, with $\Delta_k \in \mathcal{B}(\mathbb{R})$ and $a_1, \ldots, a_p \in \mathbb{C}$, such that $\phi(f) = \sum_{k=1,p} a_k \mathbb{I}_{\Delta_k}(f)$.

We denote by E the algebra of step functions carried by a compact set, and we consider the application

$$T : E \qquad\qquad\qquad \rightarrow L^2(\Omega, \mathcal{A}, dP)$$

$$\sum_{k=1,p} a_k \mathbb{I}_{\Delta_k}(f) \rightarrow \sum_{k=1,p} a_k \hat{Z}(\Delta_k),$$

(4.7)

where $\hat{Z}$ is a fixed stochastic measure. We shall note

$$\sum_{k=1,p} a_k \hat{Z}(\Delta_k) = \int_{\mathbb{R}} \phi(f) d\hat{Z}(f).$$

(4.8)

It is clear that T defines a homomorphism that maps $E \subset L^2(\mathbb{R}, \mathcal{B}(\mathbb{R}), d\mu_{\hat{Z}})$ onto $L^2(\Omega, \mathcal{A}, dP)$ since for $\phi_k(f) = \sum_{l=1,n_k} a_l^{(k)} \mathbb{I}_{\Delta_l^k}(f)$ $(k = 1, 2)$,

$$< \phi_1, \phi_2 > = \int \sum_{k,l} a_k^{(1)} a_l^{(2)*} \mathbb{I}_{\Delta_k \cap \Delta_l}(f) d\mu_{\hat{Z}}(f)$$

$$= \sum_{k,l} a_k^{(1)} a_l^{(2)*} \mu_{\hat{Z}}(\Delta_k \cap \Delta_l)$$

$$= \sum_{k,l} a_k^{(1)} a_l^{(2)*} \| \hat{Z}(\Delta_k \cap \Delta_l) \|^2$$

(4.9)

$$= < \sum_k a_k^{(1)} \hat{Z}(\Delta_k), \sum_l a_l^{(2)} \hat{Z}(\Delta_l) >$$

$$= < T(\phi_1), T(\phi_2) > .$$

Moreover, E is dense in $L^2(\mathbb{R}, \mathcal{B}(\mathbb{R}), d\mu_{\hat{Z}})$ and T is continuous. Therefore, from the continuous extension theorem of a bounded linear operator (see Appendix C or [30] p.97), T has a homomorphic extension that maps $L^2(\mathbb{R}, \mathcal{B}(\mathbb{R}), d\mu_{\hat{Z}})$ onto $L^2(\Omega, \mathcal{A}, dP)$.

To be more precise, for $\phi \in L^2(\mathbb{R}, \mathcal{B}(\mathbb{R}), d\mu_{\hat{Z}})$, there exists a sequence of functions $(\phi_n)_{n \in \mathbb{N}}$ of E that converges towards ϕ, and we set by definition

$$\int_{\mathbb{R}} \phi(f) d\hat{Z}(f) = \lim_{n \to \infty} \int_{\mathbb{R}} \phi_n(f) d\hat{Z}(f).$$

(4.10)

Moreover, the preservation of the norm is expressed by

$$\| \int_{\mathbb{R}} \phi(f) d\hat{Z}(f) \|^2 = \int_{\mathbb{R}} |\phi(f)|^2 d\mu_{\hat{Z}}(f).$$

(4.11)

We have therefore defined the stochastic integrals associated with a stochastic measure $\hat{Z}$, and we have checked that the stochastic integration establishes a homomorphism that maps $L^2(\mathbb{R}, \mathcal{B}(\mathbb{R}), d\mu_{\hat{x}})$ onto $L^2(\Omega, \mathcal{A}, dP)$ for the scalar products defined on these spaces.

For processes with discrete indices, we shall define the stochastic integral in a similar way, by replacing the domain of integration $\mathbb{R}$ by the interval $\mathcal{I} =]-1/2, 1/2]$.

4.2 Kolmogorov's Isomorphism

We denote by H_X the linear envelope of X, that is, the Hilbert subspace of $L^2(\Omega, \mathcal{A}, dP)$ generated by the linear combinations of the random variables $(X_t)_{t \in \mathbb{R}}$, or of the random variables $(X_n)_{n \in \mathbb{Z}}$ for the discrete case, and the limits of sequences of such combinations: $H_X = \overline{span}\{X_t; t \in \mathbb{R}\}$. We then have the following result:

Theorem 4.1 *The linear operator*

$$T_X : \sum_n \alpha_n e^{2i\pi t_n f} \to \sum_n \alpha_n X_{t_n}, \tag{4.12}$$

where the sum applies to a finite number of terms, extends to an isomorphism that maps $L^2(\mathbb{R}, \mathcal{B}(\mathbb{R}), d\mu_X(f))$ onto H_X, and $L^2(\mathcal{I}, \mathcal{B}(\mathcal{I}), d\mu_X(f))$ onto H_X for the discrete case.

Proof See Appendix D.

Remark In the discrete case, when the norm defined by μ_X is equivalent to the one defined by Lebesgue's measure, that is, $d\mu_X(f) = S_X(f)df$ and $\forall f \in \mathcal{I}, 0 < m \le S_X(f) \le M < \infty$, the functions $(\chi_n(f) = e^{2i\pi nf})_{n \in \mathbb{Z}}$ form a basis of $L^2(\mathcal{I}, \mathcal{B}(\mathcal{I}), S_X(f)df)$. Then, any element of H_X can be represented in a unique way as the limit of a series of the form $\sum_n \alpha_n X_n$, with $\sum_n |\alpha_n|^2 < \infty$.

4.3 Spectral Representation

It can easily be checked that for any stochastic measure $\hat{X}$ the process $X_t = \int_{\mathbb{R}} e^{2i\pi ft} d\hat{X}(f)$ is zero mean, WSS and mean square continuous. Its autocovariance function is given by $R_X(t) = \int_{\mathbb{R}} e^{2i\pi ft} d\mu_{\hat{x}}(f)$.

From Bochner's theorem, the uniqueness of the positive measure associated with $R_X(t)$ shows that the measure $\mu_{\hat{x}}$ associated with the stochastic measure $\hat{X}$ coincides with the spectral measure μ_X of the process X.

An essential result is that all mean square continuous stationary processes have representations of the form $X_t = \int_{\mathbb{R}} e^{2i\pi ft} d\hat{X}(f)$, where $\hat{X}$ is a stochastic measure:

Theorem 4.2 *For a zero mean, WSS, mean square continuous process X indexed by $\mathbb{R}$, there exists a unique stochastic measure $\hat{X}(f)$ such that $X_t = \int_{\mathbb{R}} e^{2i\pi ft} d\hat{X}(f)$. For a discrete time, zero mean, WSS process, there exists a unique stochastic measure $\hat{X}(f)$ such that $X_n = \int_{\mathcal{I}} e^{2i\pi nf} d\hat{X}(f)$.*

Proof See Appendix D.

We note that for processes indexed by $\mathbb{R}$, we saw in Chapter 2 that the mean square continuity hypothesis is equivalent to the continuity of $R_X(t)$ at 0. This property is often satisfied in practice. However, for certain processes such as a white noise process indexed by $\mathbb{R}$, this hypothesis is not satisfied.

We shall see in Chapter 6 that if the spectral representation does not exist in the sense that we have just defined it, we may still associate a stochastic measure $\hat{X}(f)$ with the process X. The difficulty with this kind of situation comes from the fact that $e^{2i\pi ft} \notin L^2(\mathcal{I}, \mathcal{B}(\mathcal{I}), d\mu_{\hat{X}}(f))$. Here, we are confronted with the same kind of problem as when we wish to compute the Fourier transform of deterministic signals that, as functions, do not have Fourier transforms, like for example sinusoidal signals. As in the case of the Fourier transform of deterministic signals, we shall see that distribution theory enables us to extend the class of stochastic processes for which we can define a spectral representation.

4.4 Sampling

Let X be a WSS process indexed by $\mathbb{R}$. A discrete process $Y = (Y_n)_{n \in \mathbb{Z}}$ obtained by the regular sampling of X is still WSS. Indeed, setting $Y_n = X_{nT}$,

$$\mathbb{E}[Y_{n+k} Y_n^*] = \mathbb{E}[X_{(n+k)T} X_{nT}^*]$$

$$= R_X(kT). \tag{4.13}$$

We assume now that X has the spectral representation $X_t = \int_{\mathbb{R}} e^{2i\pi tf} d\hat{X}(f)$. We denote by $\hat{Y}$ the spectral measure of Y, and we look for its expression as a function of that of X. As $Y_n = X_{nT}$,

$$Y_n = \int_{\mathbb{R}} e^{2i\pi nTf} d\hat{X}(f)$$

$$= \sum_{k \in \mathbb{Z}} \int_{](k-1/2)T^{-1},(k+1/2)T^{-1}]} e^{2i\pi nTf} d\hat{X}(f)$$

$$= \sum_{k \in \mathbb{Z}} \int_{\mathcal{I}} e^{2i\pi nf} d\hat{X}(\frac{f+k}{T}) \tag{4.14}$$

$$= \int_{\mathcal{I}} e^{2i\pi nf} [\sum_{k \in \mathbb{Z}} d\hat{X}(\frac{f+k}{T})].$$

Thus, the stochastic measure of Y is

$$d\hat{Y}(f) = \sum_{k \in \mathbb{Z}} d\hat{X}(\frac{f+k}{T}). \tag{4.15}$$

4. Spectral Representation of WSS Processes

Purpose We want to obtain a generalisation of the expression of purely harmonic processes $X_t = \sum_{k=1,p} \xi_k e^{2i\pi f_k t}$ in the case of WSS processes that would be of the form $X_t = \int_{\mathbb{R}} e^{2i\pi f t} d\hat{X}(f)$, where the nature of $d\hat{X}(f)$ should be specified. This representation will enable the filtering of stationary processes to be studied in a simple way in the next chapter.

4.1 Stochastic Measures and Stochastic Integrals

With a view to obtaining a frequency representation of wide sense stationary processes, we begin by defining stochastic measures.

4.1.1 Definition

A function $\hat{Z}$ defined on the Borel σ-algebra $\mathcal{B}(\mathbb{R})$ and with values in $L^2(\Omega, \mathcal{A}, dP)$ is called a stochastic measure if it verifies the following properties:

$$\forall \Delta \in \mathcal{B}(\mathbb{R}), \quad \mathbb{E}[\hat{Z}(\Delta)] = 0,$$

$$\forall (\Delta_n)_{n \in \mathbb{N}} \in \mathcal{B}(\mathbb{R})^{\mathbb{N}}, \text{ if } \Delta_n \cap \Delta_m = \emptyset \text{ for } m \neq n,$$

$$\hat{Z}(\cup_{n \in \mathbb{N}} \Delta_n) = \sum_{n \in \mathbb{N}} \hat{Z}(\Delta_n), \tag{4.1}$$

$$\forall \Delta_1, \Delta_2 \in \mathcal{B}(\mathbb{R}), \text{ if } \Delta_1 \cap \Delta_2 = \emptyset, \quad \mathbb{E}[\hat{Z}(\Delta_1)\hat{Z}(\Delta_2)^*] = 0.$$

4.1.2 Measure $\mu_{\hat{Z}}$ Associated with $\hat{Z}$

We define on $\mathcal{B}(\mathbb{R})$ the positive measure $\mu_{\hat{Z}}$ associated with the stochastic measure $\hat{Z}$ by

$$\mu_{\hat{Z}}(\Delta) = \| \hat{Z}(\Delta) \|^2 . \tag{4.2}$$

The σ-additivity property of $\mu_{\hat{Z}}$ stems directly from the fact that if $\Delta_n \cap \Delta_m = \emptyset$ for $m \neq n$,

4.5 (Spectral measure inversion formula) We consider a WSS process $X = (X_n)_{n \in \mathbb{Z}}$ and $\mathcal{X}(f) = \hat{X}(]-1/2, f])$. Parallelling the discussion of Exercise 3.3, show that for all f_1, f_2 in $]-1/2, 1/2[$,

$$\tfrac{1}{2}[\mathcal{X}(f_2) + \mathcal{X}(f_2^-)] - \tfrac{1}{2}[\mathcal{X}(f_1) + \mathcal{X}(f_1^-)]$$

$$\overset{m.q.}{=} (f_2 - f_1)X_0 + \lim_{N \to \infty} \sum_{|n| \leq N, n \neq 0} X_n \left(\frac{e^{-2i\pi n f_2} - e^{-2i\pi n f_1}}{-2i n \pi} \right).$$

$$(4.20)$$

4.6 (Parseval's relation) Let $X = (X_t)_{t \in \mathbb{R}}$ denote a mean square continuous WSS process, and $g(t)$ a continuous function such that $\hat{g}(f) \in L^2(\mathbb{R}, \mathcal{B}(\mathbb{R}), d\mu_X(f))$, where $\hat{g}(f)$ is the Fourier transform of $g(t)$. Show that

$$\int_{\mathbb{R}} g(t)^* X_t dt = \int_{\mathbb{R}} \hat{g}(f)^* d\hat{X}(f).$$

$$(4.21)$$

4.7 (Circulant covariance matrices) We consider $X = (X_n)_{n \in \mathbb{Z}}$ such that $R_X(n)$ is periodic with period N.

a) Show that the covariance matrix $T_{X,N-1}$ of size $N \times N$ with general term $[T_{X,N-1}]_{ab} = R_X(a - b)$ is circulant, that is, successive lines of $T_{X,N-1}$ are obtained by cyclic permutations.

b) Show that the eigenvalues of $T_{X,N-1}$ are the vectors

$$d(f_k) = (1/\sqrt{N})[1, e^{2i\pi f_k}, \ldots, e^{2i\pi(N-1)f_k}]^T,$$

$$(4.22)$$

where $f_k = kN^{-1}$ and $k = 0, \ldots, N-1$.

c) Calculate the spectral measure of X in terms of $X_0, \ldots, X_{n-1}$. (Hint: note that the stochastic measure of X is of the form $d\hat{X}(f) = \sum_{k=0,n-1} \xi_k e^{2i\pi n f_k}$.)

4.8 (Extension of the stochastic measure definition) Let $\mathcal{Z} = (\mathcal{Z}_t)_{t \in \mathbb{R}}$ denote a process with uncorrelated increments, that is, for all $t_1 < t_2 \leq t_3 < t_4$,

$$\mathbb{E}[(\mathcal{Z}_{t_4}^c - \mathcal{Z}_{t_3}^c)(\mathcal{Z}_{t_2}^c - \mathcal{Z}_{t_1}^c)^*] = 0,$$

$$(4.23)$$

where $\mathcal{Z}_t^c = \mathcal{Z}_t - m(t)$ and $m(t) = \mathbb{E}[\mathcal{Z}_t]$. In other words, $\mathcal{Z}^c = (\mathcal{Z}_t^c)_{t \in \mathbb{R}}$ is a process with orthogonal increments (see Exercise 4.4). Let $\hat{\mathcal{Z}}^c$ denote the stochastic measure associated with $\mathcal{Z}^c$ as in Exercise 4.1. We extend the definition of the stochastic integral as follows: let $\phi \in L^2(\mathbb{R}, \mathcal{B}(\mathbb{R}), d\hat{\mathcal{Z}}^c(t)) \cap L^1(\mathbb{R}, \mathcal{B}(\mathbb{R}), dm(t))$, where $\int_{]a,b]} dm(t) = m(b) - m(a)$. Then, we define $\int_{\mathbb{R}} \phi(t) d\hat{\mathcal{Z}}(t)$ as

$$\int_{\mathbb{R}} \phi(t) d\hat{\mathcal{Z}}(t) = \int_{\mathbb{R}} \phi(t) d\hat{\mathcal{Z}}^c(t) + \int_{\mathbb{R}} \phi(t) dm(t).$$

$$(4.24)$$

Show that with this definition we have for all ϕ_1 and ϕ_2 in $L^2(\mathbb{R}, \mathcal{B}(\mathbb{R}), d\hat{Z}^c(t))$ $\cap L^1(\mathbb{R}, \mathcal{B}(\mathbb{R}), dm(t))$,

$$\mathbb{E}[(\int_{\mathbb{R}} \phi_1(t) d\hat{Z}(t))(\int_{\mathbb{R}} \phi_2(t) d\hat{Z}(t))^*]$$

$$= \int_{\mathbb{R}} \phi_1(t)\phi_2^*(t) d\mu_{\hat{Z}^c}(t) + (\int_{\mathbb{R}} \phi_1(t) dm(t))(\int_{\mathbb{R}} \phi_2(t) dm(t))^*. \tag{4.25}$$

4.9 (Iterated integrals) We consider a stochastic measure $\hat{Z}$ and $\phi(s,t) \in L^1(\mathbb{R}^2, \mathcal{B}(\mathbb{R}^2), ds \times dt)$. In addition, we assume that $\phi(s,t) \in L^2(\mathbb{R}, \mathcal{B}(\mathbb{R}), d\mu_{\hat{Z}}(t))$ for almost every s (with respect to the Lebesgue measure).

a) Approximating $\phi(s,t)$ by a convergent sequence of functions of the form $\phi_n(s,t) = \sum_{k=1,N_n} \phi_{1,k}^{(n)}(s)\phi_{2,k}^{(n)}(t)$ where the functions $(\phi_{1,k}^{(n)}(s)\phi_{2,k}^{(n)}(t))_{k=1,N_n}$ are non-overlapping, show that $\int_{\mathbb{R}} \phi(s,t) d\hat{Z}(t)$ is measurable with respect to the variables s and ω.

b) We consider the following inequality:

$$\int_{\mathbb{R}} (\int_{\mathbb{R}} |\phi(s,t)| ds)^2 d\mu_{\hat{Z}}(t) < \infty. \tag{4.26}$$

Show that if inequality (4.26) holds, then the expression

$$X_1 = \int_{\mathbb{R}} (\int_{\mathbb{R}} \phi(s,t) ds) d\hat{Z}(t) \tag{4.27}$$

is well defined.

c) We consider the following inequality

$$\int_{\mathbb{R}} (\int_{\mathbb{R}} |\phi(s,t)|^2 d\mu_{\hat{Z}}(t))^{1/2} ds < \infty. \tag{4.28}$$

Show that if inequality (4.28) holds, then the expression

$$X_2 = \int_{\mathbb{R}} (\int_{\mathbb{R}} \phi(s,t) d\hat{Z}(t)) ds \tag{4.29}$$

is well defined.
(Hint: use the Loève mean square integrability criterion given by Equation (2.27).)

d) If conditions (4.26) and (4.28) hold, show that $P(X_1 = X_2) = 1$, which then means that the order of integration does not affect the result. The iterated integral will be denoted indifferently by the right-hand side of (4.27), of (4.29) or by $\int_{\mathbb{R}^2} \phi(s,t) ds d\hat{Z}(t)$.
(Hint: use a sequence of approximations of ϕ as in question a).)

e) As an example of an application, show that if $h(s) \in C^1([a,b])$, then we obtain the following integration by part formula

$$\int_{[a,b]}(h(t) - h(a))d\hat{Z}(t) = \int_{[a,b]}\hat{Z}([s,b])h'(s)ds$$

$$= [h(b) - h(a)]\hat{Z}([a,b]) - \int_{[a,b]}\hat{Z}([a,s[)h'(s)ds.$$

$$(4.30)$$

(Hint: let $\phi(s,t) = h'(s)\mathbb{I}_{[s,b]}(t)$.)

4.10 (General orthogonal expansion) Let $X = (X_t)_{t \in \mathbb{R}}$ denote a stochastic process, not necessarily stationary. We assume that there exists a positive measure ν and a set of functions $(\phi_t)_{t \in \mathbb{R}}$ belonging to $L^2(\mathbb{R}, \mathcal{B}(\mathbb{R}), d\nu_X(f))$ such that $\overline{span}\{\phi_t; t \in \mathbb{R}\} = L^2(\mathbb{R}, \mathcal{B}(\mathbb{R}), d\nu_X(f))$ and

$$R_X(t_1, t_2) = \int_{\mathbb{R}} \phi_{t_1}(f)\phi_{t_2}(f)d\nu_X(f).$$

$$(4.31)$$

a) Show that there exists a stochastic measure $\hat{Z}_X$ such that X has the representation

$$X_t = \int_{\mathbb{R}} \phi_t(f)d\hat{Z}_X(f).$$

$$(4.32)$$

b) Example: let $X_t = g(t)Y_t$, where $g(t)$ is a bounded continuous function and $Y = (Y_t)_{t \in \mathbb{R}}$ a WSS process. Express relations (4.31) and (4.32) in terms of ν_Y and $\hat{Y}$ respectively.

5. Filtering of WSS Processes

Purpose We wish to generalise the results obtained in the context of deterministic signal filtering in the case of WSS process filtering.

5.1 Elements of Deterministic Signal Filtering

We recall that a filter is defined as a linear homogeneous system. By homogeneous, we mean that if the input $x(t)$ yields the output $y(t)$ then, for a fixed τ, the input $x(t - \tau)$ yields the output $y(t - \tau)$. We may characterise an important class of filters from the following result (see, for example, [40] p. 118):

Theorem 5.1 *Any bounded linear form T defined on $L^\infty(\mathbb{R}, \mathcal{B}(\mathbb{R}), dt)$ may be written in the form $T : x \to \int x d\mu$, where μ is a bounded measure.*

In order to study deterministic filtering, we generally restrict ourselves to the case of bounded measures that are absolutely continuous with respect to Lebesgue's measure, that is, such that $d\mu(t) = g(t)dt$, with $g \in L^1(\mathbb{R}, \mathcal{B}(\mathbb{R}), dt)$. We now consider filters for which a bounded input signal $x(t)$ yields a bounded output signal $y(t)$. This condition is often called the *Bounded Input-Bounded Output* (BIBO) condition. We shall therefore have, for $t \in \mathbb{R}$ and from Theorem 5.1, $y(t) = \int_\mathbb{R} x(u)g_t(u)du$, where $g_t(u) \in L^1(\mathbb{R}, \mathcal{B}(\mathbb{R}), du)$.

Moreover, the homogeneity condition is expressed by

$$g_t(u) = g_0(u - t). \tag{5.1}$$

Indeed, $y(t - \tau) = \int_\mathbb{R} x(u)g_{t-\tau}(u)du$ also satisfies the relations

$$y(t - \tau) = \int_\mathbb{R} x(u - \tau)g_t(u)du$$

$$= \int_\mathbb{R} x(u)g_t(u + \tau)du, \tag{5.2}$$

hence $g_{t-\tau}(u) = g_t(u+\tau)$. Taking $t = \tau$, it results that $g_0(u) = g_t(u+t)$, that is, $g_t(u) = g_0(u - t)$. Moreover, writing $h_t(u) = g_t(-u)$, it clearly appears that the filtering corresponds to a convolution operation:

$$y(t) = \int_{\mathbb{R}} x(u)g_t(u)du$$

$$= \int_{\mathbb{R}} x(u)g_0(u-t)du \tag{5.3}$$

$$= \int_{\mathbb{R}} x(u)h_0(t-u)du.$$

Of course, as is seen in deterministic signal processing, defining filters as homogeneous linear systems does not involve the impulse response $h_0(t)$ belonging to $L^1(\mathbb{R}, \mathcal{B}(\mathbb{R}), df)$. However, we generally restrict ourselves to using filters that satisfy the BIBO condition.

5.2 Filtering of WSS Processes

We consider a zero mean, mean square continuous, WSS process X. It therefore has a spectral representation of the form $X_t = \int_{\mathbb{R}} e^{2i\pi ft} d\hat{X}(f)$. We define in $L^2(\Omega, \mathcal{A}, dP)$ the filtering operations of X from the functions $H(f)$ that belong to $L^2(\mathbb{R}, \mathcal{B}(\mathbb{R}), d\mu_X)$, by the relation

$$Y_t = \int_{\mathbb{R}} e^{2i\pi ft} H(f) d\hat{X}(f). \tag{5.4}$$

We then have $d\hat{Y}(f) = H(f)d\hat{X}(f)$. It is important to restrict oneself to functions $H(f)$ that belong to $L^2(\mathbb{R}, \mathcal{B}(\mathbb{R}), d\mu_X)$, so that the process Y_t is also a second order process, for

$$\| Y_t \|^2 = \int_{\mathbb{R}} |H(f)|^2 d\mu_X(f). \tag{5.5}$$

In the discrete case, $X_n = \int_{\mathcal{I}} e^{2i\pi nf} d\hat{X}(f)$ and the filtering operations of X are defined from the functions $H(f) \in L^2(\mathcal{I}, \mathcal{B}(\mathcal{I}), d\mu_X)$. We then have

$$Y_n = \int_{\mathcal{I}} e^{2i\pi nf} H(f) d\hat{X}(f). \tag{5.6}$$

Following the conservation of the scalar product expressed by relation (4.11), the autocovariances are expressed as a function of the spectral measures by the relations

$$R_Y(t) = \mathbb{E}[Y_{t+s}Y_s^*] = \int_{\mathbb{R}} e^{2i\pi ft} |H(f)|^2 d\mu_X(f),$$

$$R_{YX}(t) = \mathbb{E}[Y_{t+s}X_s^*] = \int_{\mathbb{R}} e^{2i\pi ft} H(f) d\mu_X(f), \tag{5.7}$$

$$R_{XY}(t) = \mathbb{E}[X_{t+s}Y_s^*] = \int_{\mathbb{R}} e^{2i\pi ft} H^*(f) d\mu_X(f).$$

Consequently, the spectral measures are linked by the relations

$$d\mu_Y(f) = |H(f)|^2 d\mu_X(f),$$

$$d\mu_{YX}(f) = H(f)d\mu_X(f), \qquad (5.8)$$

$$d\mu_{XY}(f) = H^*(f)d\mu_X(f),$$

which, in the case of absolutely continuous spectra, yields

$$S_Y(f) = |H(f)|^2 S_X(f),$$

$$S_{YX}(f) = H(f)S_X(f), \qquad (5.9)$$

$$S_{XY}(f) = H^*(f)S_X(f).$$

The equality of the Fourier transforms of the measures that appear in relations (5.8) leads to the corresponding expressions for the autocovariance functions:

$$R_Y(t) = h(t) \star h^*(-t) \star R_X(t),$$

$$R_{YX}(t) = h(t) \star R_X(t), \qquad (5.10)$$

$$R_{XY}(t) = h^*(-t) \star R_X(t),$$

where $\star$ denotes the convolution operator: $(g_1 * g_2)(t) = \int x(u)y(t-u)du$. We establish, for example, expression (5.10) for $R_Y(t)$: using Fubini's theorem,

$$R_Y(t) = \mathbb{E}[Y_{u+t} Y_u^*]$$

$$= \mathbb{E}[(\int_{\mathbb{R}} e^{2i\pi f(u+t)} H(f)d\hat{X}(f))(\int_{\mathbb{R}} e^{2i\pi fu} H(f)d\hat{X}(f))^*]$$

$$= \int_{\mathbb{R}} e^{2i\pi ft} H(f)H^*(f)d\mu_X(f)$$

$$= \int_{\mathbb{R}^3} e^{2i\pi f(t-t_1+t_2)} h(t_1)h^*(t_2)d\mu_X(f)dt_1 dt_2 \qquad (5.11)$$

$$= \int_{\mathbb{R}^2} e^{2i\pi f(t-u)}[\int_{\mathbb{R}} h(t_1)h^*(t_1-u)dt_1]d\mu_X(f)du$$

$$= \int_{\mathbb{R}^2} e^{2i\pi f(t-u)}[h(u) \star h^*(-u)]d\mu_X(f)du$$

$$= h(t) \star h^*(-t) \star R_X(t).$$

If X is a wide sense stationary process indexed by $\mathbb{Z}$, we obtain similar formulas such as

$$R_{XY}(t) = \int_{\mathcal{I}} e^{2i\pi n f} H^*(f) d\mu_X(f). \tag{5.12}$$

5.3 Comparison of the Deterministic with the Stochastic Case

We remark that if X is, for example, a discrete white noise, the corresponding stable filters, that is, those whose output defines a second order process, have frequency responses of the form $H(f) \in L^2(\mathcal{I}, \mathcal{B}(\mathcal{I}), df)$. We then have $H(f) = \sum_{n \in \mathbb{Z}} h_n e^{-2i\pi n f}$ with $\sum_{n \in \mathbb{Z}} |h_n|^2 < \infty$. In a deterministic context, stable filters are often characterised by the more restrictive condition $\sum_n |h_n| < \infty$. However, it must be noted that in the deterministic case, the stability condition is imposed over the set of bounded signals, whereas in the stochastic case we define the class of stable filters associated with each process or, to be more precise, with each spectral measure. Moreover, in the first case BIBO stability is concerned, whereas in the second it is the conservation of the finite power of the process that is concerned.

Another difference between filtering deterministic signals and filtering processes: difference equations of the form

$$Y_n + \sum_{k=1,p} a_k Y_{n-k} = \sum_{l=0,q} b_l X_{n-l} \tag{5.13}$$

will only be able to characterise a single filter with input X and output Y in the case where X and Y are WSS processes. To prove this, we can show that the spectral measure of Y is defined uniquely from that of X (see Chapter 10).

On the contrary, in the discrete case, Equation (5.13) will have several solutions. Consider, for example, the equation $Y_n = a Y_{n-1} + X_n$, where the input signal is given by $X_n = \delta_{0,n}$. It has a causal and an anti-causal solution given by

$$Y_n = 0, \text{ for } n < 0, Y_0 = 1, \text{ and } Y_n = a^n, \text{ for } n > 0,$$

$$\text{and } Y_n = -a^{-n}, \text{ for } n < 0, Y_0 = 0, \text{ and } Y_n = 0, \text{ for } n > 0, \tag{5.14}$$

respectively. The corresponding transfer functions $1 + \sum_{n>0} a^n z^{-n}$ and $-\sum_{n<0}(-a)^{-n} z^{-n}$ therefore represent two possible solutions to the difference equation $Y_n = a Y_{n-1} + X_n$. Often, we mean by the solution to Equation (5.13) that which is stable, that is, bounded and which here is the causal solution if $|a| < 1$.

5.4 Examples

5.4.1 Bandpass Filters

We consider a process X, with spectral representation $X_t = \int_{\mathbb{R}} e^{2i\pi ft} d\hat{X}(f)$. If $H(f) = \mathbb{I}_{[f_1, f_2]}(f)$, then the spectral measure of $Y_t = \int_{\mathbb{R}} e^{2i\pi ft} H(f) d\hat{X}(f)$ is given by

$$d\mu_Y(f) = \mathbb{I}_{[f_1, f_2]}(f) d\mu_X(f). \tag{5.15}$$

This type of filtering, called *bandpass filtering*, leads to a process whose mean power distribution is the same as that of X on $[f_1, f_2]$, and is equal to zero elsewhere.

We consider in particular the problem of sampling a process X. The sampled process with period T can be defined by $X_{e,t} = \sum_{n \in \mathbb{Z}} X_{nT} \delta_{nT}$. Assuming that the spectrum of X is carried by a sub-interval of $[-(2T)^{-1}, (2T)^{-1}]$, we shall see that we can find X from X_e using the filter with frequency response $H(f) = \mathbb{I}_{[-(2T)^{-1}, (2T)^{-1}]}(f)$, whose impulse response we shall denote by $h(t)$. The process Y at the output of this filter is expressed by

$$
\begin{aligned}
Y_t &= (h \star X_e)_t \\
&= \sum_{n \in \mathbb{Z}} h(t - nT) X(nT) \\
&= \sum_{n \in \mathbb{Z}} h(t - nT) \int_{\mathbb{R}} e^{2i\pi nfT} d\hat{X}(f) \\
&= \int_{\mathbb{R}} \left(\sum_{n \in \mathbb{Z}} h(t - nT) e^{2i\pi nfT} \right) d\hat{X}(f).
\end{aligned}
\tag{5.16}
$$

Now, for any value of f contained in the support of $\hat{X}$,

$$
\begin{aligned}
\sum_{n \in \mathbb{Z}} h(t - nT) e^{2i\pi nfT} &= e^{2i\pi ft} [h(u) e^{-2i\pi fu} \star \sum_{n \in \mathbb{Z}} \delta_{nT}]_{u=t} \\
&= e^{2i\pi ft} \mathcal{F}^{-1} [H(f' - f) T^{-1} \sum_{n \in \mathbb{Z}} \delta_{nT^{-1}}(f')] \\
&= e^{2i\pi ft} T^{-1}.
\end{aligned}
\tag{5.17}
$$

Hence, $Y_t = T^{-1} X_t$.

5.4.2 Differentiators

For a WSS process X, we can easily check that the necessary and sufficient mean square derivability condition, given by the existence of $\frac{\partial^2}{\partial t^2} R_X(t_1, t_2)$ at point 0, is expressed by $\int_{\mathbb{R}} f^2 d\mu_X(f) < \infty$. In this case, the function

$$h \rightarrow h^{-1}\big(e^{2i\pi f(t+h)} - e^{2i\pi ft}\big) \tag{5.18}$$

converges into $L^2(\mathbb{R}, \mathcal{B}(\mathbb{R}), d\mu_X(f))$, when h tends towards 0, towards $(2i\pi f) \times e^{2i\pi ft}$. As

$$\frac{X_{t+h} - X_t}{h} = \int_{\mathbb{R}} \frac{e^{2i\pi f(t+h)} - e^{2i\pi ft}}{h} d\hat{X}(f), \tag{5.19}$$

and $\int_{\mathbb{R}} f^2 d\mu_X(f) < \infty$, X_t is mean square derivable, with derivative

$$X_t' = \int_{\mathbb{R}} 2i\pi f e^{2i\pi ft} d\hat{X}(f). \tag{5.20}$$

The mean square derivation of a process therefore corresponds to a filtering operation, with frequency response $H(f) = 2i\pi f$.

5.4.3 Linear Partial Differential Equations

For a process indexed by $\mathbb{Z}^n$ or $\mathbb{R}^n$ that is wide sense stationary with respect to the different indices, we can generalise the notion of spectral representation directly. Thus, for such a process of the form $X = (X_{t_1,\ldots,t_n})_{t_1,\ldots,t_n \in \mathbb{R}^n}$, the autocovariance function is given by

$$R_X(t_1,\ldots,t_n) = \mathbb{E}[X_{\tau_1+t_1,\ldots,\tau_n+t_n} X^*_{\tau_1,\ldots,\tau_n}] \tag{5.21}$$

and the spectral measure of X, denoted by $d\mu_X(f_1,\ldots,f_n)$, is defined by the relation

$$R_X(t_1,\ldots,t_n) = \int_{\mathbb{R}^n} e^{2i\pi(f_1 t_1 + \cdots + f_n t_n)} d\mu_X(f_1,\ldots,f_n). \tag{5.22}$$

Similarly, we can define a spectral representation of X in the form

$$X_{t_1,\ldots,t_n} = \int_{\mathbb{R}^n} e^{2i\pi(f_1 t_1 + \cdots + f_n t_n)} d\hat{X}(f_1,\ldots,f_n), \tag{5.23}$$

when X is mean square continuous.

We now consider a four-index stationary process (three for the spatial position and one for the time), denoted by X_{xyzt}, which satisfies the wave equation:

$$\frac{\partial^2 X}{\partial x^2} + \frac{\partial^2 X}{\partial y^2} + \frac{\partial^2 X}{\partial z^2} - \frac{1}{c^2}\frac{\partial^2 X}{\partial t^2} = 0, \tag{5.24}$$

where c represents the propagation speed of the waves in the medium.

We denote by $\mu_X(f_1, f_2, f_3, f_4)$ the spectral measure of this process. By considering the study of differentiators again, we can easily check that expression (5.24) has a meaning when

$$\int f_k^4 d\mu_X(f_1, f_2, f_3, f_4) < \infty,$$ (5.25)

for $k = 1, \ldots, 4$.

Moreover, using the spectral representation of X, relation (5.24) leads to

$$\int_{\mathbb{R}^4} \left(\sum_{k=1,3} (2i\pi f_k)^2 - \left(\frac{2i\pi f_4}{c}\right)^2 \right) e^{2i\pi(f_1 x + f_2 y + f_3 z + f_4 t)} d\hat{X}(f_1, f_2, f_3, f_4) = 0.$$ (5.26)

Hence,

$$\int_{\mathbb{R}^4} \left| \sum_{k=1,3} (2\pi f_k)^2 - \left(\frac{2\pi f_4}{c}\right)^2 \right|^2 d\mu_X(f_1, f_2, f_3, f_4) = 0,$$ (5.27)

which shows that the support of the spectrum of X is carried by a cone of $\mathbb{R}^4$, of equation

$$f_4^2 = c^2 \sum_{k=1,3} f_k^2.$$ (5.28)

Exercises

5.1 We consider a WSS process $X = (X_t)_{t \in \mathbb{R}}$ and a filter with impulse response $h(t) = \sum_{k=1,p} c_k e^{-\lambda_k t} \mathbb{I}_{\mathbb{R}_+}(t)$. Calculate the autocovariance and the spectrum of the process Y at the output of this filter when X is the input, in terms of the autocovariance and of the spectrum of X respectively.

5.2 We consider a WSS process $X = (X_n)_{n \in \mathbb{Z}}$ with PSD $S_X(f)$, and Y defined by $Y_n = K^{-1} \sum_{k=0, K-1} X_{n-k}$.
a) Calculate the spectrum of Y.
b) Calculate $R_Y(n)$ when X is a white noise with variance σ^2.

5.3 Let $X = (X_t)_{t \in \mathbb{R}}$ denote a WSS process. Show that $Y_t = X_t^* \star X_{-t}$ is well defined when $\int_{\mathbb{R}} R_X(t)^2 < \infty$.

5.4 Let $X = (X_t)_{t \in \mathbb{R}}$ denote a WSS process, with autocovariance function $R_X(t)$ and PSD $S_X(f)$, that models a signal transmitted over a propagation channel. X propagates to a receiver following several paths and each path introduces a delay and an attenuation of X. Finally, a signal of the form $Y_t = \sum_{k=1,p} a_k X_{t-\tau_k}$ with $a_1, \ldots, a_p, \tau_1, \ldots, \tau_p \in \mathbb{R}$ is observed at the receiver side.

a) Calculate $R_Y(t)$ and then $S_Y(f)$. Check that the propagation channel operates as a filter. Give its frequency response and its impulse response.

b) We now assume that the coefficients $(a_k)_{k=1,p}$ are zero mean independent random variables, and independent of X. Give the expression of $R_Y(t)$.

5.5 We consider a WSS process $X = (X_t)_{t \in \mathbb{R}}$. We assume that $\int_{\mathbb{R}} f^4 d\mu_X(f) < \infty$ and that X satisfies equation $d^2X/dt^2 + k^2X = 0$. Show that $X = \xi_1 e^{-ikt} + \xi_2 e^{ikt}$, where ξ_1 and ξ_2 are uncorrelated random variables.

5.6 Let us consider a filter with transfer function

$$g_N(z) = \frac{1}{2N+1} \sum_{n=-N,N} e^{2i\pi n f_0} z^{-n}. \tag{5.29}$$

Draw the shape of $|g_N(e^{2i\pi f})|$. What is the interest of such a filter when N is large?

5.7 We consider $h(z) = \frac{z^{-1} - \alpha}{1 - \alpha^* z^{-1}}$, with $|\alpha| < 1$.

   a) Show that $h(z)$ represents the transfer function of an admissible filter for any second order WSS process.

   b) Check that two distinct processes may have the same spectrum.

5.8 We consider X and Y, two WSS indexed by $\mathbb{Z}$. We assume that X and Y are zero mean and jointly stationary. We want to approximate Y as a filtered version of X. Let $h(z) = \sum_{k=0,p} h_k z^{-k}$ denote this approximating filter. $h(z)$ is chosen as the solution to the following constrained optimisation problem:

$$\begin{cases} \min_{h \in \mathbb{C}^{p+1}} \| Y_n - [h(z)]X_n \|^2 \\ \| Y_n \|^2 = \| [h(z)]X_n \|^2, \end{cases} \tag{5.30}$$

where $h = [h_1, \ldots, h_p]^T$. Calculate $h(z)$.

5.9 (Spatial sampling of propagating waves) We consider the example of Section 5.4.3. We assume that the process $X = (X_{xyzt})_{(xyzt) \in \mathbb{R}^4}$ is observed along the x-axis, yielding the process $Y_{xt} = X_{x00t}$.

   a) Express the spectral measure of Y in terms of that of X. What subset of $\mathbb{R}^2$ supports the spectrum of Y?

   b) We assume that the support of the time spectrum of Y lies inside the frequency interval $[-f_{max}, f_{max}]$. Y is regularly sampled along the x-axis at positions nd ($n \in \mathbb{Z}$). Show that d must be chosen such that $d \leq c/(2f_{max})$ to avoid spectrum aliasing. Note that this particular application of Shannon's sampling theorem is of practical interest in the field of array processing where a signal is observed at the outputs of an array of sensors.

5.10 (Parseval's relation) Let $X = (X_t)_{t \in \mathbb{R}}$ denote a mean square continuous WSS process, and $g(t)$ a continuous function such that $\hat{g}(f) \in L^2(\mathbb{R}, \mathcal{B}(\mathbb{R}), d\mu_X(f))$, where $\hat{g}(f)$ is the Fourier transform of $g(t)$. Show that

$$\int_{\mathbb{R}} g(t-u)^* X_u \, du = \int_{\mathbb{R}} e^{2i\pi ft} \hat{g}(f)^* d\hat{X}(f). \tag{5.31}$$

(Hint: see Exercise 4.6.)

5.11 Let $X = (X_t)_{t \in \mathbb{R}}$ denote a WSS process. We assume that the sample paths of X belong to $L^1(\mathbb{R}, \mathcal{B}(\mathbb{R}), dt)$ and we define the path integrals

$$\tilde{X}(f) = \int_{\mathbb{R}} e^{-2i\pi f t} X_t dt. \tag{5.32}$$

a) Check that for all ω in Ω the function $\hat{X}_\omega : \mathcal{B}(\mathbb{R}) \to \mathbb{R}$ defined by $\hat{X}_\omega(A) = \tilde{X}(A)(\omega)$ is a measure and that

$$X_t(\omega) = \int_{\mathbb{R}} e^{-2i\pi f t} d\hat{X}_\omega(f). \tag{5.33}$$

b) Show that we have the following path integral relationship:

$$\tilde{X}(f) = \int_{\mathbb{R}} \delta_{f-u} d\hat{X}(u). \tag{5.34}$$

(Hint: calculate $\int_{\mathbb{R}} e^{2i\pi f t} \tilde{X}(f) df$.)

6. Important Particular Processes

Purpose Here, we briefly present a few families of processes that often play an important part in statistical signal processing. The presentation of Markov processes is delayed until Chapter 15, where they will play an important role.

6.1 Gaussian Processes

Definition 6.1 *Real Gaussian processes are those for which the joint distributions of vector random variables of the form* $\mathbf{X} = [X_{t_1}, \ldots, X_{t_n}]^T$ *are Gaussian.*

We recall that a real Gaussian random vector $\mathbf{X}$ of size n is characterised by a characteristic function of the form

$$\varphi_{\mathbf{X}}(u) = \exp(i.u^T m_{\mathbf{X}} - (1/2)u^T \Gamma_{\mathbf{X}} u), \tag{6.1}$$

for all $u \in \mathbb{R}^n$. We show that $m_{\mathbf{X}}$ and $\Gamma_{\mathbf{X}}$ thus represent the mean vector and the covariance matrix of $\mathbf{X}$ respectively. When $|\Gamma_{\mathbf{X}}| \neq 0$, $\mathbf{X}$ has a density distribution of the form

$$f_{\mathbf{X}}(\mathbf{x}) = (2\pi)^{-N/2} |\Gamma_{\mathbf{X}}|^{-1/2} \exp[-\frac{1}{2}(\mathbf{x} - m_{\mathbf{X}})^T \Gamma_{\mathbf{X}}^{-1}(\mathbf{x} - m_{\mathbf{X}})]. \tag{6.2}$$

For the complex case, the probability density of $\mathbf{X}$ is obtained by identifying $\mathbf{X}$ with a random vector of $\mathbb{R}^{2n}$ whose components are the real and imaginary parts of the components of $\mathbf{X}$.

Knowledge of a function $m(t)$ and of a function of the positive type $R(t_1, t_2)$ entirely characterises the distribution of a Gaussian process for which $\mathbb{E}[X_t] = m(t)$, and $\mathrm{cov}[X_{t_1} X_{t_2}] = R(t_1, t_2)$; it is easy to verify that the family of Gaussian vector distributions thus constructed defines the distribution of a Gaussian process from Kolmogorov's consistency theorem (Theorem 2.1).

The limit of a sequence of Gaussian random variables $(X_n)_{n \in \mathbb{N}}$, that converges in the space $L^2(\Omega, \mathcal{A}, dP)$, is a Gaussian random variable X. To prove this, we merely have to verify that $\mathbb{E}[X_n]$ and $\mathbb{E}[X_n^2]$ converge towards $\mathbb{E}[X] = m_X$ and $\mathbb{E}[X^2] = \sigma_X^2 + m_X^2$ respectively, then to note that the sequence of characteristic functions $\phi_{X_n}(u)$ of the random variables X_n converges towards the characteristic function $\exp(i.m_X u - \sigma_X^2 u^2/2)$ of a $\mathcal{N}(m_X, \sigma_X^2)$ distribution at any point $u \in \mathbb{R}$.

Moreover, we show that if X and Y are vectors with joint Gaussian distribution, $\mathbb{E}[X|Y]$ coincides with the orthogonal projection of X in $L^2(\Omega, \mathcal{A}, P)$ on the vector space generated by the components of Y. More precisely, denoting by $\mathcal{N}(m, \Gamma)$ a multivariate Gaussian distribution with mean vector m and covariance matrix Γ, we have the following result:

Theorem 6.1 *If (X, Y) has a Gaussian distribution, and $m_X, m_Y, \Gamma_X, \Gamma_Y,$ Γ_{XY} denote the means, covariances and cross-covariance of X and Y,*

$$\mathbb{E}[X|Y] = m_X + \Gamma_{XY}\Gamma_Y^{-1}(Y - m_Y), \tag{6.3}$$

and the distribution of X conditional to $Y = y$ is

$$\mathcal{N}(m_X + \Gamma_{XY}\Gamma_Y^{-1}(y - m_Y), \Gamma_X - \Gamma_{XY}'\Gamma_Y^{-1}\Gamma_{XY}^T). \tag{6.4}$$

Proof We note

$$Z = X - m_X - \Gamma_{XY}\Gamma_Y^{-1}(Y - m_Y). \tag{6.5}$$

It is easy to see that

$$\begin{pmatrix} Z \\ Y \end{pmatrix} = \mathcal{N}\left(\begin{bmatrix} 0 \\ m_Y \end{bmatrix}, \begin{bmatrix} \Gamma_X - \Gamma_{XY}\Gamma_Y^{-1}\Gamma_{XY}^T & 0 \\ 0 & \Gamma_Y \end{bmatrix} \right). \tag{6.6}$$

Consequently,

$$\mathbb{E}[X|Y] = \mathbb{E}[Z|Y] + m_X + \Gamma_{XY}\Gamma_Y^{-1}\mathbb{E}[(Y - m_Y)|Y]$$

$$= \mathbb{E}[Z] + m_X + \Gamma_{XY}\Gamma_Y^{-1}(Y - m_Y) \tag{6.7}$$

$$= m_X + \Gamma_{XY}\Gamma_Y^{-1}(Y - m_Y).$$

It is also clear that

$$\mathbb{E}[e^{iu^T X}|Y]$$

$$= \mathbb{E}[e^{iu^T Z}|Y]\exp[iu^T(m_X + \Gamma_{XY}\Gamma_Y^{-1}(Y - m_Y))]$$

$$= \exp[iu^T(m_X + \Gamma_{XY}\Gamma_Y^{-1}(Y - m_Y)) - \frac{1}{2}u^T(\Gamma_X - \Gamma_{XY}\Gamma_Y^{-1}\Gamma_{XY}^T)u], \tag{6.8}$$

which completes the proof. $\square$

6.2 Poisson Processes

Poisson processes are very useful for describing random phenomena like people arriving at different times at a counter, calls arriving at a telephone network, or the number of times (assumed to be independent) that a machine breaks down. Studying Poisson processes is important when faced with the correct sizing of machines or services in charge of managing these phenomena.

Definition 6.2 *A Poisson process $X = (X_t)_{t \in \mathbb{R}_+}$ of intensity λ is a process with independent increments, that is, if $t_1 < t_2 < t_3 < t_4$ then $X_{t_2} - X_{t_1}$ and $X_{t_4} - X_{t_3}$ are independent, and such that if $t_1 < t_2$ the distribution of $X_{t_2} - X_{t_1}$ is a Poisson distribution of parameter $\lambda(t_2 - t_1)$, that is, $P(X_{t_2} - X_{t_1} = k) = (1/k!)(\lambda(t_2 - t_1))^k e^{-\lambda(t_2 - t_1)}$.*

To show that Poisson processes are likely to model the above-mentioned phenomena, we denote by T_n the random instant when the nth event occurs and $\Delta_n = T_{n+1} - T_n$ the time interval between T_n and T_{n+1}. The memoryless nature of the occurrence of the events is expressed by

$$P(\Delta_n > t + h / \Delta_n > t) = P(\Delta_n > h). \tag{6.9}$$

We can easily deduce that Δ_n has an exponential distribution, whose intensity we shall denote by λ. Indeed, from relation (6.9) $P(\Delta_n > t + h) = P(\Delta_n > t)P(\Delta_n > h)$. Deriving this relation with respect to h and then fixing $h = 0$, it results that the probability density of Δ_n satisfies the differential equation

$$f_{\Delta_n}(t) = f_{\Delta_n}(0) \int_t^\infty f_{\Delta_n}(u)du, \tag{6.10}$$

whose solution on $\mathbb{R}_+$ is

$$f_{\Delta_n}(t) = \lambda \exp(-\lambda t) \mathbb{I}_{\mathbb{R}_+}(t), \tag{6.11}$$

with $\lambda = f_{\Delta_n}(0)$.

Consequently, as $T_n = \Delta_1 + \ldots + \Delta_n$ and the random variables $(\Delta_k)_{k \in \mathbb{N}}$ form a sequence of independent random variables with the same distributions, the density of the distribution of T_n is given by the expression

$$f_{T_n}(t) = \frac{\lambda^n t^{n-1} e^{-\lambda t}}{(n-1)!} \mathbb{I}_{\mathbb{R}_+}(t), \tag{6.12}$$

which is easily obtained by induction. This distribution is called Erlang's distribution of order n and of parameter λ.

If we now denote by N_t the number of events that have occurred up to instant t, it is clear that N_t is distributed according to a Poisson distribution of parameter λt:

$$`P(N_t = n) = P(T_n \leq t < T_{n+1})$$

$$= P(T_n \leq t) - P(T_{n+1} \leq t)$$

$$= \frac{\lambda^n}{(n-1)!} [\int_0^t e^{-\lambda u} u^{n-1} du - \int_0^t \lambda e^{-\lambda u} \frac{u^n}{n} du] \qquad (6.13)$$

$$= e^{-\lambda t} \frac{(\lambda t)^n}{n!}.$$

Moreover, it is easy to see that N_t is a process with independent increments. N is therefore a Poisson process.

Additional results concerning Poisson processes may be found in [2].

6.3 White Noise

Definition 6.3 *A white noise process is a WSS process that has an absolutely continuous spectral measure with respect to Lebesgue's measure, and whose PSD is a constant function.*

In the case of a process indexed by $\mathbb{Z}$, we easily verify that this means that the process is made up of a sequence of uncorrelated random variables, with the same mean and the same variance.

The continuous case is trickier to deal with. Indeed, the PSD of a white noise process is then a constant function on $\mathbb{R}$. Therefore, the integral of the spectral density is not defined, which means that for a white noise process U we have $\| U_t \|^2 = +\infty$; U is therefore not a second order process!

For such a process, we could, however, define the autocovariance function in the sense of the distribution theory, since a constant function equal to σ^2 represents the Fourier transform of the distribution $\sigma^2 \delta_0$, where δ_0 represents the Dirac distribution.

In order to understand the problem better, we present a first attempt at constructing a white noise process. For this, we consider a process X with orthogonal increments, that is, such that

$$\mathbb{E}[(X_{t_3} - X_{t_2})(X_{t_1} - X_{t_0})^*] = 0, \text{ for } t_0 < t_1 < t_2 < t_3. \qquad (6.14)$$

We assume, moreover, that for $t_2 \geq t_1$, $\| X_{t_2} - X_{t_1} \|^2 = \sigma^2(t_2 - t_1)$. We shall give an example of such a process later. The processes $U^{(n)}$ defined by

$$U_t^{(n)} = h_n^{-1}(X_{t+h_n} - X_t), \qquad (6.15)$$

have autocovariance functions of the form

$$R_{U^{(n)}}(t) = \sigma^2 \left(\frac{h_n - |t|}{h_n^2} \right) \mathbb{1}_{[-h_n, h_n]}(t). \qquad (6.16)$$

We assume now that $\lim_{n\to\infty} h_n = 0$. Then, $R_{U^{(n)}}(t)$ converges towards the Dirac distribution $\sigma^2\delta_0$. Clearly, the sequence of spectral densities of the processes $U^{(n)}$, given by

$$S_{U^{(n)}}(f) = \sigma^2 \left(\frac{\sin \pi f h_n}{\pi f h_n}\right)^2, \tag{6.17}$$

converges uniformly on any interval $[-A, A]$ towards a PSD equal to σ^2.

Unfortunately, we cannot define a white noise process that would be the derivative in $L^2(\Omega, \mathcal{A}, dP)$ of the process with orthogonal increments X. Indeed, $R_{U^{(n)}}(0)$ does not have a finite limit when n tends towards $+\infty$, and X is therefore not mean square derivable.

6.3.1 Generalised Processes

In deterministic signal processing we introduce the theory of distributions to generalise the Fourier transform for classes of functions that are not integrable with respect to Lebesgue's measure, such as sinusoidal functions. In light of the above remarks, it is easier to understand that, like for deterministic signal processing, we have to introduce the notion of generalised processes to rigorously describe white noise processes in the continuous case.

Mean Square and Weak Convergence

Definition 6.4 *We say that a sequence of processes* $(X^{(n)})_{n\in\mathbb{N}}$ *defined on a set of indices T mean square converges towards X if*

$$\forall t \in T, \quad \lim_{n\to\infty} X_t^{(n)} \stackrel{m.s.}{=} X_t. \tag{6.18}$$

In many cases, such a limit does not exist, whereas integral expressions of the type

$$Y_t^{(n)} = \int_T g(t, \tau) X_\tau^{(n)} d\tau \tag{6.19}$$

have a mean square limit, when n tends towards $+\infty$, for large classes of functions g.

To specify this point, we consider the example of the previous section where we saw that the processes $U^{(n)}$ do not mean square converge. We now note that the relations $W([a, b[) = X_b - X_a$, for $a < b$, define a stochastic measure W, and that $d\mu_W(f) = \sigma^2 df$, since $\| W([a, b[) \|^2 = \sigma^2(b - a)$. We then have the following result:

Theorem 6.2 *We consider the process* $U^{(n,h)}$ *defined by*

$$U_t^{(n,h)} = \int_T h(t - \tau) U_\tau^{(n)} d\tau, \tag{6.20}$$

where h is continuous. Then, for all $t \in T$ the sequence of random variables $(U_t^{(n,h)})_{n \in \mathbb{N}}$. mean square converges towards

$$U_t^h = \int_T h(t - \tau)dW(\tau),$$ (6.21)

where the stochastic measure W is defined by $W([a, b[) = X_b - X_a$ $(a < b)$.

Proof

$$\| U_t^{(n,h)} - U_t^h \|^2$$
$$= \sigma^2 \int_T \int_T h(t - \tau)h^*(t - \tau')\frac{h_n - |\tau - \tau'|}{h_n^2}\mathbb{I}_{[-h_n,h_n]}(\tau - \tau')d\tau d\tau'$$

$$+ \sigma^2 \int_T |h(t - \tau)|^2 d\tau$$

$$- 2\mathbb{E}[\int_T \int_T h(t - \tau)h^*(t - \tau')U_\tau^{(n)}d\tau dW^*(\tau')].$$ (6.22)

Writing $u = h_n^{-1}(\tau - \tau')$, the first integral can also be written as

$$\sigma^2 \int_T h(t - \tau) \left(\int_{[-1,1]} h^*(t - \tau + h_n u)(1 - |u|)du \right) d\tau.$$ (6.23)

As h is continuous, the integral in brackets converges uniformly towards $h^*(t - \tau)$ when n tends towards $+\infty$. Expression (6.23) therefore converges towards $\sigma^2 \int_T |h(t - \tau)|^2 d\tau$.

Concerning the third term of (6.22), since

$$U_\tau^{(n)} = \int_u \frac{1}{h_n}\mathbb{I}_{[\tau,\tau+h_n[}(u)dW(u),$$ (6.24)

we obtain

$$\mathbb{E}[\int_T \int_T h(t - \tau)h^*(t - \tau')U_\tau^{(n)}d\tau dW^*(\tau')]$$

$$= \int_\tau h(t - \tau)\mathbb{E}[\int_{\tau'} h^*(t - \tau')\int_u \frac{1}{h_n}\mathbb{I}_{[\tau,\tau+h_n[}(u)dW^*(\tau')dW(u)]d\tau$$

$$= \int_\tau h(t - \tau) \left(\int_{\tau'} h^*(t - \tau')\frac{1}{h_n}\mathbb{I}_{[\tau,\tau+h_n[}(\tau')\sigma^2 d\tau' \right) d\tau$$

$$= \sigma^2 \int_T |h(t - \tau)|^2 d\tau.$$ (6.25)

It thus seems that $\| U_t^{(n,h)} - U_t^h \|^2$ tends towards 0 when n tends towards $+\infty$, which completes the proof. $\square$

U^h may be interpreted as the output of a filter with impulse response $h(t)$ and with a white noise input process. We are then led to introduce the following definition:

Definition 6.5 *We shall say that a sequence $(X^{(n)})_{n \in \mathbb{N}}$ of random processes converges weakly (in the mean square sense) towards a process X relative to a certain class Φ of functions, called test functions, if*

$$\forall \phi \in \Phi, \quad \lim_{n \to \infty} \int_T \phi(t) X_t^{(n)} dt \stackrel{m.s.}{=} \int_T \phi(t) X_t dt. \tag{6.26}$$

Φ will, for example, be the set $\mathcal{D}$ of the functions of class C^∞ with compact supports.

This definition is not yet sufficient for, as we have just seen, X does not necessarily exist as a second order process. The notion of a generalised process allows this situation to be taken into account.

Generalised Processes In this paragraph, we shall merely recall the definition of distributions, and briefly present the definition of generalised processes, insofar as this notion will not be essential for the rest of this book, where we mainly envisage discrete time processes. We begin by recalling the definition of a distribution.

Definition 6.6 *A distribution u is a linear form defined on the set $\mathcal{D}$ of functions of $C^\infty(\mathbb{R})$ with compact support $(u : \phi \in \mathcal{D} \to\, < u, \phi > \, = u(\phi) \in \mathbb{R})$ that satisfies the following property: for any compact set K of $\mathbb{R}$, there exists an integer p and a constant C such that*

$$\forall \phi \in C^\infty(K), \; |u(\phi)| < C \sup_{x \in K, \alpha \leq p} |\frac{d^\alpha \phi(x)}{dx^\alpha}|. \tag{6.27}$$

The vector space of the distributions is denoted by $\mathcal{D}'$.

We shall say that a sequence $(u_k)_{k \in \mathbb{N}}$ of distributions converges (in $\mathcal{D}'$) towards u if

$$\forall \phi \in \mathcal{D}, \quad \lim_{k \to \infty} u_k(\phi) = u(\phi). \tag{6.28}$$

One main interest of distributions lies in the fact that if for a sequence $(u_k)_{k \in \mathbb{N}}$ of $\mathcal{D}'$, and for any function $\phi \in \mathcal{D}$, $u_k(\phi)$ converges towards a limit denoted by $u(\phi)$, then u defines an element of $\mathcal{D}'$.

Definition 6.7 *We call a generalised process $(X_t)_{t \in T}$ an application that maps $T \times \Omega$ onto $\mathcal{D}' : \forall (t, \omega) \in T \times \Omega$, $X_t(\omega)$ is a distribution of $\mathcal{D}'$ and $\forall \phi \in \mathcal{D}$ the relation $Y_t^\phi(\omega) = \, < X_t(\omega), \phi >$ defines a random process $Y^\phi = (Y_t^\phi)_{t \in T}$.*

In particular, any mean square continuous process X may be seen as a generalised process: with any function $\phi \in \mathcal{D} \subset L^2(\mathbb{R}, \mathcal{B}(\mathbb{R}), d\mu_X(f))$ a process Y^ϕ can be associated, defined by

$$Y_t^\phi = \int_\mathbb{R} \phi(t - \tau) X_\tau d\tau$$

$$= \int_\mathbb{R} e^{2i\pi ft} \hat{\phi}(f) d\hat{X}(f), \tag{6.29}$$

where $\hat{\phi}(f)$ here represents the Fourier transform of $\phi(t)$. But it is particularly for processes that are not mean square continuous that the notion of a generalised process is interesting. In particular, if we take the example again where X represents a process with orthogonal increments and W the stochastic measure associated with it, like in Theorem 6.2, we shall be able to characterise a spectral representation of X from knowledge of the outputs of the filters whose impulse response belongs to $\mathcal{D}$ and with input X. We already know that the processes Y^ϕ obtained by filtering X are given by

$$Y_t^\phi = \int_\mathbb{R} \phi(t - \tau) dW(\tau). \tag{6.30}$$

We can also show that there exists a second stochastic measure, denoted by $\hat{W}$ and such that

$$Y_t^\phi = \int_\mathbb{R} e^{2i\pi ft} \hat{\phi}(f) d\hat{W}(f). \tag{6.31}$$

Indeed, defining $\hat{W}$ by

$$\hat{W}([a, b]) = \int_\mathbb{R} \mathcal{F}^{-1}(\mathbb{1}_{[a,b]}(t)) dW(t), \tag{6.32}$$

using the linearity of the integral and of the Fourier transform, and Kolmogorov's isomorphism, it is easy to verify that $\hat{W}$ is a stochastic measure. Using the approximation of $e^{2i\pi ft} \hat{\phi}(f)$ by a sequence of step functions, and going to the limit, we obtain

$$Y_t^\phi = \int_\mathbb{R} \phi(t - \tau) dW(\tau)$$

$$= \int_\mathbb{R} \mathcal{F}^{-1}(e^{2i\pi ft} \hat{\phi}(f)) dW(t) \tag{6.33}$$

$$= \int_\mathbb{R} e^{2i\pi ft} \hat{\phi}(f) d\hat{W}(f).$$

It is clear that if X is a white noise process, that is, if $\| dW(t) \|^2 = \sigma^2 dt$, the previous representation can be generalised for all the functions ϕ of $L^2(\mathbb{R}, \mathcal{B}(\mathbb{R}), dt)$.

We shall not deal further with these notions, but it was important to highlight certain difficulties linked with idealised representations of physical phenomena, such as white noise. Finally, we remark that the notion of distribution only reintroduces into a mathematical framework the idea that, in practice, the observation of a signal is always limited in time or in frequency, a limitation taken into account by the distribution theory through the notion of test functions. For further details about distributions and generalised processes, refer, for example, to [31] or [34] and [9] respectively. For a detailed study of white noise, see, for example, [21].

6.3.2 Brownian Motion

We shall now investigate the case of Gaussian white noise processes, which are simultaneously white noise processes and Gaussian processes. These processes are widely used to model noise phenomena or to study Gaussian processes, which may be seen as the output of a filter with a white noise input.

Constructing such a process does not pose a problem in the discrete case. In the case of processes indexed by $\mathbb{R}$, we can obtain a Gaussian white noise process as the derivation, in the sense of generalised processes, of a Brownian motion.

Definition 6.8 *A Brownian motion (also called a Wiener process) is a process B with orthogonal increments such that the random variables $B_t - B_{t'}$ $(t' < t)$ have zero mean Gaussian distributions of variance $\sigma^2(t - t')$.*

In fact, this definition characterises a simple model of the trajectory of a particle in a liquid for which we assume, in addition to the continuity of the trajectory, that the particle undergoes many impacts and has negligible inertia (the process has orthogonal increments) and that the properties of the medium, assumed to be homogeneous, do not evolve over time (the distribution of $B_{t+\tau} - B_{t'+\tau}$ does not depend on τ).

As $\| B_{t+h} - B_t \|^2 = \sigma^2 h$, it is clear that the trajectories of B are mean square continuous, but nowhere mean square differentiable.

We define the Gaussian white noise in the same way as when dealing with any white noise distribution (see Theorem 6.2), starting from the fact that the relations $W([a, b[) = B(b) - B(a)$ $(a < b)$ define a stochastic measure W. The generalised process Y, characterised by the relations

$$Y_t^h : h \in \mathcal{D} \to \int_{\mathbb{R}} h(t - \tau)dW(\tau), \tag{6.34}$$

$\forall h \in \mathcal{D}$, defines a Gaussian white noise process: Y^h is a zero mean Gaussian process, since it may be written as the limit of a sequence of Gaussian processes, h being the limit of a sequence of step functions. Integrals of the form $\int_{\mathbb{R}} h(u)dW(u)$ are called *Wiener integrals*. The autocovariance function of Y^h is

$$R_{Y^h}(t) = \sigma^2 h(t) \star h^*(-t), \tag{6.35}$$

and its PSD $S_{Y^h}(f) = \sigma^2 |H(f)|^2$, where $H(f)$ is the Fourier transform of h. These results show that Y^h may be seen as the output of a filter with impulse response h, and with a Gaussian input process that has a constant PSD.

6.4 Cyclostationary Processes

A process X is said to be strict sense cyclostationary, or strict sense periodically correlated, of period T, if the distribution of the random vector $[X_{t_1}, \ldots, X_{t_n}]^T$ is the same as that of $[X_{t_1+kT}, \ldots, X_{t_n+kT}]^T$, $\forall n, k \in \mathbb{N}$, and $\forall t_1, \ldots, t_n \in \mathbb{R}$. It is said to be second order cyclostationary if we have simply $\mathbb{E}[X_{t+kT}] = \mathbb{E}[X_t]$ and $R_X(t_1 + kT, t_2 + kT) = R_X(t_1, t_2)$, $\forall k$.

Example (linear modulations in digital communications) In many situations, the processes observed in digital transmissions may be represented in the form

$$X_t = \sum_{k \in \mathbb{Z}} A_k \mathbb{1}_{[0,T[}(t - kT) + B_t, \tag{6.36}$$

where the random variables A_k ($k \in \mathbb{Z}$) are independent random variables that represent the transmitted symbols and B_t a white noise process, independent of the random variables A_k, that corrupts the transmission. The random variables A_k take their values in a finite set $\{\alpha_1, \ldots, \alpha_M\}$ with the respective probabilities $p_1, \ldots, p_M$ ($\sum_{i=1,M} p_i = 1$). We show that X_t is second order cyclostationary, of period T:

$$\mathbb{E}[X_t] = \mathbb{E}[A_k]$$

$$= \sum_{i=1,M} \alpha_i p_i, \quad \text{with } kT \le t < (k+1)T. \tag{6.37}$$

If $kT \le t < (k+1)T$ and $lT \le t + \tau < (l+1)T$,

$$\mathbb{E}[X_{t+\tau} X_t^*] = \mathbb{E}[A_k]\mathbb{E}[A_l^*]$$

$$= \left(\sum_{i=1,M} \alpha_i p_i\right)^2 \qquad \text{if } k \ne l,$$

$$\mathbb{E}[X_{t+\tau} X_t^*] = \mathbb{E}[|A_k|^2] + \| B_t \|^2 \delta_{0,\tau} \tag{6.38}$$

$$= \sum_{i=1,M} \alpha_i^2 p_i + \| B_t \|^2 \delta_{0,\tau} \quad \text{if } k = l.$$

For $0 < \alpha < T$, $\mathbb{E}[X_{t+\tau+\alpha} X_{t+\alpha}^*]$ may therefore differ from $\mathbb{E}[X_{t+\tau} X_t^*]$: take, for example, $0 < t < T - \alpha$ and $T - \alpha - t < \tau < T - t$. But for $\alpha = T$ these two terms are identical, which proves the result.

In order to be able to use the results related to stationary processes in a cyclostationary context, we can refer to the following result:

Theorem 6.3 *If X is strict cyclostationary of period T, the process $Y_t = X_{t-\phi}$, where ϕ is a random variable with uniform distribution on $[0,T]$, is strict sense stationary. The distribution functions F_X and F_Y, expressed by*

$$F_X(x_1, \ldots, x_n; t_1, \ldots, t_n) = P(X_{t_1} \leq x_1, \ldots, X_{t_n} \leq x_n), \quad (6.39)$$

are then linked by the relation

$$F_Y(y_1, \ldots, y_n; t_1, \ldots, t_n) = \frac{1}{T} \int_0^T F_X(y_1, \ldots, y_n; t_1 - \tau, \ldots, t_n - \tau) d\tau. \quad (6.40)$$

If X is second order cyclostationary, Y is then second order stationary, and

$$\mathbb{E}[Y_t] = \frac{1}{T} \int_0^T \mathbb{E}[X_\tau] d\tau \quad (6.41)$$

and $R_Y(t) = \frac{1}{T} \int_0^T R_X(t + \tau, \tau) d\tau.$

Proof We denote by A_c the event

$$(Y_{t_1+c} \leq y_1, \ldots, Y_{t_n+c} \leq y_n) = (X_{t_1+c-\phi} \leq y_1, \ldots, X_{t_n+c-\phi} \leq y_n), \quad (6.42)$$

where c is a constant. We have $P(A_c) = \mathbb{E}[\mathbb{1}_{A_c}] = \mathbb{E}[\mathbb{E}[\mathbb{1}_{A_c}|\phi]]$, and because of the periodicity of the distribution of the vector $[X_{t_1}, \ldots, X_{t_n}]^T$,

$$\mathbb{E}[\mathbb{E}[\mathbb{1}_{A_c}|\phi]] = \frac{1}{T} \int_0^T P[(X_{t_1+c-\phi} \leq y_1, \ldots, X_{t_n+c-\phi} \leq y_n)|\phi = \varphi] d\varphi$$

$$= \frac{1}{T} \int_0^T F_X(y_1, \ldots, y_n; t_1 + c - \varphi, \ldots, t_n + c - \varphi) d\varphi$$

$$= \frac{1}{T} \int_0^T F_X(y_1, \ldots, y_n; t_1 - \tau, \ldots, t_n - \tau) d\tau. \quad (6.43)$$

Hence, $P(A_c)$ does not depend on the value of c and

$$P(A_c) = F_Y(y_1, \ldots, y_n, t_1, \ldots, t_n). \quad (6.44)$$

Relation (6.40) is therefore satisfied and Y is stationary in the strict sense.

When X is second order cyclostationary, we similarly show that Y is second order stationary, and we obtain expressions (6.41) by using the relations

$$\mathbb{E}[Y_t] = \mathbb{E}[\mathbb{E}[X_{t-\phi}|\phi]] \quad (6.45)$$

and $\mathbb{E}[Y_{t+\tau}Y_\tau^*] = \mathbb{E}[\mathbb{E}[(X_{t+\tau-\phi}X_{\tau-\phi}^*)|\phi]].\ \square$

For more details about cyclostationary processes, see, for instance, [46]

6.5 Circular Processes

A process X is circular if, for all $\theta \in \mathbb{R}$, the process Z defined by $Z_n = X_n e^{i\theta}$ has the same distribution as X. We say that X is circular up to the p-th order if only the moments of Z up to order p are identical to the corresponding moments of X. We then show that:

Theorem 6.4 *A complex random process X is circular up to the p-th order if and only if all its moments*

$$\mathbb{E}[\prod_{\sum a_m = a} X_m^{a_m} \prod_{\sum b_n = b} (X_n^*)^{b_n}] \tag{6.46}$$

are zero when $a \neq b$ and $a + b \leq p$.

Proof We write $Z = X e^{i\theta}$. The hypothesis of p-th order circularity is characterised by the relations

$$\mathbb{E}[\prod_{\sum a_m = a} Z_m^{a_m} \prod_{\sum b_n = b} (Z_n^*)^{b_n}]$$

$$= e^{i(a-b)\theta} \mathbb{E}[\prod_{\sum a_m = a} X_m^{a_m} \prod_{\sum b_n = b} (X_n^*)^{b_n}] \tag{6.47}$$

$$= \mathbb{E}[\prod_{\sum a_m = a} X_m^{a_m} \prod_{\sum b_n = b} (X_n^*)^{b_n}],$$

for $a + b \leq p$. Relations (6.47) are satisfied for any value of θ if and only if

$$\mathbb{E}[\prod_{\sum a_m = a} X_m^{a_m} \prod_{\sum b_n = b} (X_n^*)^{b_n}] = 0 \tag{6.48}$$

when $a \neq b$ and $a + b \leq p$. $\square$

We also indicate the following result, for second order circular random variables:

Theorem 6.5 *The complex random variable $X = X_1 + iX_2$, where X_1 and X_2 are random variables with real values, is second order circular if and only if $\mathbb{E}[X] = 0$, $\| X_1 \| = \| X_2 \|$ and $\mathbb{E}[X_1 X_2] = 0$.*

Proof From Theorem 6.4, X is second order circular if and only if $\mathbb{E}[X] = 0$, and $\mathbb{E}[X^2] = 0$. The second relation is expressed by

$$\mathbb{E}[X^2] = (\| X_1 \|^2 - \| X_2 \|^2) + 2i\mathbb{E}[X_1 X_2] = 0, \tag{6.49}$$

that is, $\| X_1 \| = \| X_2 \|$ and $\mathbb{E}[X_1 X_2] = 0$. $\square$

We note a direct consequence of this result: if $X = X_1 + iX_2$ is a Gaussian circular random variable of variance σ_X^2, we then easily verify that its probability density may be expressed in the form of a function of the complex vector variable $x = x_1 + ix_2$:

$$f_X(x) = \frac{1}{\pi \sigma_X^2} \exp(\frac{-|x|^2}{\sigma_X^2}).$$
(6.50)

Concerning Gaussian circular vectors $X = [X_1, \ldots, X_n]^T = X_r + iX_i$, we easily check that they are characterised by the relations $\mathbb{E}[X] = 0$, $\mathbb{E}[X_r X_r^T] = \mathbb{E}[X_i X_i^T]$, and $\mathbb{E}[X_r X_i^T] = -\mathbb{E}[X_i X_r^T]$. It is then easy to establish (see Exercice 6.1) that the probability density of X may be expressed as a function of the complex vector variable $x = x_r + ix_i$:

$$f_X(x) = \frac{1}{\pi^N |C_X|} \exp(-x^H C_X^{-1} x),$$
(6.51)

where n is the size of X and $C_X = 2\mathbb{E}[X_r X_r^T]$.

The Fourier transform of the processes satisfies the following circularity property, presented here for a process indexed by $\mathbb{Z}$:

Theorem 6.6 *The Fourier transform of a WSS zero mean process* $X = (X_n)_{n \in \mathbb{Z}}$, *with discrete index and defined by*

$$Y_f = \frac{1}{\sqrt{n}} \sum_{l=1,n} X_l e^{-2i\pi fl/n},$$
(6.52)

converges towards a second order circular random process $(Y_f)_{f \in \mathcal{I}}$.

Proof The proof can be derived directly from Theorem 6.5, and the convergence properties of the Fourier transform, presented in the chapter relating to non-parametric spectral estimation. $\square$

Here are some other properties of circular complex vectors:

Theorem 6.7 *Circularity is preserved by linear transforms. The linear combination of circular independent vectors is another circular vector. If X is a circular vector, its characteristic function $\Phi_X(u)$ is a function of $|u|$.*

Proof The first property is immediate and the second one follows from Theorem 6.4. Finally, to establish the third property, we let $X = X_1 + iX_2 = |X|(\cos \alpha + i \sin \alpha)$, and $u = u_1 + iu_2 = |u|(\cos \theta + i \sin \theta)$. By taking into account the independence of the distribution of X with respect to the value of α, the characteristic function of X at point u is given by

$$\Phi_X(u) = \mathbb{E}[e^{i(u_1 X_1 + u_2 X_2)}]$$

$$= \mathbb{E}[e^{i|u||X| \cos(\theta - \alpha)}]$$
(6.53)

$$= \mathbb{E}[e^{i|u||X|}]. \square$$

6.6 Multivariate Processes

Multivariate processes are processes with values in a product space, for instance the space $\mathbb{C}^k$. All the results concerning the spectral properties of WSS scalar processes presented in this book can be generalised for the vector case (see, for instance, [15],[19]).

In signal processing the use of multivariate processes can be useful, in particular, to describe a signal observed simultaneously on several sensors. Moreover, the order k linear differential equations, or difference equations in the discrete case, which bring into play scalar processes, may be studied as first order linear differential equations involving multivariate processes of dimension k.

In this section, we consider multivariate processes indexed by $\mathbb{Z}$. Note that similar results could be obtained by dealing with processes indexed by $\mathbb{R}$. Let $X = (X_n)_{n \in \mathbb{Z}}$ be a multivariate process of size p, that is, a process such that X_n is a complex random vector of size p for any n. We assume that X is a WSS process, that is, its mean is of the form $m_X = \mathbb{E}[X_m]$ and its autocovariance of the form $R_X(n) = \mathbb{E}[X_{m+n} X_m^H] - m_X m_X^H$ ($\forall m \in \mathbb{Z}$).

Also, we assume that X is a second order process, that is $\| R_X(0) \| < \infty$, where the matrix norm is defined by $\| M \| = tr(MM^H)^{1/2}$. Note that it is equivalent to assume that the scalar components of X are second order processes.

Bochner's theorem First, let us define positive Hermitian measures. μ is said to be a positive Hermitian measure of size $p \times p$ defined on $\mathcal{I}$ if, for any $\Delta \in \mathcal{B}(\mathcal{I})$, $\mu(\Delta)$ is a $p \times p$ positive Hermitian matrix. In order to show that WSS multivariate processes have spectra, just like WSS scalar processes, we recall the following result that generalises Bochner's theorem [12]: a sequence of matrices $(R(n))_{n \in \mathbb{Z}}$, with $R(-n) = R(n)^H$, represents the Fourier coefficients of a positive Hermitian measure μ, that is,

$$R(n) = \int_{\mathcal{I}} e^{2i\pi nf} d\mu(f), \qquad (6.54)$$

if and only if for any $k \in \mathbb{N}$ the size $(k+1)p \times (k+1)p$ block Toeplitz Hermitian matrix denoted by T_k, with block (i, j) equal to $R(i - j)$, is positive. Then, μ is unique. From this, we see readily that any multivariate WSS process $X = (X_n)_{n \in \mathbb{Z}}$ has a matrix spectral measure μ_X. When the scalar measures $[d\mu_X(f)]_{ij}$ are absolutely continuous with respect to the Lebesgue's measure, $d\mu_X(f)$ can be written as $d\mu_X(f) = S_X(f)df$, where $S_X(f)$ is called the power spectrum density matrix of X.

Filtering Assume now that $h(z)$ represents a matrix transfer function with p inputs and q outputs, that is $[h(z)]_{ij}$ ($1 \leq i \leq q, 1 \leq j \leq p$) is a scalar transfer function. In other words, $h(z) = \sum_{k \in \mathbb{Z}} h_k z^{-k}$, where h_k are matrices of size $q \times p$. We say that $h(z)$ is an admissible filter for X if the process $Y = (Y_n)_{n \in \mathbb{Z}}$, defined by

$$Y_n = [h(z)]Y_n = \sum_{k \in \mathbb{Z}} h_k Y_{n-k}, \qquad (6.55)$$

is a second order process. This means that we must have $\| R_Y(0) \| < \infty$, which amounts to $\| \sum_{k,l \in \mathbb{Z}} h_k R_X(l-k) h_l^H \| < \infty$ or equivalently

$$\| \int_I h(e^{2i\pi f}) d\mu_X(f) h(e^{2i\pi f})^H \, df \| < \infty. \qquad (6.56)$$

Clearly, the spectral matrix of Y is

$$d\mu_Y(h) = h(e^{2i\pi f}) d\mu_X(f) h(e^{2i\pi f})^H. \qquad (6.57)$$

Exercises

6.1 Let $X = (X_t)_{t \in \mathbb{R}}$ denote a complex valued Gaussian process, and note $\mathbf{X} = [X_{t_1}, \dots, X_{t_n}]^T$.

a) Show that the probability density function of $\mathbf{X}$ can be written as a function of the complex vector $\mathbf{x} \in \mathbb{C}^n$:

$$f_\mathbf{X}(\mathbf{x}) = \frac{1}{\pi^n \sqrt{|C|}} \exp(-\frac{1}{2}[\mathbf{x}^H \mathbf{x}^T] C^{-1} [\mathbf{x}^T \mathbf{x}^H]^T), \qquad (6.58)$$

where $C = \mathbb{E}\big[[\mathbf{X}^T \mathbf{X}^H]^T [\mathbf{X}^H \mathbf{X}^T]\big]$.

b) When X is a circular process, check that we obtain formula (6.51).

6.2 Let $X = (X_t)_{t \in \mathbb{R}}$ be a zero mean WSS process and define

$$\hat{R}_{X,T}(t) = (2T)^{-1} \int_{-T+|t|/2}^{T-|t|/2} X_{u+t/2} X_{u-t/2} du. \qquad (6.59)$$

a) Calculate $\mathbb{E}[\hat{R}_{X,T}(t)]$.

b) If X is Gaussian, show that the variance of $\hat{R}_{X,T}(t)$ is

$$\frac{1}{2T} \int_{-T+|t|/2}^{T-|t|/2} \left(1 - \frac{|t|+|u|}{2T} \right) (R_X^2(u) + R_X(u+t) R_X(u-t)) du. \qquad (6.60)$$

6.3 (Processes with orthogonal increments) We consider $X = (X_t)_{t \in \mathbb{R}}$, a zero mean process of $L^2(\Omega, \mathcal{A}, P)$ with uncorrelated increments, that is $\mathbb{E}[(X_{t_4} - X_{t_3})(X_{t_2} - X_{t_1})] = 0$ for $t_1 \le t_2 < t_3 \le t_4$. We also assume that $\| X_{t_2} - X_{t_1} \|^2 = \sigma^2 |t_2 - t_1|$. Show that the random variables $X_t - X_s$ are Gaussian.

(Hint: consider the characteristic function of

$$X_t - X_s = \sum_{k=0, N-1} (X_{s+(k+1)(t-s)/N} - X_{s+k(t-s)/N}), \qquad (6.61)$$

and let N tend to $+\infty$.)

6.4 (Poisson white noise) Let $N = (N_t)_{t \in \mathbb{R}_+}$ represent a Poisson process: $P(N_t = n) = e^{-\lambda t}(\lambda t)^n/n!$. Let $T = (T_n)_{n \in \mathbb{N}^*}$ denote the corresponding instants of the events counted by N. We consider N, the function defined on $\mathcal{B}(\mathbb{R})$ by $N(A) = \sum_{n \in \mathbb{N}^*} \mathbb{I}_A(T_n)$.

a) Check that N is a generalised stochastic measure (see Exercise 4.4) and that for two functions g and h in $L^1(\mathbb{R}, \mathcal{B}(\mathbb{R}), dt) \cap L^2(\mathbb{R}, \mathcal{B}(\mathbb{R}), dt)$, we have

$$\mathbb{E}[(\int_{\mathbb{R}} g(t) dN(t))(\int_{\mathbb{R}} h(t) dN(t))^*] = \lambda \int_{\mathbb{R}} g(t) h^*(t) dt$$

$$+ (\lambda \int_{\mathbb{R}} g(t) dt)(\lambda \int_{\mathbb{R}} h(t) dt)^*. \tag{6.62}$$

b) We assume that N represents the number of particles arriving at a sensor. The nth particle generates an impulse of amplitude ξ_n, where ξ_n are independent random variables with the same distribution. Since the sensor is not perfect, the impulses are filtered by it. Let $h(t)$ denote the corresponding impulse response. Finally, the observed process is of the form

$$Y_t = \sum_{n \in \mathbb{N}} \xi_n h(t - T_n). \tag{6.63}$$

Express Y_t in an integral form and calculate the autocovariance and the spectrum of Y.

6.5 Using Kolmogorov's theorem, show that the definition of the distribution of a Brownian motion presented in Section 6.3.2 is consistent with Kolmogorov's theorem.

6.6 Let W denote a Brownian motion.
a) Calculate the autocovariance of W.
b) Calculate the Karhunen-Loeve expansion of W (see Exercise 3.2).

6.7 (Brownian increments) We consider $X_t = h^{-1}(W_{t+h} - W_t)$, where $W = (W_t)_{t \in \mathbb{R}}$ is a Brownian motion. Show that

$$R_X(t) = h^{-1}(1 - h^{-1}|t|) \mathbb{I}_{[-h,h]}(t), \tag{6.64}$$

and that

$$S_X(f) = \left(\frac{\sin(\pi h f)}{\pi h f}\right)^2. \tag{6.65}$$

6.8 Let $W = (W_t)_{t \in \mathbb{R}_+}$ denote a Brownian motion, with $W_t \sim \mathcal{N}(0, t)$. Letting $t = t_0^{(n)} < \ldots < t_n^{(n)} = t + T$, we define $\Delta W_k^{(n)} = W_{t_k^{(n)}} - W_{t_{k-1}^{(n)}}$, $S_n = \sum_{k=1,n}(\Delta W_k^{(n)})^2$ and $M_n = \max_{k=1,N} t_k^{(n)} - t_{k-1}^{(n)}$. Show that

$$\lim_{M_n \to 0} S_n \stackrel{m.q.}{=} T. \tag{6.66}$$

6.9 (White noise and martingales) A stochastic process $M = (M_t)_{t \in \mathbb{R}_+}$ is called a martingale with respect to a non-decreasing sequence of σ-algebra $\mathcal{M} = (\mathcal{M}_t)_{t \in \mathbb{R}_+}$, that is σ-algebra such that $\mathcal{M}_s \subset \mathcal{M}_t$ for $s \leq t$, if $\forall t$, M_t is $\mathcal{M}_t$-measurable, $\mathbb{E}[|M_t|] < \infty$, and $\forall s \leq t$, $\mathbb{E}[M_t | \mathcal{M}_s] = M_s$. Prove that a Brownian motion is a martingale.

6.10 (Wiener process as a Markov process) Prove that a Wiener process is a Markov process, that is $P(W_t \in A | W_{\tau_1}, \ldots, W_{\tau_n}) = P(W_t \in A | W_{\tau_n})$, for any $n \in \mathbb{N}^*$ and any $0 \leq \tau_1 \leq \ldots \leq \tau_n \leq t$.

6.11 (Spectra of some digital codes) The PSD of a cyclostationary process $X = (X_t)_{t \in \mathbb{R}}$ with period T is defined as the PSD of the stationary process $Y = (X_{t+\phi})_{t \in \mathbb{R}}$, where ϕ is a random variable with uniform distribution on $[0, T[$.

Let $A = (A_n)_{n \in \mathbb{Z}}$ be a stationary sequence of real-valued random symbols and $h(t)$ a known waveform, with $h(t) = 0$ for $t \notin [0, T[$. $X_t = \sum_{k \in \mathbb{Z}} A_k h(t - kT)$ represents the signal used to transmit the symbols A_n.

a) Prove that the spectrum of X is given by

$$d\mu_X(f) = \left[T^{-1} \sum_{k \in \mathbb{Z}} R_A(k) e^{-2\pi k f T} \right] |H(f)|^2 df$$

$$+ T^{-2} |\mathbb{E}[A_n]|^2 \sum_{k=-\infty,\infty} |H(\tfrac{k}{T})|^2 \delta_{k/T}. \tag{6.67}$$

(Remark: since $\mathbb{E}[A_n]$ may not be equal to zero, it has to be accounted for in the calculation of the spectrum.)

b) Calculate the PSD for the following codes

- Non-Return to Zero (NRZ) code: the random variables A_k $(k \in \mathbb{Z})$ are independent, $P(A_k = \pm 1) = 1/2$ and $h(t) = \mathbb{1}_{[0,T[}(t)$;
- M-ary code: A_k are independent,

$$P(A_k = \pm 1) = P(A_k = \pm 3) = \ldots = P(A_k = \pm M - 1) = 1/M \quad (6.68)$$

and $h(t) = \mathbb{1}_{[0,T[}(t)$;
- Return to Zero (RZ) code: the random variables A_k are independent, $P(A_k = \pm 1) = 1/2$ and $h(t) = \mathbb{1}_{[0,\lambda T[}(t)$ $(\lambda \in]0, T[)$;
- biphase code: the random variables A_k are independent, $P(A_k = \pm 1) = 1/2$ and $h(t) = \mathbb{1}_{[0,T/2[}(t) - \mathbb{1}_{[T/2,T[}(t)$;
- Alternate Marked Inversion (AMI): the random variables A_k are dependent: if a sequence of independent 0 and 1 symbols with the same probability has to be transmitted, 0 symbols are encoded by $A_k = 0$, and 1 symbols are encoded by $A_k = 1$ and $A_k = -1$ alternately. $h(t) = \mathbb{1}_{[0,T/2[}(t) - \mathbb{1}_{[T/2,T[}(t)$.

c) What is the practical interest of these codes?

6.12 (Cyclic autocovariance and spectral measure) Let $X = (X_n)_{n \in \mathbb{Z}}$ represent a zero mean cyclostationary process of period T. Show that $\forall m, n \in \mathbb{Z}$ $R_X(m+n, m)$ can be written in the form

$$R_X(m+n, m) = \sum_{k=0, T-1} R_k(n) e^{2i\pi km/T}, \tag{6.69}$$

and express $R_k(n)$ in terms of R_X. The T sequences $(R_k(n))_{n \in \mathbb{Z}}$, $k = 0, T-1$, are called *cyclocovariance sequences*. What does the sequence $(R_0(n))_{n \in \mathbb{Z}}$ correspond to?

6.13 Let $X = (X_n)_{n \in \mathbb{Z}}$ denote a WSS process with PSD $S_X(f)$, and $Y_n = [h(z)]X_n$. Calculate the autocovariance matrix function and the spectrum of the vector process Z defined by $Z_n = [X_n, Y_n]^T$.

6.14 Let $X = (X_n)_{n \in \mathbb{Z}}$ denote a multivariate process of size p, and $v = [v_0^T, \dots, v_k^T]^T$ any vector of size $(k+1)p$. We note $v(z)$ the polynomial vector transfer function of size p defined by $v(z) = \sum_{l=0,k} v_k z^k$. Show that if all the block Toeplitz matrices $T_{X,k}$ of size $(k+1)p \times (k+1)p$ with block (i, j) equal to $R_X(i-j)$ are strictly positive (that is all their eigenvalues are greater than 0), then $\| [v(z)]^H X_n \| = 0$ if and only if $v = 0$.

7. Non-linear Transforms of Processes

Purpose The transforms undergone by the autocovariance functions of processes in non-linear systems cannot be expressed using formulas as simple as those obtained in the context of filtering. In fact, studying these relations must generally be done case by case. Here, we present some simple examples of memoryless non-linear transforms that we encounter commonly in the fields of electronics and telecommunications.

7.1 Square Law Detector and Hard Limiter

In this paragraph, we denote by X a zero mean, real-valued, WSS process indexed by $\mathbb{R}$.

7.1.1 Square Law Detector

A square law detector can be used, for example, to measure the power of received signals. The output of a square law detector with input signal X_t is of the form $Y_t = X_t^2$. If $f_{X_t}(x)$ is the probability density function of X_t, then clearly that of Y_t is given by

$$f_{Y_t}(y) = \frac{1}{2\sqrt{y}}(f_{X_t}(\sqrt{y}) + f_{X_t}(-\sqrt{y})). \tag{7.1}$$

If, moreover, X is Gaussian and zero mean, it appears that

$$R_Y(t) = \mathbb{E}[X_{\tau+t}^2 X_\tau^2] - \mathbb{E}[X_{\tau+t}^2]\mathbb{E}[X_\tau^2]$$

$$= 2R_X^2(t), \tag{7.2}$$

because, for Gaussian real random variables X_1, X_2, X_3, X_4 (see Chapter 14 dealing with higher order statistics),

$$\mathbb{E}[X_1 X_2 X_3 X_4] = \mathbb{E}[X_1 X_2]\mathbb{E}[X_3 X_4] + \mathbb{E}[X_1 X_3]\mathbb{E}[X_2 X_4]$$

$$+ \mathbb{E}[X_1 X_4]\mathbb{E}[X_2 X_3]. \tag{7.3}$$

We now remark that in the case of a real-valued, zero mean, WSS Gaussian process indexed by $\mathbb{R}$, Slutsky's theorem (Theorem 2.6) is expressed by

$$\lim_{T \to \infty} \frac{1}{2T} \int_{[-T,T]} R_X(t) dt = 0, \tag{7.4}$$

and $\forall l \in \mathbb{R}$,

$$\lim_{T \to \infty} \frac{1}{2T} \int_{[-T,T]} \mathbb{E}[X_t X_{t+l} X_l X_0] dt - R_X(l)^2$$

$$= \lim_{T \to \infty} \frac{1}{2T} \int_{[-T,T]} (R_X(t+l) R_X(t-l) + R_X(t)^2) dt \tag{7.5}$$

$$= 0.$$

From Cauchy-Schwarz's inequality, this latter condition is satisfied when $\lim_{T \to \infty} \frac{1}{2T} \int_{[-T,T]} R_X^2(t) dt = 0$.

Under these ergodicity conditions, the estimator

$$U_T = \frac{1}{2T} \int_{[-T,T]} X_t^2 dt \tag{7.6}$$

of the power of the process converges in the mean square sense towards $R_X(0)$ when T tends towards $+\infty$.

7.1.2 Hard Limiter

We now consider the hard limiter whose output is given by $Y_t = sign(X_t)$. Typically, it represents the output of an operational amplifier without feedback.

It is clear that $P(Y_t = 1) = P(X_t > 0) = 1 - P(Y_t = -1)$ and that, if X is zero mean,

$$R_Y(t) = P(X_{t+\tau} X_\tau > 0) - P(X_{t+\tau} X_\tau < 0). \tag{7.7}$$

Theorem 7.1 *If X is a WSS Gaussian process, the autocovariance function of the output of the hard limiter is given by*

$$R_Y(t) = \frac{2}{\pi} \arcsin\left(\frac{R_X(t)}{R_X(0)}\right). \tag{7.8}$$

Proof We begin by computing $P(X_{t+\tau} X_\tau > 0)$. The random vector $[X_{t+\tau}, X_\tau]^T$ is distributed according to a zero mean Gaussian distribution with covariance matrix denoted by Σ, with coefficients $\Sigma_{11} = \Sigma_{22} = R_X(0)$ and $\Sigma_{12} = \Sigma_{21} = R_X(t)$. We can easily transform this vector into a Gaussian random vector, denoted by $[Y_1, Y_2]^T$, with zero mean and identity covariance matrix, with the transform

$$X_\tau = \sqrt{\frac{R_X(0)}{2}}(Y_1\sqrt{1+\rho} + Y_2\sqrt{1-\rho}),$$

$$X_{t+\tau} = \sqrt{\frac{R_X(0)}{2}}(Y_1\sqrt{1+\rho} - Y_2\sqrt{1-\rho}),$$

(7.9)

where $\rho = R_X(t)R_X(0)^{-1}$. The condition $X_{t+\tau}X_\tau > 0$ is also expressed by $Y_1^2(1+\rho) - Y_2^2(1-\rho) > 0$. We denote by $f_{Y_1}(y_1)$ and $f_{Y_2}(y_2)$ the probability density functions of Y_1 and of Y_2. Considering the change of the variables $Y_1 = r\cos\theta$, and $Y_2 = r\sin\theta$, we obtain

$$P(X_{t+\tau}X_\tau > 0) = \int_{Y_1^2(1+\rho)-Y_2^2(1-\rho)>0} f_{Y_1}(y_1)f_{Y_2}(y_2)dy_1\,dy_2$$

$$= \int_{\{\tan^2(\theta)<\frac{1+\rho}{1-\rho};\theta\in[-\pi,\pi]\}} \frac{d\theta}{2\pi}$$

$$= \frac{2}{\pi}\int_{\{\cos(2\theta)>-\rho;\theta\in[0,\pi/2]\}} d\theta,$$

(7.10)

$$= \frac{1}{\pi}\arccos(-\rho)$$

$$= \frac{1}{2} + \frac{1}{\pi}\arcsin(\rho),$$

for

$$\tan^2(\theta) < \frac{1+\rho}{1-\rho} \Leftrightarrow (1-\rho)\sin^2\theta < (1+\rho)\cos^2\theta$$

(7.11)

$$\Leftrightarrow \sin^2\theta - \cos^2\theta = -\cos(2\theta) < \rho$$

and $\arccos x + \arcsin x = \pi/2$. Finally, since $R_Y(t) = 2P(X_{t+\tau}X_\tau > 0) - 1$,

$$R_Y(t) = \frac{2}{\pi}\arcsin\left(\frac{R_X(t)}{R_X(0)}\right). \quad \Box$$

(7.12)

7.1.3 Bussgang's Theorem

The following result, valid for memoryless transforms, may be useful to simplify the practical calculation of the autocovariance of a process.

Theorem 7.2 (Bussgang) *Let X be a zero mean, WSS Gaussian process, and g a memoryless transform. Then, $R_X(t)$ is proportional to $R_{XY}(t)$, where $Y_t = g(X_t)$.*

Proof

$$R_{XY}(t) = \mathbb{E}[X_{t+\tau} g(X_\tau)]$$

$$= \mathbb{E}[\mathbb{E}[X_{t+\tau} g(X_\tau)|X_\tau]] \tag{7.13}$$

$$= \mathbb{E}[g(X_\tau)\mathbb{E}[X_{t+\tau}|X_\tau]].$$

But, as X is Gaussian and zero mean,

$$\mathbb{E}[X_{t+\tau}|X_\tau] = \mathbb{E}[X_{t+\tau} X_\tau](\mathbb{E}[X_\tau X_\tau])^{-1} X_\tau$$

$$\tag{7.14}$$

$$= R_X(t) R_X(0)^{-1} X_\tau.$$

Hence, $R_{XY}(t) = R_X(0)^{-1} R_{g(X),X}(0) R_X(t)$, which proves the result. $\square$

Thus, by taking again, for example, the case of the hard limiter, that is, for $g(x) = sign(x)$,

$$R_X(t) = \frac{R_X(0)}{\mathbb{E}[|X_t|]} \mathbb{E}[X_{t+\tau} sign(X_\tau)]. \tag{7.15}$$

This formula can be used in practice to estimate $R_X(t)$ with a low computational cost.

7.2 Amplitude Modulation

We consider a process Y, obtained by modulating the amplitude of a sinusoidal signal with a zero mean WSS process X: $Y_t = X_t \times \cos(2\pi f_0 t)$. Then,

$$Y_t = \int_{\mathbb{R}} \frac{1}{2}[e^{2i\pi(f-f_0)t} + e^{2i\pi(f+f_0)t}]d\hat{X}(f)$$

$$\tag{7.16}$$

$$= \int_{\mathbb{R}} e^{2i\pi ft}[\frac{d\hat{X}(f+f_0) + d\hat{X}(f-f_0)}{2}]$$

and $d\hat{Y}(f) = (1/2)[d\hat{X}(f+f_0) + d\hat{X}(f-f_0)]$. We assume that the support of μ_X lies in an interval $[-f_c, f_c]$ such that $f_0 > f_c$. The supports of $d\hat{X}(f+f_0)$ and of $d\hat{X}(f-f_0)$ are then non-overlapping and

$$d\mu_Y(f) = \frac{1}{4}[d\mu_X(f+f_0) + d\mu_X(f-f_0)]. \tag{7.17}$$

Up to one factor, the spectrum of Y is therefore obtained by a translation of the spectrum of X around the frequencies $-f_0$ and $+f_0$.

7.2.1 Phase and Quadrature Modulation

If $X_{1,t}$ and $X_{2,t}$ are real-valued, zero mean, WSS processes, we can define the phase and quadrature modulation of these processes by

$$Y_{1,t} = X_{1,t} \cos(2\pi f_0 t) - X_{2,t} \sin(2\pi f_0 t). \qquad (7.18)$$

Y_1 is zero mean, and

$$\mathbb{E}[Y_{t+\tau} Y_\tau] = \tfrac{1}{2}[R_{X_1}(t) + R_{X_2}(t)] \cos(2\pi f_0 t)$$

$$+ \tfrac{1}{2}[R_{X_1}(t) - R_{X_2}(t)] \cos(4\pi f_0 \tau + 2\pi f_0 t) \qquad (7.19)$$

$$- R_{X_1 X_2}(t) \sin(4\pi f_0 \tau + 2\pi f_0 t).$$

Hence, Y_1 is WSS if and only if $R_{X_1}(t) = R_{X_2}(t)$ and $R_{X_1 X_2}(t) = 0$. In what follows, we shall assume that this condition is satisfied.

Phase and quadrature modulation enables two signals to be transmitted simultaneously in the same frequency band. The two signals thus modulated may then be separated simply. We assume that the processes X_1 and X_2 have spectra carried by $[-f_c, f_c]$. In this case, the spectrum of Y is carried by

$$[-f_c - f_0, f_c - f_0] \cup [-f_c + f_0, f_c + f_0]. \qquad (7.20)$$

In practice, the frequencies f_0 and f_c are generally known and $f_0 \gg f_c$. X_1 and X_2 can then be recovered from Y by the transforms

$$X_{1,t} = 2F_{f_c}[\cos(2\pi f_0 t) Y_t]$$

$$\qquad (7.21)$$

$$X_{2,t} = 2F_{f_c}[-\sin(2\pi f_0 t) Y_t],$$

where F_{f_c} represents the filtering operation with frequency response $\mathbb{I}_{[-f_c, f_c]}$ (f). Transform (7.21) is known as phase and quadrature demodulation.

In addition, we note that we can introduce a complex representation of the phase and quadrature modulated signals by noting

$$Y_2(t) = X_{1,t} \sin(2\pi f_0 t) + X_{2,t} \cos(2\pi f_0 t),$$

$$\qquad (7.22)$$

$$\text{and} \quad Z_t \quad = Y_{1,t} + iY_{2,t} = (X_{1,t} + iX_{2,t}) e^{2i\pi f_0 t}.$$

7.2.2 Analytic Representation and SSB Modulation

We now introduce the notion of an analytic process associated with a process. Hilbert's filter is defined by the frequency response $H(f) = -i.sign(f)$. Hilbert's transform of the process X is therefore given by

$$\tilde{X}_t = \int_{\mathbb{R}} -i.sign(f) e^{2i\pi f t} d\hat{X}(f). \qquad (7.23)$$

The stochastic measure of $\tilde{X}$ is given by $-i.sign(f)d\hat{X}(f)$. As X is real-valued, it is clear that $d\hat{X}(f) = d\hat{X}(-f)^*$. This property is again satisfied by the stochastic measure of $\tilde{X}$; $\tilde{X}$ is therefore also a real-valued process, for

$$\tilde{X}_t = 2\mathcal{R}e[\int_{\mathbb{R}_+} e^{2i\pi ft}d\hat{\tilde{X}}(f)]. \tag{7.24}$$

The process defined by $Z_t = X_t + i\tilde{X}_t$ then has the remarkable property of having a spectral support contained in $\mathbb{R}_+$:

$$d\mu_Z(f) = 4 \times \mathbb{1}_{\mathbb{R}_+}(f)d\mu_X(f), \tag{7.25}$$

for $d\hat{Z}(f) = d\hat{X}(f) + id\hat{\tilde{X}}(f) = [1 + sign(f)]d\hat{X}(f)$.

In addition, X can easily be obtained from Z: $Z_t = X_t + i\tilde{X}_t$ and X_t and $\tilde{X}_t$ are real, therefore $X_t = (Z_t + Z_t^*)/2$. The process $(1/2)Z_t$ is called the *analytical* part of X: $(1/2)d\hat{Z}(f) = \mathbb{1}_{\mathbb{R}_+}(f)d\hat{X}(f)$.

If the support of $S_X(f)$ is bounded, it appears that $S_Z(f)$ occupies a support twice as small. This property is exploited in transmissions in order to limit the spectral bandwidth of transmitted signals. Indeed, we sometimes transmit the signal $\mathcal{R}e[Z_t\cos(2\pi f_0 t)]$ rather than $X_t\cos(2\pi f_0 t)$. This amounts to performing the phase and quadrature modulation of $(X_{1,t}, X_{2,t}) = (X_t, \tilde{X}_t)$. This technique is known as Single Side Band (SSB) modulation.

7.2.3 Rice's Representation

We may wonder whether, for a given real stationary process Y_t, we can obtain a representation of type (7.18). For a fixed f_0, we can put $W_t = Y_t e^{-2i\pi f_0 t} = X_{1,t} + iX_{2,t}$. We then have

$$Y_t = \mathcal{R}e[W_t e^{2i\pi f_0 t}]$$
$$= X_{1,t}\cos(2\pi f_0 t) - X_{2,t}\sin(2\pi f_0 t). \tag{7.26}$$

The desired representation therefore exists, but it is not unique. For example, for any fixed value of ϕ, we can define $X_{1,t} + iX_{2,t} = Y_t e^{-2i\pi f_0 t + i\phi}$. We now consider the particular representation defined by

$$Z_t = X_{1,t} + iX_{2,t} = (Y_t + i\tilde{Y}_t)e^{-2i\pi f_0 t}, \tag{7.27}$$

where $\tilde{Y}$ is the Hilbert transform of Y. The corresponding representation (7.18) is called Rice's representation. The process Z_t is called the complex envelope of Y. Its stochastic measure and spectral measure are given by

$$d\hat{Z}(f) = 2 \times \mathbb{1}_{\mathbb{R}_+}(f + f_0)d\hat{Y}(f + f_0)$$
$$\text{and } d\mu_Z(f) = 4 \times \mathbb{1}_{\mathbb{R}_+}(f + f_0)d\mu_Y(f + f_0). \tag{7.28}$$

When f_0 does not represent a known frequency, we often consider the representation obtained for the value of f_0 that minimises $\int_{\mathbb{R}}(f-f_0)^2 d\mu_Y(f)$, that is,

$$f_0 = \frac{1}{\int_{\mathbb{R}} d\mu_Y(f)} \int_{\mathbb{R}} f d\mu_Y(f). \tag{7.29}$$

Now we return to phase and quadrature demodulation. The demodulation of the process Y, whose spectrum is carried by $[-f_c - f_0, f_c - f_0] \cup [-f_c + f_0, f_c + f_0]$, yields Rice's representation of Y. Indeed, the stochastic measure of $X_1 + iX_2$, where X_1 and X_2 are given by (7.21) is of the form

$$d\hat{X}_1(f) + id\hat{X}_2(f) = 2 \times \mathbb{I}_{[-f_c, f_c]}(f)d\hat{Y}(f + f_0)$$

$$= 2 \times \mathbb{I}_{\mathbb{R}^+}(f + f_0)d\hat{Y}(f + f_0). \tag{7.30}$$

Hence, from (7.27) and (7.28), $X_{1,t} + iX_{2,t} = e^{-2i\pi f_0 t}(Y_t + i\tilde{Y}_t)$. It will be of interest to graphically represent the spectral supports of the processes and filters brought into play for a better understanding of the relations studied in this paragraph.

7.2.4 Demodulation in the Presence of Noise

In a practical context, it is often necessary to study how the demodulation operation behaves with respect to the presence of an additive white noise B, with constant power spectral density equal to σ^2. We shall denote by $B_{1,t} = 2F_{f_c}[\cos(2\pi f_0 t)B_t]$ and $B_{2,t} = 2F_{f_c}[-\sin(2\pi f_0 t)B_t]$ the demodulated phase and quadrature noise. The demodulation of the process $Y + B$ then leads to $X_1 + B_1$ and $X_2 + B_2$, whose spectral measures are $d\mu_{X_1}(f) + 2 \times \mathbb{I}_{[-f_c, f_c]}(f)\sigma^2 df$ and $d\mu_{X_2}(f) + 2 \times \mathbb{I}_{[-f_c, f_c]}(f)\sigma^2 df$ respectively. We shall note the factor 2 that appears in the expression of the PSD of the demodulated noise.

This factor 2 might lead us to believe that the demodulation operation introduces a degradation of the Signal to Noise Ratio (SNR), that is, of the ratio between the power of the signals of interest and that of the noise. This is not the case. Indeed, a quick calculation shows that for the transmitted signals, the total SNR for the process $Y + B$ in the bandwidth of interest $[-f_c - f_0, f_c - f_0] \cup [-f_c + f_0, f_c + f_0]$ is equal to

$$SNR_{\text{mod}} = \frac{\frac{1}{2}[\int_{\mathbb{R}} d\mu_{X_1}(f) + \int_{\mathbb{R}} d\mu_{X_2}(f)]}{(4f_c)\sigma^2}, \tag{7.31}$$

and that after demodulation, it becomes, in the bandwidth of interest $[-f_c, f_c]$,

$$SNR_{\text{demod}} = \frac{\int_{\mathbb{R}} d\mu_{X_1}(f) + \int_{\mathbb{R}} d\mu_{X_2}(f)}{(2f_c) \times 2\sigma^2 + (2f_c) \times 2\sigma^2}. \tag{7.32}$$

The SNR is therefore not modified by the demodulation operation.
Moreover, we remark that B_1 and B_2 are uncorrelated since

$$d\mu_{B_1,B_2}(f) = 4\mathbb{1}_{[-f_c,f_c]}(f)\mathbb{E}[(\tfrac{d\hat{B}(f-f_0)+d\hat{B}(f+f_0)}{2})(\tfrac{d\hat{B}(f-f_0)-d\hat{B}(f+f_0)}{2i})^*]$$

$$= -i\mathbb{1}_{[-f_c,f_c]}(f)(\sigma^2 df - \sigma^2 df) = 0.$$

$$(7.33)$$

Remark In the case where the non-linear transforms applied to a process are not instantaneous, we often describe them by means of state space models. These models will be introduced in Chapter 9 (see Equations 9.31).

Exercises

7.1 We consider a WSS process $X = (X_t)_{t\in\mathbb{R}}$, and we define $Y = (Y_t)_{t\in\mathbb{R}}$ by $Y_t = X_{h(t)}$. What condition about $h(t)$ ensures that Y is a WSS process?

7.2 We consider a white Gaussian noise $V = (V_n)_{n\in\mathbb{Z}}$. Calculate the spectrum $Y = (Y_n)_{n\in\mathbb{Z}}$, where $Y_n = sign(bV_n + V_{n-1})$.

7.3 Let $X = (X_n)_{n\in\mathbb{Z}}$ denote a complex valued WSS circular Gaussian process with autocovariance function $R_X(n)$. We define $Y = (Y_n)_{n\in\mathbb{Z}}$ by

$$Y_n = \sum_{m=0,p}\left(\sum_{k=1,r} a_{m,k}X_{n-m}^k\right). \qquad (7.34)$$

Calculate the autocovariance function of Y.

7.4 We consider a real-valued Gaussian WSS process $X = (X_t)_{t\in\mathbb{R}}$, and we define $Y = (Y_t)_{t\in\mathbb{R}}$ by $Y_t = \sin(X_t)$. Calculate the mean and autocovariance functions of Y.

7.5 (Square law detector) The signal $\mathbb{1}_{[0,T]}(t)\cos(2\pi ft)$ is transmitted to a receiver over an unknown propagation channel with impulse response $h(t) = \sum_{k=1,p} A_k e^{i\phi_k}\delta_{\tau_k}$.
  a) Give the expression of the signal at the receiver side, denoted by Y_t.
  b) In order to estimate the amplitudes A_k and the time delays τ_k, Y_t goes through a quadratic receiver that performs phase and quadrature demodulation followed by filtering with impulse response $\mathbb{1}_{[0,T]}(t)$ and square summation; the output is

$$Z_t = (\mathbb{1}_{[0,T]}(t) \star (Y_t \times \cos(2\pi ft))^2 + (\mathbb{1}_{[0,T]}(t) \star (Y_t \times \sin(2\pi ft))^2. \qquad (7.35)$$

We assume that $f \gg T^{-1}$ and $T < (1/2)|\tau_k - \tau_l|$ for $k \neq l$. What is the expression of Z_t in terms of the channel parameters? How can the channel parameters $(A_k, \tau_k)_{k=1,p}$ be recovered from Z_t?

7.6 (Phase modulation) Let $X = (X_t)_{t\in\mathbb{R}}$ denote a strict sense stationary, real-valued, process, and $Y = (Y_t)_{t\in\mathbb{R}}$ defined by

$$Y_t = \exp(2i\pi f_0 t + i\alpha X_t + i\phi), \tag{7.36}$$

where α is a constant and ϕ a random variable independent of X and with uniform distribution on $[0, 2\pi]$. Calculate $R_Y(t)$ when X is a Gaussian process with autocovariance function $R_X(t)$.

7.7 (Maximum detection) Let $X = (X_n)_{n\in\mathbb{Z}}$ denote a sequence of zero mean independent random variables with the same distribution, and $Y^{(N)} = (Y_n^{(N)})_{n\in\mathbb{Z}}$ defined by $Y_n^{(N)} = \max\{X_n, X_{n-1}, \ldots, X_{n-N+1}\}$.

a) Calculate the distribution of $Y_n^{(N)}$ in terms of that of X_n.

b) Show that $Y^{(N)}$ is stationary.

7.8 (Soft limiter) Let $X = (X_t)_{t\in\mathbb{R}}$ denote a WSS process and $Z = (Z_t)_{t\in\mathbb{R}}$ defined by

$$Z_t = \max\{-1, X_t\}\mathbb{1}_{\mathbb{R}_-}(X_t) + \min\{1, X_t\}\mathbb{1}_{\mathbb{R}_+}(X_t). \tag{7.37}$$

a) Calculate the distribution of Z_t in terms of that of X_t.

b) Calculate the autocovariance function of Z when X is a harmonic process of the form $X_t = \xi e^{2i\pi ft}$ and when X is a Gaussian process with autocovariance function $R_X(t)$.

7.9 (Doppler effect) We consider a moving source emitting the signal $X_t = e^{2i\pi ft}$ from a point r with speed vector v. The signal travels at speed c to a receiver located at point 0. We shall note $u = \| r \|^{-1} r$.

a) Show that up to one amplitude factor and a propagation time delay the receiver observes the signal $y(t) = e^{2i\pi f(1+u^T v/c)t}$.

b) More generally, if X is a mean square continuous WSS process delivered by the source, show that the signal received at the receiver end is a stationary process Y defined by

$$Y_t = \int_\mathbb{R} e^{2i\pi f(1+u^T v/c)t} d\hat{X}(f). \tag{7.38}$$

c) Calculate $R_Y(t)$ and $d\mu_Y(f)$ in terms of $R_X(t)$ and of $d\mu_X(f)$ respectively.

7.10 (Quantisation) We consider a stationary process $X = (X_n)_{n\in\mathbb{Z}}$. In order to perform the digital processing of X, it is quantised in the following way: we define $Y = (Y_n)_{n\in\mathbb{Z}}$ by

$$\begin{cases} Y_n = m^{-1}(k+1/2) & \text{if } k/m \le X_n \le (k+1)/m \\ & \text{and } k = -N, -N+1, \ldots, N-1, \\ \\ Y_n = m^{-1}(N-1/2) & \text{if } X_n \ge N/m \\ \\ Y_n = m^{-1}(-N+1/2) & \text{if } X_n \le -N/m. \end{cases} \tag{7.39}$$

a) Calculate the quantisation error variance defined by $\| X_n - Y_n \|^2$ when $X_n \sim \mathcal{U}_{[-N/m, N/m]}$.

b) More generally, we define K values denoted by $(X_{q,k})_{k=1,K}$ and $K-1$ thresholds $(S_k)_{k=1,K-1}$ such that $Y_n = X_{q,1}$ if $X_n < S_1$, $Y_n = X_{q,k}$ if $S_k \leq X_n < S_{k+1}$ for $1 < k < K-2$, and $Y_n = X_{q,K}$ if $X_n > S_{K-1}$. Express $\| X_n - Y_n \|^2$ in terms of the $X_{q,K}$, of S_k, and of the distribution of X.

c) Explain how it is possible to search for a (local) minimum for $\| X_n - Y_n \|^2$ by means of an alternate iterative search that involves optimising the choice of the thresholds S_k for fixed values of $X_{q,k}$ and then optimising the choice of the $X_{q,K}$ for fixed values of the thresholds. This technique for designing a quantiser is called the *K-mean algorithm*.

7.11 (Spread spectrum communications) Let $X = (X_t)_{t \in \mathbb{R}}$, with

$$X_t = \sum_{k \in \mathbb{Z}} A_k \mathbb{1}_{[0, T_s[}(t - kT_s + \phi). \tag{7.40}$$

The coefficients A_k are independent random variables with $P(A_k = 1) = P(A_k = -1) = 1/2$ and ϕ is a random variable independent of A_k ($k \in \mathbb{Z}$), with uniform distribution on $[0, T_s]$.

a) Give the expression of the spectrum of X.

b) We now consider a sequence of independent random variables C_m ($m = 0, \ldots, N-1$), with $P(C_m = 1) = P(C_m = -1) = 1/2$, and we note

$$s(t) = \frac{1}{\sqrt{N}} \sum_{m=0, N-1} C_m \mathbb{1}_{[0, T_c[}(t - mT_c), \tag{7.41}$$

with $T_c = T_s/N$. Calculate the spectrum of $Y = (Y_t)_{t \in \mathbb{R}}$ defined by

$$Y_t = \sum_{k \in \mathbb{Z}} A_k \mathbb{1}_{[0, T_s[}(t - kT_s + \phi)s(t - kT_s + \phi), \tag{7.42}$$

and compare it to that of X.

c) If we assume that the receiver knows the sequence $(C_m)_{m=1,N}$ used for spectrum spreading at the transmitter side, show that X can be recovered from Y.
(Hint: consider $Y_t \times (\sum_{k \in \mathbb{Z}} s(t - kT_s + \phi))$.)

d) Let $(C_m^{(1)})_{m=0,N-1}$ and $(C_m^{(2)})_{m=0,N-1}$ denote two sequences and $s^{(1)}(t)$ and $s^{(2)}(t)$ the corresponding signals calculated as in Equation (7.41). We assume that $\sum_{m=0,N-1} C_m^{(1)} C_m^{(2)} = 0$. In this case, show that it is possible to simultaneously transmit signals of the form $X_t^{(1)} = \sum_{k \in \mathbb{Z}} A_k^{(1)} \mathbb{1}_{[0, T_s[}(t - kT_s + \phi)$ and $X_t^{(2)} = \sum_{k \in \mathbb{Z}} A_k^{(2)} \mathbb{1}_{[0, T_s[}(t - kT_s + \phi)$ in the same spectrum bandwidth and then recover them without any cross interference residual term.

8. Linear Prediction of WSS Processes

Purpose We wish to approach, in the sense of the minimum error variance criterion, the variable X_n of a WSS process $X = (X_n)_{n \in \mathbb{Z}}$ by means of a linear combination of the random variables $(X_{n-k})_{k \geq 1}$. Some important applications like system modelling, filter synthesis, or signal compression, use the results of linear prediction theory.

8.1 Definitions

8.1.1 Conditional Expectation and Linear Prediction

At this point of our presentation, it is perhaps useful to recall some basic elements concerning conditional expectation.

Let X and Y denote a random variable and a random vector of size n of $L^2(\Omega, \mathcal{A}, dP)$ respectively. We sometimes have to evaluate X from knowledge of vector Y. To do this, we can try to approach X by a random variable of the form $\phi(Y)$. A possible and fairly natural choice involves selecting this function such that the random variable $\phi(Y)$ is as close as possible to X in $L^2(\Omega, \mathcal{A}, dP)$.

Let $\hat{\phi}$ be the function that realises this optimum:

$$\hat{\phi}(Y) = \arg \min_{\phi(Y) \in L^2(\Omega, \mathcal{A}, dP)} \| X - \phi(Y) \|^2,$$
$$\hat{\phi} \quad = \arg \min_{\phi \in L^2(\mathbb{R}^n, \mathcal{B}(\mathbb{R}^n), dP_Y)} \| X - \phi(Y) \|^2 .$$

(8.1)

We now recall the following definition:

Definition 8.1 *Let X and Y denote a random variable and a random vector of $L^2(\Omega, \mathcal{A}, dP)$. We call the expectation of X conditional to $Y = [Y_1, \ldots, Y_n]^T$, and we denote it by $\mathbb{E}[X|Y]$, the projection in $L^2(\Omega, \mathcal{A}, dP)$ of X on the set $L^2(\Omega, \sigma(Y), dP)$ of random variables of $L^2(\Omega, \mathcal{A}, dP)$ measurable for the σ-algebra $\sigma(Y)$ generated by the family $\{Y_i^{-1}(B); i = 1, \ldots, n, \text{ and } B \in \mathcal{B}(\mathbb{R})\}$. $\sigma(Y)$ is called the σ-algebra generated by Y.*

Remark In fact, the notion of conditional expectation is generally introduced in a different way. In particular, the problem is addressed in the space $L^1(\Omega, \mathcal{A}, dP)$ and we define the expectation of random variables, or events, conditionally to σ-algebras. The definition that we have just presented may be seen as an equivalent definition, valid for the subset $L^2(\Omega, \mathcal{A}, dP)$ of $L^1(\Omega, \mathcal{A}, dP)$.

It is easy to check that the set of elements of $L^2(\Omega, \sigma(Y), dP)$ coincides with the set of real random variables of the form $Z = \phi(Y)$, where $\phi \in L^2(\mathbb{R}^n, \mathcal{B}(\mathbb{R}^n), dP_Y)$. Consequently,

$$\mathbb{E}[X|Y] = \hat{\phi}(Y). \tag{8.2}$$

More generally, when $Y = (Y_n)_{n \in \mathbb{Z}}$ is a random process, $\mathbb{E}[X|Y]$ represents the projection in $L^2(\Omega, \mathcal{A}, dP)$ of X onto the Hilbert subspace $L^2(\Omega, \sigma(Y), dP)$. $L^2(\Omega, \sigma(Y), dP)$ is made up of random variables measurable for the σ-algebra $\sigma(Y)$ generated by the family $\{Y_n^{-1}(B); n \in \mathbb{Z} \text{ and } B \in \mathcal{B}(\mathbb{R})\}$. The generalisation of these definitions in the case where X is a random vector is immediate.

In the context of predicting a process X, we are more particularly interested in evaluating conditional expectations such as $\mathbb{E}[X_n | (X_{n-k})_{k=1,p}]$ or $\mathbb{E}[X_n | (X_{n-k})_{k \geq 1}]$. Unfortunately, if the knowledge of the process X is limited to that of its means and autocovariance functions, computing these values is not possible, except in the case where X is a Gaussian process. Moreover, knowing the distribution of X does not guarantee that we can explicitly compute the function $\hat{\phi}$ such that $\hat{\phi}((X_{n-k})_{k=1,p}) = \mathbb{E}[X_n | (X_{n-k})_{k=1,p}]$.

We shall therefore restrict ourselves, in the context of linear prediction, to seeking an estimator $\hat{X}_n$ of X_n in the form of a linear combination of the past values of the process: $\hat{X}_n = \sum_{k \geq 1} a_k X_{n-k}$. More precisely, we shall take for $\hat{X}_n$ the projection of X_n on the space $H_{X,n-1} = \overline{span}\{X_{n-k}; k \geq 1\}$, where $\overline{span}\{.\}$ designates the Hilbert space generated by the finite linear combinations of the variables under consideration, and the limits in $L^2(\Omega, \mathcal{A}, dP)$ of such sequences of linear combinations.

We shall denote this projection by $X_n/H_{X,n-1}$. We notice that the sets $H_{X,n}$ are nested: $H_{X,n} \subset H_{X,n+k}, \forall k > 0$. In fact, $X_n/H_{X,n-1}$ is called the *one step linear predictor*, $X_n/H_{X,n-p}$ representing the p step linear predictor.

Finite past linear prediction simply involves computing the projection of X_n on the linear space generated by a finite number $X_{n-1}, \ldots, X_{n-p}$ of random variables of the past of X, and denoted by $H_{X,n-1,n-p}$. Computing this projection amounts to solving a system of linear equations and is of great practical interest, in particular, when we only know a finite number of autocovariance coefficients of X.

8.1.2 Innovation Process

The innovation process of X is the process I defined by

$$I_n = X_n - X_n/H_{X,n-1}. \tag{8.3}$$

We shall see that this process, which at every instant represents the part of the process that cannot be predicted from its past (in the sense of linear prediction), plays a very important role.

We verify that I is a white noise process and that I and X are jointly stationary. Clearly, $I_n \in H_{X,n}$ and $I_n \perp H_{X,n-1}$. Therefore, the random variables I_n are uncorrelated. Moreover, the stationarity of X implies that of I. Finally, as $I_n \in H_{X,n}$, I_n is jointly stationary with X. We note that if X is not WSS, we may again define I in the same way. The random variables I_n is again uncorrelated, but then I is no longer a stationary process.

To model certain physical systems, it can be of great practical interest to represent their output as that of a causal filter with a white noise input. The results of linear prediction theory make it possible to define under what conditions this approach is justified.

In this chapter, we are therefore particularly interested in the conditions under which X can be seen as the output of a causal filter with a white noise input. We shall then see that this white noise necessarily corresponds (up to one factor) to the innovation process.

In what follows, we shall use the notion of normalised innovation. Normalised innovation is defined by $\nu_n = \| I_n \|^{-1} I_n$ if $I_n \neq 0$, and by $\nu_n = 0$ if $I_n = 0$. Moreover, we shall note the innovation processes I_X and ν_X, instead of I and ν, in the case where there is any ambiguity over the process X under consideration.

8.1.3 Regular and Singular Processes

We shall note

$$H_{X,+\infty} = \cup_{n \in \mathbb{Z}} H_{X,n},$$
$$\text{and } H_{X,-\infty} = \cap_{n \in \mathbb{Z}} H_{X,n}. \tag{8.4}$$

We say that X is regular if $H_{X,-\infty} = \{0\}$, and singular (or deterministic) if $H_{X,+\infty} = H_{X,-\infty}$. If a process is both regular and singular, it is zero, for then $H_{X,+\infty} = \{0\}$. We shall avoid using the term 'deterministic' because the notion of a deterministic process unfortunately has a non-stochastic flavour.

8.1.4 Examples

White noise A white noise process is a regular process.

To show this result, we consider a white noise process W and a random variable V of $H_{W,-\infty}$. It has to be shown that necessarily $V = 0$. As for any n, V belongs to $H_{W,n}$, and can be written in the form

$$V = \sum_{k \geq 0} \alpha_k W_{n-k}. \tag{8.5}$$

The random variables W_k ($k \in \mathbb{Z}$) are orthogonal, therefore $\| V \|^2 = \| W_n \|^2 \sum_{k \geq 0} |\alpha_k|^2$, and V is zero if and only if all the coefficients α_k are equal to zero. But, as we also have $V \in H_{W,n-1}$ and since $\mathbb{E}[W_k W_l^*] = 0$ for $k \neq l$, it is clear that $V - V/H_{W,n-1} = \alpha_0 W_n = 0$. Therefore, $\alpha_0 = 0$, and by induction it results that all the coefficients α_k are equal to zero, which leads to $V = 0$.

Harmonic Processes A harmonic process X defined by $X_n = \sum_{k=1,K} \xi_k \times e^{2i\pi n f_k}$ is singular. To show this, we consider the filter with frequency response

$$H(f) = \prod_{k=1,K} (1 - e^{-2i\pi(f-f_k)})$$

$$= 1 - \sum_{n=1,K} c_n e^{-2i\pi n f}.$$

$$(8.6)$$

Clearly,

$$X_n - \sum_{n=1,K} c_n X_{n-k} = 0. \qquad (8.7)$$

Thus, $X_n = \sum_{n=1,K} c_n X_{n-k}$, and $X_n \in H_{X,n-1}$. Similarly, we can check that $X_{n-1} \in H_{X,n-2}$. Consequently, $X_n \in H_{X,n-2}$, and by induction we obtain $X_n \in H_{X,-\infty}$. As this property is satisfied for any n, $H_{X,+\infty} \subset H_{X,-\infty}$ and finally $H_{X,-\infty} = H_{X,+\infty}$.

8.2 Wold's Decomposition Theorem

We shall see that a WSS process X can be represented as the sum of a regular process and of a singular process. In fact, this result is connected to Radon-Nikodym's theorem (see, for example, [37] p.117). This theorem indicates, in particular, that a bounded measure μ_X can be written as the sum of an absolutely continuous measure with respect to Lebesgue's measure, with density $S_X(f)$, and of a measure μ_x^s, carried by a set of measure zero with respect to Lebesgue's measure:

$$d\mu_X(f) = S_X(f)df + d\mu_X^s(f). \qquad (8.8)$$

Representation (8.8) is called Lebesgue's decomposition of μ_X.

Theorem 8.1 (Wold's decomposition)

$$H_{X,n} = H_{\nu,n} \oplus H_{X,-\infty}. \qquad (8.9)$$

Proof It is clear that $H_{X,n} = span\{\nu_n\} \oplus H_{X,n-1}$. Thus, if a random variable V of $H_{X,n}$ is orthogonal to ν_n, it belongs to $H_{X,n-1}$. Similarly, $H_{X,n-1} = span\{\nu_{n-1}\} \oplus H_{X,n-2}$. Therefore, if V is orthogonal to ν_n and to ν_{n-1}, it belongs to $H_{X,n-2}$. Then, the result is obtained by induction. $\square$

An important consequence of Wold's decomposition theorem is that since the spaces $H_{\nu,n}$ and $H_{X,-\infty}$ are orthogonal and their sum is equal to $H_{X,n}$, we have $X_n = Y_n + Z_n$, where $Y_n = X_n/H_{\nu,n}$ and $Z_n = X_n/H_{X,-\infty}$. The link between Wold's decomposition of $H_{X,n}$ and Lebesgue's decomposition of μ_X then appears in the following result:

Theorem 8.2 *Wold's decomposition leads to the representation of X in the form $X_n = Y_n + Z_n$, with $Y_n = X_n/H_{\nu,n}$ and $Z_n = X_n/H_{X,-\infty}$. Y and Z are regular and singular respectively.*

If Lebesgue's decomposition of μ_X is given by $d\mu_X(f) = S_X(f)df + d\mu_X^s(f)$, the innovation I of X verifies

$$\| I_n \|^2 = \exp\left(\int_I \log S_X(f)df \right). \tag{8.10}$$

Singular processes are characterised by the fact that $I_n = 0$. If $I_n \neq 0$, the spectral measures of Y and of Z are then given by $d\mu_Y(f) = S_X(f)df$, and $d\mu_Z(f) = d\mu_X^s(f)$ respectively.

Proof See Appendix E.

Formula (8.10) is known as the Kolmogorov-Szegö formula.

Corollary 8.3 *A non-zero process X is regular if and only if $d\mu_X(f) = S_X(f)df$, with $\int_I \log S_X(f)df > -\infty$.*

Proof If X is regular and non-zero, it is not singular and, consequently, $I_n \neq 0$. Therefore, $\int_I \log S_X(f)df > -\infty$. Moreover, as $H_{X,-\infty} = \{0\}$, $Z_n = X_n/H_{X,-\infty} = 0$. Thus, $X_n = Y_n + Z_n = Y_n$, and $d\mu_X(f) = d\mu_Y(f) = S_X(f)df$.

Conversely, from Theorem 8.2, if $\int_I \log S_X(f)df > -\infty$, the innovation I of X is non-zero, and $d\mu_Y(f) = S_X(f)df = d\mu_X(f)$. As here $\mu_Z = \mu_X^s = 0$, it is clear that $Z = 0$. Therefore, $X = Y$ and X is regular. $X \neq 0$ for $S_X(f)$ is not the null function ($\int_I \log S_X(f)df > -\infty$). □

From the above, we deduce that regular processes are those that can be represented as the output of a causal filter (and therefore realisable in practice) with a white noise input, as stated by the following result:

Theorem 8.4 *X is regular if and only if $H_{X,n} = H_{I,n}$, where I is the innovation of X. I is then the only white noise B, up to one factor, such that $H_{X,n} = H_{B,n}$.*

Proof If X is regular, $X = Y$ from the proof of the above-mentioned corollary. Now, $Y_n \in H_{\nu,n}$ and $H_{\nu,n} = H_{I,n}$ (ν is the normalised innovation). Therefore, $X_n \in H_{I,n}$ and $H_{X,n} \subset H_{I,n}$. Moreover, $I_n \in H_{X,n}$, therefore, $H_{I,n} \subset H_{X,n}$ and finally $H_{I,n} = H_{X,n}$.

Conversely, if $H_{X,n} = H_{I,n}$, $H_{X,-\infty} = H_{I,-\infty} = \{0\}$, for I is a white noise process. X is therefore regular.

If $H_{X,n} = H_{B,n}$, as X_n is of the form $X_n = \sum_{k \geq 0} g_k B_{n-k}$, it results that

$$I_n = X_n - X_n/H_{X,n-1}$$

$$= X_n - X_n/H_{B,n-1} \tag{8.11}$$

$$= g_0 B_n,$$

and $I = g_0 B$. $\square$

Remark We can check that Wold's decomposition, the fact that the processes Y and Z are regular and singular respectively, or that the first part of Theorem 8.4, do not require the hypothesis that X is stationary to be made.

8.3 Finite Past Linear Prediction

In what follows, we shall note $H_{X,n_1,n_2} = span\{X_{n_1}, X_{n_1+1}, \ldots, X_{n_2}\}$ ($n_1 \leq n_2$).

Often, in practice, the random variables X_{n-k} are only observed for a limited set of values of k: $k = 1, \ldots, p$. In these conditions, the problem of finite past linear prediction leads to us looking for

$$X_n/H_{X,n-p,n-1} = \sum_{k=1,p} a_k X_{n-k}, \tag{8.12}$$

that is, the coefficients $\alpha_k = a_k$ that minimise the criterion $\| X_n - \sum_{k=1,p} a_k X_{n-k} \|^2$. We shall denote by σ_p^2 the minimum of this criterion, which represents the prediction error variance.

It is then easy to verify that the values a_k of the coefficients α_k that realise the minimum are given by

$$a = T_{p-1}^{-1} r_p, \tag{8.13}$$

where $a = [a_1, \ldots, a_p]^T$, $r_p = [R_X(1), \ldots, R_X(p)]^T$, and T_{p-1} is the Toeplitz Hermitian matrix of size p and with general term $[T_{p-1}]_{i,j} = R_X(i - j)$, $(i, j) = 1, p$. We recall that a Toeplitz matrix T is characterised by the fact that all the terms in any parallel to the diagonal have the same values, that is, $[T]_{i,j}$ only depends on $i - j$.

To justify formula (8.13), we notice that from the projection theorem, the coefficients a_k are characterised by the relations

$$\| (X_n - \sum_{k=1,p} a_k X_{n-k}) X_{n-l}^* \|^2 = R_X(l) - \sum_{k=1,p} a_k R_X(l - k)$$

$$= 0, \quad \text{for } l = 1, p, \tag{8.14}$$

and the equations of this system are written in the matrix form $r_p - T_{p-1} a = 0$, that is, $a = T_{p-1}^{-1} r_p$ if $|T_{p-1}| \neq 0$. The relations

$$R_X(l) - \sum_{k=1,p} a_k R_X(l-k) = 0, \quad l = 1, p, \qquad (8.15)$$

are called the *Yule-Walker equations*, and the transfer function filter $a(z) = 1 - \sum_{k=1,p} a_k z^{-k}$ is often called the order p Prediction Error Filter (PEF).

If the matrix T_{p-1} is not invertible, we show (see the proof of Theorem I.3, in Appendix I.2) that the minimum of the criterion is equal to 0. It then appears that

$$\| I_n \|^2 = \| X_n - X_n/H_{X,n-1} \|^2$$

$$\leq \| X_n - X_n/H_{X,n-p,n-1} \|^2 \qquad (8.16)$$

$$\leq 0.$$

Hence, $I_n = 0$ and the process X is singular. In what follows, we shall assume that X is not singular.

Theorem 8.5 *The variance of finite past prediction error is given by*

$$\sigma_p^2 = \| X_n - \sum_{k=1,p} a_k X_{n-k} \|^2$$

$$= R_X(0) - r_p^H T_{p-1}^{-1} r_p \qquad (8.17)$$

$$= \frac{|T_p|}{|T_{p-1}|}.$$

Proof As $a = T_{p-1}^{-1} r_p$, it results from relations (8.14) that

$$\sigma_p^2 = \mathbb{E}[(X_n - \sum_{k=1,p} a_k X_{n-k})(X_n - \sum_{k=1,p} a_k X_{n-k})^*]$$

$$= \mathbb{E}[(X_n - \sum_{k=1,p} a_k X_{n-k}) X_n^*]$$

$$= R_X(0) - \sum_{k=1,p} a_k R_X(-k) \qquad (8.18)$$

$$= R_X(0) - r_p^H a$$

$$= R_X(0) - r_p^H T_{p-1}^{-1} r_p.$$

Moreover,

$$T_p = \begin{bmatrix} R_X(0) & r_p^H \\ r_p & T_{p-1} \end{bmatrix}, \qquad (8.19)$$

and developing the determinant of T_p following the first line yields

$$|T_p| = R_X(0)|T_{p-1}| - \sum_{k=1,p} R_X(-k)\Delta_k, \qquad (8.20)$$

where Δ_k is the determinant of a matrix corresponding to T_{p-1} whose k^{th} column is replaced by r_p. Up to the factor $|T_{p-1}|$, the coefficients Δ_k are therefore the components v_k of the vector $v = [v_1, \ldots, v_p]^T$, that is, the solution of the linear system of equations $T_{p-1}v = r_p$. Therefore, $\Delta_k = a_k|T_{p-1}|$ and

$$| T_p| = R_X(0)| T_{p-1}| - \sum_{k=1,p} R_X(-k)a_k|T_{p-1}|$$

$$= \sigma_p^2| T_{p-1}|,$$

(8.21)

which completes the proof. $\square$

Remark When the process X is not stationary, the problem of finite past linear prediction again amounts to solving a linear system of equations. But, in this case, the coefficients $R_X(i,j) = \mathbb{E}[X_i X_j^*]$ of T_p and of r_p do not depend only on $i-j$. Therefore, T_{p-1} no longer has a Toeplitz structure. The Toeplitz structure of the stationary case is very satisfactory since it enables us to envisage fast algorithms, in particular to invert T_{p-1}, such as the Levinson algorithm, which will be presented in Chapter 11.

By assuming that X is regular, and even more simply if $\int_{\mathbb{R}} \log S_X(f)df > -\infty$, we shall show the convergence of the sequence of the prediction error variances σ_p^2 towards $\| I_n \|^2$:

Theorem 8.6 *If X is a WSS process such that $\int_{\mathbb{R}} \log S_X(f)df > -\infty$, then $\lim_{p \to \infty} \sigma_p^2 = \| I_n \|^2$, and*

$$\log \| I_n \|^2 = \lim_{p \to \infty} \frac{1}{p} \log |T_p|$$

$$= \int_{\mathcal{I}} \log S_X(f)df.$$

(8.22)

Proof First, we notice that

$$X_n - X_n/H_{X,n-p,n-1} = I_n + (X_n/H_{X,n-1} - X_n/H_{X,n-p,n-1})$$

$$= I_n + X_n/F_{n-1-p},$$

(8.23)

where F_{n-1-p} represents the subspace of $H_{X,n-1-p}$ such that $H_{X,n-1} = H_{X,n-p,n-1} \oplus F_{X,n-1-p}$. The previous relation then leads to the inequalities

$$\| X_n - X_n/H_{X,n-1} \| \leq \| X_n - X_n/H_{X,n-p,n-1} \|$$

$$\leq \| I_n \| + \| X_n/F_{n-1-p} \|,$$

(8.24)

that is,

$$\| I_n \|^2 \leq \sigma_p^2$$

$$\leq (\| I_n \| + \| X_n/F_{n-1-p} \|)^2$$

$$\leq (\| I_n \| + \| X_n/H_{X,n-1-p} \|)^2.$$

(8.25)

But, as $H_{X,-\infty} = \{0\}$, $\lim_{p\to\infty} X_n / H_{X,n-1-p} = 0$ and, therefore, $\lim_{p\to\infty} \sigma_p^2$ $= \parallel I_n \parallel^2$.

The convergence of a sequence u_n towards a limit l implies the convergence of $n^{-1} \sum_{k=1,n} u_k$ towards the same value l (Cesaro's mean). Therefore, by taking into account the fact that $\sigma_p^2 = |T_p||T_{p-1}|^{-1}$,

$$\log \parallel I_n \parallel^2 = \lim_{p\to\infty} \log \sigma_p^2$$

$$= \lim_{p\to\infty} \frac{1}{p} \sum_{n=1,p} \log \sigma_n^2 \tag{8.26}$$

$$= \lim_{p\to\infty} \frac{1}{p} \log |T_p|.$$

The second equality of (8.22) is obtained from the fact that $\int_{\mathcal{I}} \log S_X(f) df > -\infty$. $\square$

In Chapter 11, we shall present other results concerning the finite past linear prediction of stationary processes.

8.4 Causal Factorisation of a PSD

8.4.1 Causal Factorisation

For a regular process $X = (X_n)_{n\in\mathbb{Z}}$, the problem of linear prediction amounts to that of looking for the representation of X as a function of its normalised innovation ν in the form $X_n = \sum_{k\geq 0} h_k \nu_{n-k}$, since then $X_n / H_{X,n-p} = \sum_{k\geq p} h_k \nu_{n-k}$.

We write $h(e^{2i\pi f}) = \sum_{k\geq 0} h_k e^{-2i\pi k f}$, and we give the properties of the filter thus defined. As

$$X_n = \int_{\mathcal{I}} e^{2i\pi n f} h(e^{2i\pi f}) d\hat{\nu}(f), \tag{8.27}$$

it is clear that $S_X(f) = |h(e^{2i\pi f})|^2$. A causal transfer function $h(z)$ such that $S_X(f) = |h(e^{2i\pi f})|^2$ is called a *causal factorisation* of $S_X(f)$. We notice that $h(z)$ is not unique. Thus, for example, $\left(\frac{e^{-2i\pi f} - \alpha^*}{1 - \alpha e^{-2i\pi f}}\right) h(e^{2i\pi f})$ where $|\alpha| < 1$, is another example of a causal factorisation of $S_X(f)$.

Theorem 8.7 *All the causal factorisations of $S_X(f)$ are of the form $\theta(e^{2i\pi f})$ $\times h(e^{2i\pi f})$, where $h(z)$ is the transfer function of the representation of X as a function of its normalised innovation, and $\theta(z)$ a causal transfer function with modulus equal to 1.*

Proof If $h'(e^{2i\pi f}) = \sum_{k\geq 0} h'_k e^{-2i\pi k f}$ is a causal factorisation of $S_X(f)$, then the process defined by

$$Y_n = \sum_{k\geq 0} h'_k \nu_{n-k}, \qquad (8.28)$$

where ν is the normalised innovation of X, is such that $Y_n \in H_{\nu,n}$. Therefore, $Y_n \in H_{X,n}$. Consequently, there exists a causal filter with frequency response $\theta(e^{2i\pi f}) = \sum_{k\geq 0} \theta_k e^{-2i\pi k f}$ such that

$$Y_n = \sum_{k\geq 0} \theta_k X_{n-k}. \qquad (8.29)$$

The representation of X as a function of its normalised innovation is given by the filter with frequency response $h(e^{2i\pi f}) = \sum_{k\geq 0} h_k e^{-2i\pi k f}$:

$$X_n = \sum_{k\geq 0} h_k \nu_{n-k}. \qquad (8.30)$$

Finally, relations (8.28), (8.29), and (8.30) lead to

$$d\hat{Y}(f) = h'(e^{2i\pi f})d\hat{\nu}(f)$$

$$= \theta(e^{2i\pi f})d\hat{X}(f) \qquad (8.31)$$

$$= \theta(e^{2i\pi f})h(e^{2i\pi f})d\hat{\nu}(f),$$

and therefore $h'(e^{2i\pi f}) = \theta(e^{2i\pi f})h(e^{2i\pi f})$. Moreover, $h(e^{2i\pi f})$ and $h'(e^{2i\pi f})$ are causal factorisations of $S_X(f)$ and the relation

$$|h'(e^{2i\pi f})|^2 = |\theta(e^{2i\pi f})|^2 |h(e^{2i\pi f})|^2, \qquad (8.32)$$

yields $|\theta(e^{2i\pi f})| = 1$ for almost every value of f since $\int_{\mathcal{I}} \log S_X(f)df > -\infty$, which completes the proof. $\square$

8.4.2 Minimum-phase Causal Factorisation

It is possible to find the causal filter h such that $X_n = [h(z)]\nu_n$ among the set of all causal factorisations of $S_X(f)$, thanks to the following result:

Theorem 8.8 *The causal filter $h(e^{2i\pi f})$ of representation of a regular process as a function of its normalised innovation is the only causal factorisation $g(e^{2i\pi f}) = \sum_{k\geq 0} g_k e^{-2i\pi k f}$ of $S_X(f)$ (up to a modulus 1 factor) that satisfies*

$$\log |g_0|^2 = \int_{\mathcal{I}} \log S_X(f)df. \qquad (8.33)$$

$h(e^{2i\pi f})$ *is called the minimum-phase causal factorisation of X. It is also called the innovation filter of X.*

Proof Let $g(e^{2i\pi f})$ be a causal factorisation of $S_X(f)$ and E the set of points where $g(e^{2i\pi f})$ is equal to zero. As $\int_I \log|g(e^{2i\pi f})|^2 df > -\infty$, E is a zero measure set with respect to Lebesgue's measure. Consequently, the relation $d\hat{W}(f) = d\hat{X}(f)g^{-1}(e^{2i\pi f})$, defined for $f \notin E$, characterises the stochastic measure of a white noise W with variance equal to 1. This white noise is unique, for if two white noise processes W_1 and W_2 verify this definition,

$$\| W_{1,n} - W_{2,n} \| = \| \int_E e^{2i\pi n f} d\hat{W}_1(f) - \int_E e^{2i\pi n f} d\hat{W}_2(f) \|$$

$$\tag{8.34}$$

$$\leq 2(\int_E df)^{1/2}.$$

But, $\int_E df = 0$. Hence, $W_{1,n} = W_{2,n}$.

We then have $d\hat{X}(f) = g(e^{2i\pi f})d\hat{W}(f)$, and it is clear that $H_{X,n-1} \subset H_{W,n-1}$. Therefore,

$$\| X_n - X_n/H_{W,n-1} \|^2 \leq \| X_n - X_n/H_{X,n-1} \|^2,$$

$$\tag{8.35}$$

which can also be written as

$$|g_0|^2 \leq \exp(\int_I \log S_X(f)df).$$

$$\tag{8.36}$$

Since $H_{X,n-1} \subset H_{W,n-1}$, we again have

$$\| X_n - X_n/H_{X,n-1} \|^2 = \| X_n - X_n/H_{W,n-1} \|^2$$

$$+ \| X_n/H_{W,n-1} - X_n/H_{X,n-1} \|^2.$$

$$\tag{8.37}$$

Therefore, in case of equality in relation (8.35), we have $X_n/H_{W,n-1} = X_n/H_{X,n-1}$, and

$$I_n = h_0\nu_n$$

$$= X_n - X_n/H_{W,n-1}$$

$$\tag{8.38}$$

$$= g_0 W_n.$$

As $I_n \neq 0$, $h_0 \neq 0$ and $g_0 \neq 0$, it is clear that $H_{W,n} = H_{\nu,n} = H_{X,n}$. Hence, from Theorem 8.4, $W = \alpha\nu$ with $|\alpha| = 1$, for $\| W_n^2 \|^2 = \| \nu_n^2 \|^2 = 1$. Consequently,

$$d\hat{X}(f) = g(e^{2i\pi f})d\hat{W}(f)$$

$$= \alpha g(e^{2i\pi f})d\hat{\nu}(f)$$

$$\tag{8.39}$$

$$= h(e^{2i\pi f})d\hat{\nu}(f),$$

and $h = \alpha g$. $\square$

Definition 8.2 *We call a minimum-phase causal filter a filter whose frequency response $h(e^{2i\pi f})$ is a minimum-phase causal factorisation of the spectrum with density $|h(e^{2i\pi f})|^2$. We show that for such a filter the knowledge of the frequency response modulus entirely determines that of the phase. Also, $\log|h(e^{2i\pi f})|$ and the phase of the filter are related through the discrete Hilbert transform (see Appendix W).*

We shall now show how, in certain cases, the causal factorisation associated with the innovation may be explicitly computed from $S_X(f)$. Let X denote a regular process whose PSD $S_X(f)$ is continuous, piecewise derivable, with bounded derivative and satisfies for any f

$$0 < m \le S_X(f) \le M < \infty. \tag{8.40}$$

Then, from Dirichlet's criterion, $\log S_X(f)$ coincides everywhere with its Fourier series development (see Appendix F).

We now note this development:

$$\log S_X(f) = \sum_{k \in \mathbb{Z}} b_k e^{-2i\pi k f}. \tag{8.41}$$

We then have the following result:

Theorem 8.9 *If $\log S_X(f) = \sum_{k \in \mathbb{Z}} b_k e^{-2i\pi k f}$, the minimum-phase causal factorisation of $S_X(f)$ is given by*

$$h(e^{2i\pi f}) = \exp\left(\frac{b_0}{2} + \sum_{k \ge 1} b_k e^{-2i\pi k f}\right). \tag{8.42}$$

Proof The frequency response

$$
\begin{aligned}
h(e^{2i\pi f}) &= \sum_{k \ge 0} h_k e^{-2i\pi f k} \\
&= \exp\left(\frac{b_0}{2} + \sum_{k \ge 1} b_k e^{-2i\pi k f}\right)
\end{aligned}
\tag{8.43}
$$

is causal, for e^z has a series expansion $\sum_{k \in \mathbb{N}} z^k/k!$. Moreover, as $b_{-k} = b_k^*$ ($S_X(f)$ is a real function), it is clear that

$$
\begin{aligned}
h(e^{2i\pi f})[h(e^{2i\pi f})]^* &= \exp\left(\sum_{k \in \mathbb{Z}} b_k e^{-2i\pi k f}\right) \\
&= S_X(f).
\end{aligned}
\tag{8.44}
$$

Finally, as $h_0 = \exp(b_0/2)$, Theorem 8.8 and the immediate relation $b_0 = \int_I \log S_X(f) df$ complete the proof. $\square$

8.5 The Continuous Case

For continuous time processes, the problem of linear prediction may be addressed in the same way. We note, however, that Paley and Wiener have shown that $S_X(f)$ has a factorisation of the form $|H(f)|^2$, where $H(f)$ and $H^{-1}(f)$ are causal if and only if (see, for example, [41] p.158, or [6] p.215)

$$\int_{\mathbb{R}} \frac{|\log S_X(f)|}{1 + f^2} df < \infty. \tag{8.45}$$

For the continuous case, this condition is similar to the regularity condition $\int_{\mathcal{I}} \log S_X(f)\, df > -\infty$ of the discrete case. When the Paley-Wiener condition is satisfied, we shall therefore speak of X as a regular continuous time process.

Exercises

8.1 What is the minimum-phase causal factorisation of the PSD $S(f) = 2 + \cos(2\pi f)$?

8.2 Let X denote a WSS process with spectrum $d\mu_X(f)$. Let $u = [u_0, \dots, u_N]^T$ and $u(z) = \sum_{n=0,N} u_n z^{-n}$. Show that

$$\min_{\|u\|=1} |u(e^{2i\pi f})|^2 d\mu_X(f) \quad \text{and} \quad \max_{\|u\|=1} |u(e^{2i\pi f})|^2 d\mu_X(f) \tag{8.46}$$

are obtained when u is equal to the smallest and to the largest eigenvalue of the Toeplitz matrix $T_{X,N}$ of size $(N+1) \times (N+1)$ defined by $[T_{X,N}]_{ab} = R_X(a-b)$.

8.3 We are looking for the solution to the following prediction problem

$$\min_{\{c_k, d_k\}_{k=1,p}} \left\| X_n - \sum_{k=1,p} c_k X_{n-k} - \sum_{l=1,p} d_l X_{n-l}^* \right\|^2. \tag{8.47}$$

a) Calculate the values of the coefficients $\{c_k, d_k\}_{k=1,p}$ that achieve the minimum of (8.47).

b) If X is a circular process, show that $d_k = 0$ for $k = 1, \dots, p$.

8.4 Let $X = (X_n)_{n \in \mathbb{Z}}$ be a circular WSS process.

a) Calculate the coefficients of the polynomial $B_p(z) = \sum_{k=0,p} b_{k,p} z^{-k}$ for which $\| [B_p(z)] X_n \|^2 - 2 \mathcal{R}e[b_{0,p}]$ is minimum.

b) How are the coefficients of $B_p(z)$ related to those of the prediction error filter?

8.5 Let $X = (X_n)_{n \in \mathbb{Z}}$ denote a WSS process with a spectral measure of the form $d\mu_X(f) = \sigma^2 df + \delta_{f_0}$.

a) Show that X satisfies a difference equation of the form

$$X_n - \alpha X_{n-1} = \beta_0 V_n - \beta_1 V_{n-1}, \tag{8.48}$$

where $V = (V_n)_{n \in \mathbb{Z}}$ is a white noise with unit variance: $\| V_n \| = 1$. Calculate α, β_0 and β_1.

b) Show that X is not a regular process and check that Equation (8.48) also has a regular solution.

8.6 (Linear interpolation) We consider a WSS process $X = (X_n)_{n \in \mathbb{Z}}$. Solve the following linear interpolation problem:

$$\min_{(a_k, b_k)_{k=1,p}} \| X_n - \sum_{k=1,p} a_k X_{n-k} - \sum_{l=1,p} b_l X_{n+l} \| . \tag{8.49}$$

8.7 (Asymptotic error variance of linear interpolation) We want to calculate the asymptotic minimum error variance of the linear interpolation problem (8.49) when p tends to $+\infty$. We shall denote this limit by σ_I^2.

a) Check that σ_I^2 exists.

b) Let us note $e^{2i\pi n f} = g(f) + g_\perp(f)$, where $g(f)$ represents the projection in $L^2(\mathcal{I}, \mathcal{B}(\mathcal{I}), d\mu_X(f)df)$ of $e^{2i\pi n f}$ on $\overline{span}\{\chi_k(f) = e^{2i\pi k f}; k \in \mathbb{Z}, k \neq n\}$. Show that $\sigma_I^2 = \| g_\perp(f) \|$.

c) Show that

$$\int_{\mathcal{I}} |g_\perp(f)|^2 d\mu_X(f) df = \int_{\mathcal{I}} g_\perp(f) e^{-2i\pi n f} d\mu_X(f) df. \tag{8.50}$$

d) We assume that X has a PSD denoted by $S_X(f)$. Show that $g_\perp(f)S_X(f) = ce^{2i\pi n f}$, where c is a constant.

e) Using (8.50), calculate c and show that $\sigma_I^2 = (\int_{\mathcal{I}} S_X^{-1}(f)df)^{-1}$.

f) Calculate the projection of X_n on $\overline{span}\{X_k; k \in \mathbb{Z}, k \neq n\}$.

8.8 (Jensen's formula) We consider a function $g(z)$, holomorphic in the open unit disk $\mathbb{D}$, with $|g(z)| > 0$ for all z in $\mathbb{D}$.

a) Show that $\phi(x, y) = \log(|g(x + iy)|)$ is a harmonic function in $\mathbb{D}$, that is $\frac{\partial^2 \phi}{\partial x^2} + \frac{\partial^2 \phi}{\partial y^2} = 0$, and check the mean relation

$$\phi(0, 0) = \int_{\mathcal{I}} \phi(\cos(2\pi f), \sin(2\pi f))df. \tag{8.51}$$

b) Show that $\int_{\mathcal{I}} \log |1 - e^{2i\pi f}| df = 0$.

c) We now consider a function $g(z)$ holomorphic in $\mathbb{D}$ and continuous on $\overline{D}$. Let $\alpha_1, \ldots, \alpha_m$ denote the zeroes of $g(z)$ that lie in $\mathbb{D}$ (some of same being possibly equal). Show that

$$|g(0)| = (\prod_{k=1,N} |\alpha_k|) \exp(\int_{\mathcal{I}} \log |g(e^{2i\pi f})| df).$$ (8.52)

(Hint: consider the function

$$h(z) = g(z) \prod_{k=1,n} \frac{1 - \alpha_k^* z}{\alpha_k - z} \prod_{k=n+1,m} \frac{\alpha_k}{\alpha_k - z},$$ (8.53)

where $\alpha_{n+1}, \ldots, \alpha_m$ denote the zeroes of $g(z)$ that lie on the unit circle.)

d) Relate this result to the minimum-phase factorisation of $|g(e^{2i\pi f})|^2$.

8.9 (Linear prediction of multivariate processes) Let $X = (X_n)_{n \in \mathbb{Z}}$ denote a WSS multivariate process of size p. Let $A_k(z) = \sum_{l=0,k} A_{k,l} z^{-l}$, where the coefficients $A_{k,l}$ are matrices of size $p \times p$. Furthermore, we assume that all the block Toeplitz matrices $(T_{X,k})_{k \geq 0}$ of size $(k+1)p \times (k+1)p$ with block (i, j) equal to $R_X(i - j)$ are strictly positive.

a) We are looking for the matrix polynomial $A_k(z)$ that minimises the criterion

$$K(M) = \int_{\mathcal{I}} [M(e^{2i\pi f})]^H d\mu_X(f) M(e^{2i\pi f}) - (M_{k,0} + M_{k,0}^H),$$

$$= \mathbf{M}^H T_{X,k} \mathbf{M} - (M_{k,0} + M_{k,0}^H),$$ (8.54)

where $M(z)$ is of the form $M(z) = \sum_{l=0,k} M_l z^{-l}$, and $\mathbf{M} = [M_0^H, \ldots, M_k^H]^H$. By $A_k(z)$ minimising K, we mean that $K(A_k) \leq K(M)$ for any $M(z)$ in the sense of the positive Hermitian matrix inequality. Show that $A_k(z) = U_k(z) T_{X,k}^{-1} U_k^T(\infty)$, where $U_k(z) = [I_p, z^{-1} I_p, \ldots, z^{-k} I_p]$.

(Hint: introduce the matrix $\mathbf{N} = T_{X,k}^{1/2}[M_0^H, \ldots, M_k^H]^H - T_{X,k}^{-1/2} U_k^T(\infty)$.)

b) Show that the matrix $A_{k,0}$ is Hermitian, strictly positive, and that $A_{k+1,0} \geq A_{k,0}$.

c) Show that $A(z)$ also achieves the minimisation of $Tr(L^H K(M)L)$, where L is any full rank matrix of size $p \times p$.

8.10 (Paley-Wiener condition) In this exercise, we want to study under which condition a WSS process $X = (X_t)_{t \in \mathbb{Z}}$ can be seen as the output of a causal filter with input white noise. This amounts to searching for a representation of the PSD of X of the form $S_X(f) = |g(f)|^2$, where $g(f) \in \overline{span}\{\chi_\tau = e^{2i\pi f\tau}; \tau \leq 0\}$ as well as $g^{-1}(f)$.

a) First, let us define the function $S_X^d(f)$ on $\mathcal{I}$ by $S_X^d(u) = S_X(f)$, where $u = \pi^{-1} \arctan f$. Show that S_X^d is the PSD of a regular process indexed by $\mathbb{Z}$ if and only if

$$\int_{\mathbb{R}} \frac{S_X(f)}{1 + f^2} df > -\infty.$$ (8.55)

b) In the following questions, we are going to prove that X is regular if and only if $S_X^d(u)$ is the PSD of a regular process. Prove that this can be done by showing that

$$\overline{span}\{\chi_n(\pi^{-1}\arctan f); -n \in \mathbb{N}\} = \overline{span}\{\log \chi_\tau(f); \tau \leq 0\}. \qquad (8.56)$$

c) Noting that

$$\chi_{-1}(u) = \exp(-2i.\arctan f) = \frac{1-if}{1+if} = -1 + 4\pi \int_{\mathbb{R}_+} e^{-2i\pi ft} e^{2\pi t} dt, \qquad (8.57)$$

prove that $\overline{span}\{\chi_n(\pi^{-1}\arctan f); -n \in \mathbb{N}\} \subset \overline{span}\{\chi_\tau(f); \tau \leq 0\}$.

d) In order to prove that $\overline{span}\{\chi_\tau(f); \tau \leq 0\} \subset \overline{span}\{\chi_n(\pi^{-1}\arctan f);$ $-n \in \mathbb{N}\}$, check first that $\exp(2\pi t(z-1)(z+1)^{-1})$ belongs to $\overline{span}\{\chi_n(u);$ $-n \in \mathbb{N}\}$, where $z = |z|e^{2i\pi u}$, $|z| > 1$ and $t < 0$. Then, letting $|z|$ tend to 1, check that $e^{2i\pi ft} \in \{\chi_n(\pi^{-1}\arctan f); -n \in \mathbb{N}\}$, when $t < 0$.

e) What can be concluded from the above?

9. Particular Filtering Techniques

Purpose In many problems, we want to evaluate a process X from partial knowledge of a process Y. According to whether knowledge about Y at instant n is limited to that of $H_{Y,n}$, or to that of $H_{Y,0:n} = span\{Y_0, \ldots, Y_n\}$, we may opt to evaluate X_n by $X_n/H_{Y,n}$, or $X_n/H_{Y,0:n}$. The resolution of these problems is known as *Wiener filtering* and *Kalman filtering* respectively. We next indicate how Kalman's recursive filtering can be generalised to computing recursively the distribution of X_n conditional to $\{Y_0, \ldots, Y_n\}$ for systems that may not be linear. We finish this part with the presentation of the *matched filter* that enables detection of a known deterministic signal in the presence of noise.

9.1 Wiener Filter

We assume that X and Y are two stationary and jointly stationary processes. Computing $X_n/H_{Y,n}$ amounts, in practice, to expressing X_n as the output of a certain causal filter with input Y. Moreover, we assume that the process Y is regular.

We first consider the process

$$\tilde{X}_n = \int_{\mathcal{I}} e^{2i\pi nf} \left(\frac{S_{XY}(f)}{S_Y(f)} \right) d\hat{Y}(f). \qquad (9.1)$$

It is clear that $\tilde{X}_n = X_n/H_Y$: it is sufficient to notice that $\tilde{X}_n \in H_Y$, and to verify that $\forall m, n \in \mathbb{N}$, $\mathbb{E}[(X_n - \tilde{X}_n)Y_m^*] = 0$. Unfortunately, the filter with frequency response $S_{XY}(f)S_Y^{-1}(f)$ is not causal and $\tilde{X}_n \neq X_n/H_{Y,n}$.

The idea is then to note that if ν represents the normalised innovation process of Y,

$$X_n/H_{Y,n} = X_n/H_{\nu,n}$$

$$= (X_n/H_\nu)/H_{\nu,n} \qquad (9.2)$$

$$= (X_n/H_Y)/H_{\nu,n}.$$

The equality $X_n/H_{\nu,n} = (X_n/H_\nu)/H_{\nu,n}$ stems from the fact that ν is a white noise and therefore that H_ν is the orthogonal sum of $H_{\nu,n}$ and of $\overline{span}\{\nu_m; m > n\}$.

As Y is regular, we may write $S_Y(f) = |h(e^{2i\pi f})|^2$, where $h(e^{2i\pi f}) = \sum_{k=0,\infty} h_k e^{-2i\pi kf}$ is the minimum-phase causal factorisation of $S_Y(f)$: $Y_n = \sum_{k=0,\infty} h_k \nu_{n-k}$.

We then have the relations

$$X_n/H_\nu = \int_\mathcal{I} e^{2i\pi nf} \left(\frac{S_{XY}(f)}{S_Y(f)} \right) d\hat{Y}(f)$$

$$= \int_\mathcal{I} e^{2i\pi nf} \left(\frac{S_{XY}(f)}{[h(e^{2i\pi f})]^*} \right) d\hat{\nu}(f),$$

(9.3)

and

$$X_n/H_{\nu,n} = \int_\mathcal{I} e^{2i\pi nf} \left[\frac{S_{XY}(f)}{[h(e^{2i\pi f})]^*} \right]_+ d\hat{\nu}(f)$$

$$= \int_\mathcal{I} e^{2i\pi nf} \left[\frac{S_{XY}(f)}{[h(e^{2i\pi f})]^*} \right]_+ \frac{1}{h(e^{2i\pi f})} d\hat{Y}(f),$$

(9.4)

where $[.]_+$ represents the frequency response of the causal part of the filter:

$$\left[\sum_{k=-\infty,\infty} g_k e^{-2i\pi kf} \right]_+ = \sum_{k=0,\infty} g_k e^{-2i\pi kf}.$$

(9.5)

Finally, we have the following result:

Theorem 9.1 *If Y is a regular process, $X_n/H_{Y,n}$ is the output of a causal filter with input Y and whose frequency response is $\left[\frac{S_{XY}(f)}{[h(e^{2i\pi f})]^*} \right]_+ \frac{1}{h(e^{2i\pi f})}$, where $h(e^{2i\pi f})$ is the minimum-phase causal factorisation of $S_Y(f)$.*

Example We look for the Wiener filter when $Y_n = X_n + W_n$, where X is a process defined by $X_n = z_0 X_{n-1} + V_n$, with $|z_0| < 1$, and V and W are uncorrelated white noise processes with respective variances σ_V^2 and σ_W^2. The PSD and cross-spectra of X and Y are given by

$$S_X(f) = \left(\frac{\sigma_V^2}{(1 - z_0 z^{-1})(1 - z_0^* z)} \right)_{z=e^{2i\pi f}},$$

$$S_Y(f) = S_X(f) + \sigma_W^2 = \left(\frac{\sigma_V^2 + \sigma_W^2(1 - z_0 z^{-1} - z_0^* z + |z_0|^2)}{1 - z_0 z^{-1} - z_0^* z + |z_0|^2} \right)_{z=e^{2i\pi f}},$$

$$S_{XY}(f) = S_X(f).$$

(9.6)

$S_Y(f)$ can also be written as

$$S_Y(f) = \left(\frac{\sigma_W^2 z_0^* z^2 - (\sigma_V^2 + \sigma_W^2(1 + |z_0|^2))z + \sigma_W^2 z_0}{z_0^* z^2 - (1 + |z_0|^2)z + z_0} \right)_{z=e^{2i\pi f}}. \tag{9.7}$$

We note that the modulus of the product of the roots of the numerator is equal to 1 and that the root with modulus smaller than 1 is

$$z_1 = \frac{\sigma_V^2 + \sigma_W^2(1 + |z_0|^2) - [(\sigma_V^2 + \sigma_W^2(1 + |z_0|^2))^2 - 4|z_0|^2 \sigma_W^4]^{1/2}}{2 z_0^* \sigma_W^2}. \tag{9.8}$$

Consequently,

$$S_Y(f) = \sigma_W^2 \left(\frac{(z - z_1)(z - z_1^{*-1})}{(z - z_0)(z - z_0^{*-1})} \right)_{z=e^{2i\pi f}}$$

$$= \frac{\sigma_W^2 z_0^*}{z_1^*} \left(\frac{(1 - z_1 z^{-1})(1 - z_1^* z)}{(1 - z_0 z^{-1})(1 - z_0^* z)} \right)_{z=e^{2i\pi f}}, \tag{9.9}$$

and the minimum-phase causal factorisation of $S_Y(f)$ is given by the transfer function

$$h(z) = \alpha^{1/2}(1 - z_1 z^{-1})(1 - z_0 z^{-1})^{-1}, \tag{9.10}$$

where $\alpha = (z_1^*)^{-1} \sigma_W^2 z_0^*$. The transfer function of the Wiener filter is therefore given by

$$G(z) = \frac{1}{\alpha} \left[\frac{\sigma_V^2}{(1 - z_0 z^{-1})(1 - z_0^* z)} (\frac{1 - z_0^* z}{1 - z_1^* z}) \right]_+ \frac{1 - z_0 z^{-1}}{1 - z_1 z^{-1}}$$

$$= \frac{\sigma_V^2}{\alpha} \left[\frac{-z(z_1^*)^{-1}}{(z - z_0)(z - (z_1^*)^{-1})} \right]_+ \frac{1 - z_0 z^{-1}}{1 - z_1 z^{-1}}$$

$$= \frac{\sigma_V^2}{\alpha} \left[\frac{z(z_1^*)^{-1}}{(z_1^*)^{-1} - z_0} (\frac{1}{z - z_0} - \frac{1}{z - (z_1^*)^{-1}}) \right]_+ \frac{1 - z_0 z^{-1}}{1 - z_1 z^{-1}}$$

$$= \left(\frac{\sigma_V^2}{\sigma_W^2(z_0^*(z_1^*)^{-1} - |z_0|^2)} \right) \frac{1}{1 - z_1 z^{-1}}. \tag{9.11}$$

9.2 Kalman Filter

Performing Wiener filtering assumes perfect knowledge of the autocovariance and cross-covariance functions of the processes X and Y that are brought into play. We now assume that we only have knowledge of the variables $\{Y_0, \ldots, Y_n\}$, and that we wish to evaluate $X_n / H_{Y,0:n}$. The Kalman filter provides an iterative technique to solve this problem for a large class of processes. In particular, unlike Wiener filtering, the second order stationarity hypothesis about the processes does not need to be satisfied here.

In many problems where we wish to compute $X_n/H_{Y,0:n}$, X is a vector process that is, in fact, a Markov chain characterised by the relations

$$X_{n+1} = F_n X_n + V_n, \quad n > 0, \tag{9.12}$$

where V is a white noise process such that H_V is orthogonal to X_0. The covariances of X_0 and of V_n (and consequently of X_n) are assumed to be known. In addition, we often assume that knowledge of the vector X_n is only accessible through that of a process Y, linked to X by a linear relation of the form

$$Y_n = H_n X_n + W_n, \tag{9.13}$$

where W is a white noise process such that $W_n \perp H_{X,n}$, and $W_n \perp H_V$. In the particular case where F_n and H_n are constants, we easily check that X and Y are stationary.

The covariance matrices of X_0, V_n and W_n will be denoted by $Q_{X,0}$, $Q_{V,n}$ and $Q_{W,n}$ respectively. These values, as well as F_n and H_n, are assumed to be known. By noting $\hat{X}_{m+n,n} = X_{m+n}/H_{Y,0:n}$ we show that $\hat{X}_{n,n}$ can be computed iteratively in the following way:

Theorem 9.2

$$\begin{cases} \hat{X}_{0,-1} = 0 \\[2mm] \hat{X}_{n,n} = \hat{X}_{n,n-1} + A_n(Y_n - H_n \hat{X}_{n,n-1}) \\[2mm] \hat{X}_{n+1,n} = F_n \hat{X}_{n,n}, \end{cases} \tag{9.14}$$

where the matrices A_n are obtained using the relations

$$\begin{cases} B_0 = Q_{X,0} \\[2mm] A_n = B_n H_n^H (H_n B_n H_n^H + Q_{W,n})^{-1} \\[2mm] B_n = \mathbb{E}[(X_n - \hat{X}_{n,n-1})(X_n - \hat{X}_{n,n-1})^H] \\[1mm] \quad\ = F_{n-1} C_{n-1} F_{n-1}^H + Q_{V,n} \\[2mm] C_n = \mathbb{E}[(X_n - \hat{X}_{n,n})(X_n - \hat{X}_{n,n})^H] \\[1mm] \quad\ = (I - A_n H_n) B_n (I - A_n H_n)^H + A_n Q_{W,n} A_n^H. \end{cases} \tag{9.15}$$

The matrix A_n is called the *Kalman gain*. We note that $\hat{X}_{n,n}$ is calculated from $\hat{X}_{n,n-1}$ by adding to it the linear prediction error $Y_n - Y_n/H_{Y,0:n-1}$, affected by the Kalman gain.

In practice, we begin by computing $A_0 = B_0 H_0^H (H_0 B_0 H_0^H + Q_{W,0})^{-1}$, $\hat{X}_{0,0} = A_0 Y_0$, $\hat{X}_{1,0} = F_0 \hat{X}_{0,0}$, then C_0. At each iteration, from C_{n-1}, and from $\hat{X}_{n,n-1}$, we compute successively B_n, then A_n, $\hat{X}_{n,n}$, $\hat{X}_{n+1,n}$, and C_n.

Proof It is clear that:

$$\hat{X}_{n+1,n} = F_n X_n / H_{Y,0,n} + V_n / H_{Y,0,n}$$

$$= F_n X_n / H_{Y,0,n} \tag{9.16}$$

$$= F_n \hat{X}_{n,n}.$$

In addition,

$$H_{Y,0:n} = span\{Y_0, \ldots, Y_n\}$$

$$= span\{Y_0, \ldots, Y_{n-1}, Y_n - H_n \hat{X}_{n,n-1}\}, \tag{9.17}$$

and,

$$(Y_n - H_n \hat{X}_{n,n-1}) / H_{Y,0,n-1}$$

$$= H_n(X_n - \hat{X}_{n,n-1}) / H_{Y,0,n-1} + W_n / H_{Y,0,n-1} \tag{9.18}$$

$$= 0.$$

Hence, $H_{Y,0:n} = H_{Y,0:n-1} \oplus span\{Y_n - H_n \hat{X}_{n,n-1}\}$. Consequently,

$$\hat{X}_{n,n} = X_n / H_{Y,0,n-1} + X_n / span\{Y_n - H_n \hat{X}_{n,n-1}\}$$

$$= \hat{X}_{n,n-1} + A_n(Y_n - H_n \hat{X}_{n,n-1}) \tag{9.19}$$

In order to compute A_n, we now recall that for a random vector $Z = [Z_1, Z_2]^T$, we see from the relation $\mathbb{E}[(Z_1 - (Z_1/Z_2))Z_2^H] = 0$ that Z_1/Z_2 is of the form AZ_2:

$$Z_1/Z_2 = \mathbb{E}[Z_1 Z_2^H]\mathbb{E}[Z_2 Z_2^H]^{-1} Z_2. \tag{9.20}$$

Letting $Z_1 = X_n$, and

$$Z_2 = Y_n - H_n \hat{X}_{n,n-1}$$

$$= H_n(X_n - \hat{X}_{n,n-1}) + W_n \tag{9.21}$$

we obtain

$$\mathbb{E}[Z_1 Z_2^H] = \mathbb{E}[X_n(X_n - \hat{X}_{n,n-1})^H]H_n^H$$

$$= \mathbb{E}[(X_n - \hat{X}_{n,n-1})(X_n - \hat{X}_{n,n-1})^H]H_n^H \tag{9.22}$$

$$= B_n H_n^H,$$

and

$$\mathbb{E}[Z_2 Z_2^H] = H_n \mathbb{E}[(X_n - \hat{X}_{n,n-1})(X_n - \hat{X}_{n,n-1})^H] H_n^H + \mathbb{E}[W_n W_n^H]$$

$$= H_n B_n H_n^H + Q_{W,n}.$$

(9.23)

Therefore,

$$A_n = \mathbb{E}[Z_1 Z_2^H] \mathbb{E}[Z_2 Z_2^H]^{-1}$$

(9.24)

$$= B_n H_n^H (H_n B_n H_n^H + Q_{W,n})^{-1}.$$

Moreover,

$$B_n = \mathbb{E}[(F_{n-1}(X_{n-1} - \hat{X}_{n-1,n-1}) + V_{n-1})$$

$$\times (F_{n-1}(X_{n-1} - \hat{X}_{n-1,n-1}) + V_{n-1})^H]$$

(9.25)

$$= F_{n-1} C_{n-1} F_{n-1}^H + Q_{V,n-1}.$$

Finally, by noticing that

$$X_n - \hat{X}_{n,n} = (X_n - \hat{X}_{n,n-1}) - A_n(Y_n - H_n \hat{X}_{n,n-1})$$

$$= (I - A_n H_n)(X_n - \hat{X}_{n,n-1}) - A_n W_n,$$

(9.26)

we have

$$C_n = \mathbb{E}[(X_n - \hat{X}_{n,n})(X_n - \hat{X}_{n,n})^H]$$

(9.27)

$$= (I - A_n H_n) B_n (I - A_n H_n) + A_n Q_{W,n} A_n^H. \quad \square$$

Example A radar measures the position and the speed of a mobile. This measure is corrupted by errors. We denote the process that represents these measures by $Y = [Y_1, Y_2]^T$, and $W = [W_1, W_2]^T$ the error process on the measures. We assume that W is a white noise with known covariance matrix. We can try to improve the estimation of the position and of the speed of the mobile by considering a state space representation for which the state vector is given by $X = [X_1, X_2, X_3]^T$, where X_1, X_2, and X_3 represent the position, the speed and the acceleration of the mobile respectively. We denote the period between two observations by Δ. When this period is fairly short, we may use the approximate expressions

$$X_{2,n-1} = \frac{X_{1,n} - X_{1,n-1}}{\Delta} \text{ and } X_{3,n-1} = \frac{X_{2,n} - X_{2,n-1}}{\Delta}.$$

(9.28)

Moreover, a simple acceleration model is given by

$$X_{3,n} = \rho X_{3,n-1} + V_n, \tag{9.29}$$

where $0 < \rho < 1$, and where V is a white noise process with variance σ_V^2. We then obtain the state space model

$$\begin{cases} X_n = \begin{bmatrix} 1 & \Delta & 0 \\ 0 & 1 & \Delta \\ 0 & 0 & \rho \end{bmatrix} X_{n-1} + \begin{bmatrix} 0 \\ 0 \\ 1 \end{bmatrix} V_n \\ \\ Y_n = \begin{bmatrix} 1 & 0 & 0 \\ 0 & 1 & 0 \end{bmatrix} X_n + W_n. \end{cases} \tag{9.30}$$

We note that the model is robust with respect to the choice of ρ and of the variance of V. However, if the behaviour of the mobile suddenly changes, we may be led to update these parameters.

The study of the Kalman filter for continuous time state space models can be found in [5] Chapter 6 or [4], for instance.

9.3 Generalisation of Kalman Filter

When the distributions of the processes brought into play are available, it may be of interest to consider the previous problem again and look not for $X_n / H_{Y,0:n}$ but for $\mathbb{E}[X_n | Y_0, \dots, Y_n]$. We note that in the Gaussian case both problems are equivalent. For the sake of conciseness we shall denote $\{Y_0, \dots, Y_n\}$ by $Y_{0:n}$.

We may also generalise the problem by considering systems of equations that are no longer linear but of the form

$$\begin{cases} X_{n+1} = F_n(X_n, V_n) \\ \\ Y_n = G_n(X_n, W_n), \end{cases} \tag{9.31}$$

where the variables of the sequence $X_0, (V_0, W_0), \dots, (V_n, W_n), \dots$, are independent ($V_k$ and W_k may be dependent). The first equation of the model is often called the *state equation* and the second the *observation equation*.

In fact, for this more general problem, we can obtain a recursive formulation for the estimation of the state vector, like for Kalman filtering, (see, for example, [7]). Although this represents a digression concerning the main scope of this chapter, which mainly considers problems involving only the second order moments of processes, it is interesting to point out here these general results that can be derived simply.

We note, however, that getting recursive expressions for the estimated state space vector does not mean that its use is always simple, since it implies computing integral expressions. In certain cases, considering a linearised approximation of equations (9.31) and using a Kalman filter enables us to get round the difficulty. Alternatively, we may also look for a numerical solution to the problem using simulation techniques. We shall come back to this approach in Chapter 15.

We assume here that the probability distributions considered are absolutely continuous, the case of more general probability distributions being treated in a similar way. We begin by giving the following definition:

Definition 9.1 *Computing the conditional probability density function* $p($ $x_{n+k}|y_{0:n})$ *is called filtering if* $k = 0$, *smoothing if* $k < 0$, *and prediction if* $k > 0$.

Two classical techniques for estimating the value taken by X_{n+k} from the knowledge of the values $y_0, \ldots, y_n$ taken by $Y_0, \ldots, Y_n$ are given by computing

$$\mathbb{E}[X_{n+k}|Y_{0:n} = y_{0:n}] = \int_{x_{n+k}} x_{n+k} p(x_{n+k}|y_{0:n}) dx_{n+k} \qquad (9.32)$$

and by maximising the probability density function $p(x_{n+k}|y_{0:n})$, this latter criterion being called the Maximum A Posteriori (MAP) criterion. In both cases we are led to compute $p(x_{n+k}|y_{0:n})$, which can be done recursively by exploiting the following property:

Theorem 9.3 *The sequence* $X_0, (X_1, Y_0), \ldots, (X_{n+1}, Y_n), \ldots$ *is a Markov process, as well as* $(X_n)_{n \in \mathbb{N}}$.

Proof By writing $\mathcal{V} = \{v; F_n(x_n, v) \in A\}$ and $\mathcal{W} = \{w; G_n(x_n, w) \in B\}$, it results that

$$P(X_{n+1} \in A, Y_n \in B|(x_n, y_{n-1}), \ldots, (x_1, y_0), x_0)$$

$$= P(V_n \in \mathcal{V}, W_n \in \mathcal{W}) \qquad (9.33)$$

$$= P(X_{n+1} \in A, Y_n \in B|X_n = x_n).$$

Hence,

$$p(x_{n+1}, y_n|(x_n, y_{n-1}), \ldots, (x_1, y_0), x_0) = p(x_{n+1}, y_n|x_n)$$
$$\qquad (9.34)$$
$$= p(x_{n+1}, y_n|x_n, y_{n-1})$$

and (X_{n+1}, Y_n) is a Markov process. Moreover,

$$p(x_{n+1}|x_{0:n}) = \int_{y_n,y_{n-1}} p(x_{n+1}, y_n, y_{n-1}|x_{0:n}) dy_n \, dy_{n-1}$$

$$= \int_{y_n,y_{n-1}} p(x_{n+1}, y_n|y_{n-1}, x_{0:n}) p(y_{n-1}) dy_n \, dy_{n-1}$$

$$= \int_{y_n,y_{n-1}} p(x_{n+1}, y_n|y_{n-1}, x_n) p(y_{n-1}) dy_n \, dy_{n-1} \qquad (9.35)$$

$$= \int_{y_n,y_{n-1}} p(x_{n+1}, y_n, y_{n-1}|x_n) dy_n \, dy_{n-1}$$

$$= p(x_{n+1}|x_n).$$

X is therefore a Markov process. $\square$

To obtain the recurrence equations for filtering, we define

$$\sigma_{n+k|n}(x_{n+k}) = p(x_{n+k}, y_{0:n}),$$

$$\pi_{n+k|n}(x_{n+k}) = p(x_{n+k}|y_{0:n}), \qquad (9.36)$$

that we can note simply $\sigma_{n+k|n}$ and $\pi_{n+k|n}$. $\sigma_{n+1|n}$ is called the *non-normalised one-step predictor*, and $\pi_{n+1|n}$ the *normalised one-step predictor*. They are obtained recursively in the following way:

$$\sigma_{n+1|n} = \int_{x_n} p(x_{n+1}, x_n, y_{0:n}) dx_n$$

$$= \int_{x_n} p(x_{n+1}, y_n|x_n) \sigma_{n|n-1}(x_n) dx_n$$

$$\pi_{n+1|n} = \frac{\sigma_{n+1|n}}{p(y_{0:n})} \qquad (9.37)$$

$$= \frac{\sigma_{n+1|n}}{\int_{x_{n+1}} \sigma_{n+1|n}(x_{n+1}) dx_{n+1}}.$$

$\sigma_{n|n}$ and $\pi_{n|n}$ are called *non-normalised* and *normalised* filters at instant n. Clearly, these filters are given by

$$\sigma_{n|n} = p(y_n|x_n, y_{0:n-1}) p(x_n, y_{0:n-1})$$

$$= p(y_n|x_n) \sigma_{n|n-1} \qquad (9.38)$$

$$\pi_{n|n} = \frac{\sigma_{n|n}}{\int_{x_n} \sigma_{n|n}(x_n) dx_n}.$$

With this formulation, calculating the filter $\sigma_{n|n}$ requires the prior evaluation of the predictor $\sigma_{n|n-1}$. But, the recurrence can also be performed directly on $\sigma_{n|n}$:

$$\sigma_{n+1|n+1} = p(y_{n+1}|x_{n+1})p(x_{n+1}, y_{0:n})$$

$$= p(y_{n+1}|x_{n+1}) \int_{x_n} p(x_{n+1}, x_n, y_{0:n}) dx_n$$

$$= p(y_{n+1}|x_{n+1}) \int_{x_n} p(x_{n+1}|x_n, y_{0:n})\sigma_{n|n}(x_n) dx_n \qquad (9.39)$$

$$= p(y_{n+1}|x_{n+1}) \int_{x_n} \frac{p(x_{n+1}, y_n|x_n)}{p(y_n|x_n)} \sigma_{n|n}(x_n) dx_n.$$

Recurrences (9.37) and (9.39) are initialised by $\sigma_{0|-1} = p(x_0)$, and $\sigma_{0|0} = p(y_0|x_0)p(x_0)$ respectively, assuming that the distribution of X_0 is known.

We notice that if V_n and W_n are independent, formulas (9.37) and (9.39) can be simplified, for then

$$p(x_{n+1}, y_n|x_n) = p(x_{n+1}|x_n)p(y_n|x_n). \qquad (9.40)$$

Using the same approach, we easily obtain the l-step prediction, and smoothing equations. For prediction, we clearly have

$$\sigma_{n+l|n} = \int_{x_{n+1}} p(x_{n+l}|x_{n+1})\sigma_{n+1|n}(x_{n+1}) dx_{n+1}, \qquad (9.41)$$

and

$$p(x_{n+l}|x_{n+1})$$

$$= \int_{x_{n+2}, \dots, x_{n+l-1}} p(x_{n+l}|x_{n+l-1})p(x_{n+l-1}|x_{n+l-2}) \cdots \qquad (9.42)$$

$$\cdots p(x_{n+2}|x_{n+1}) dx_{n+2} \dots dx_{n+l-1}.$$

For smoothing, letting $0 \le k \le n$, we have

$$p(x_k, y_{0:n})$$

$$= \int_{x_0, \dots, x_{k-1}} p(x_0)p(x_1, y_0|x_0) \dots p(x_k, y_{k-1}|x_{k-1}) dx_0 \dots dx_{k-1}$$

$$\times \int_{x_{k+1}, \dots, x_n} p(x_{k+1}, y_k|x_k) \dots p(x_n, y_{n-1}|x_{n-1})p(y_n|x_n) dx_{k+1} \dots dx_n.$$
$$(9.43)$$

To obtain the recurrence equations for smoothing, we conventionally note $\sigma_{k|k-1}(x_k) = \alpha_k(x_k)$:

$$\alpha_k(x_k) = p(x_k, y_{0,k-1})$$

$$= \int_{x_0, \dots, x_{k-1}} p(x_0)p(x_1, y_0|x_0) \dots p(x_k, y_{k-1}|x_{k-1}) dx_0 \dots dx_{k-1},$$
$$(9.44)$$

and

$$\beta_k(x_k) = p(y_{k,n}|x_k)$$

$$= \int_{x_{k+1},\dots,x_n} p(x_{k+1}, y_k|x_k)\dots p(y_n|x_n)dx_{k+1}\dots dx_n.$$

(9.45)

$\alpha_k(x_k)$ and $\beta_k(x_k)$, noted simply as α_k and β_k, are called the *forward filter* and the *backward filter* respectively.

From relation (9.43),

$$p(x_k, y_{0:n}) = \alpha_k(x_k)\beta_k(x_k),$$

(9.46)

and we can evaluate $p(x_k, y_{0:n})$ from the following recurrence equations on α_k and on β_k:

$$\alpha_{l+1} = \int_{x_l} p(x_{l+1}, y_l|x_l)\alpha_l(x_l)dx_l,$$

(9.47)

and $\quad \beta_l = \int_{x_{l+1}} p(x_{l+1}, y_l|x_l)\beta_{l+1}(x_{l+1})dx_{l+1}.$

Remark All the above results are easily extended to the case where the probability measures brought into play are not absolutely continuous with respect to Lebesgue's measure.

Viterbi algorithm In some problems, we wish to estimate the values taken by the sequence $X_0, \dots, X_n$, where X is an L state Markov chain, from the observation $y_0, \dots, y_n$. This type of problem appears conventionally in applications like digital transmissions or pattern recognition (for example, in speech recognition). The criterion generally considered in order to estimate the values taken by $x_{0:n}$ is the maximisation of $P(x_{0:n}|y_{0:n})$. The Viterbi algorithm described in Appendix G allows us to perform this maximisation without having to test individually each of the $(n+1)^L$ possible configurations of $x_{0:n}$ (see, for example, [8]) and requires a computational burden of about nL^2 operations only.

9.4 Matched Filter

Often, in applications such as radar or digital transmissions, we want to detect the presence of a known signal $g(t)$ at the input of a receiver. In fact, $g(t)$ is generally known only up to one time delay or one amplitude factor. It is this delay (in the case of radar detection) or this amplitude (in the case of transmissions), which represents the information of interest.

We assume that we are in the presence of an additive noise B corrupting the observation. This observation is therefore a process of the form

$$X_t = g(t) + B_t.$$

(9.48)

To detect the presence of the signal $g(t)$, we observe the output of a filter with impulse response $h(t)$ and frequency response $H(f)$ and with input X_t. We choose $H(f)$ such that the Signal to Noise Ratio, denoted by SNR, is maximal at a given time t_0.

The output of the filter is given by

$$Y_t = \int_{\mathbb{R}} h(u)g(t-u)du + \int_{\mathbb{R}} e^{2i\pi ft} H(f)d\hat{B}(f)$$

$$= \int_{\mathbb{R}} e^{2i\pi ft} H(f)G(f)df + \int_{\mathbb{R}} e^{2i\pi ft} H(f)d\hat{B}(f) \qquad (9.49)$$

$$= Y_{1,t} + Y_{2,t},$$

where $G(f)$ is the Fourier transform of $g(t)$. At time t, the SNR is then given by

$$SNR_t = \frac{\| Y_{1,t} \|^2}{\| Y_{2,t} \|^2}$$

$$= \frac{|\int_{\mathbb{R}} e^{2i\pi ft} H(f)G(f)df|^2}{\int_{\mathbb{R}} |H(f)|^2 d\mu_B(f)}. \qquad (9.50)$$

We shall assume that B has a PSD, denoted by $S_B(f)$. We then have the following result:

Theorem 9.4 *The maximisation, at instant $t = t_0$, of the Signal to Noise Ratio SNR_t of the process $g(t) + B_t$ is obtained for the filter with frequency response*

$$H(f) = Ke^{-2i\pi ft_0}G^*(f)S_B^{-1}(f), \qquad (9.51)$$

where K is any constant factor. At instant $t = t_0$, the SNR is then given by

$$SNR_{t_0} = \int_{\mathbb{R}} \frac{|G(f)|^2}{S_B(f)}df. \qquad (9.52)$$

Proof Using Cauchy-Schwarz's inequality, we have

$$|\int_{\mathbb{R}} e^{2i\pi ft} H(f)G(f)df|^2$$

$$= |\int_{\mathbb{R}} [e^{2i\pi ft} H(f)S_B^{1/2}(f)] \times [G(f)S_B^{-1/2}(f)]df|^2 \qquad (9.53)$$

$$\leq \int_{\mathbb{R}} |H(f)|^2 S_B(f)df \int_{\mathbb{R}} |G(f)|^2 S_B^{-1}(f)df,$$

with equality at $t = t_0$ if and only if

$$e^{2i\pi f t_0} H(f) S_B^{1/2}(f) = K G^*(f) S_B^{-1/2}(f), \tag{9.54}$$

where K is a constant, that is

$$H(f) = K e^{-2i\pi f t_0} G^*(f) S_B^{-1}(f). \tag{9.55}$$

The corresponding value of SNR_{t_0} is obtained immediately. $\square$

In practice, the position of a peak at the output of the matched filter enables us to estimate a time delay and the amplitude of this peak enables us to estimate the amplitude of the received signal.

Exercises

9.1 (Wiener filter with correlated noise and data) We consider the example in Section 9.1, where $X_n/H_{Y,n}$ is searched for, with $Y_n = X_n + W_n$ and $X_n = z_0 X_{n-1} + V_n$. Here, we assume that $V_n = W_n$. Calculate the corresponding Wiener filter.

9.2 (Wiener filter for an MA(1) process) We consider the example in Section 9.1, but replacing the AR model for X by the MA model given by $X_n = b_0 V_n + b_1 V_{n-1}$, where V is a unit variance white noise. Calculate the Wiener filter that yields $X_n/H_{Y,n}$.

9.3 We consider $Y = AX + V$, where X and V are uncorrelated random vectors, with respective sizes p and q ($q \geq p$). Q will represent the covariance matrix of V.

a) We estimate X from Y by $\hat{X} = MY$. Show that $\| X - \hat{X} \|$ is minimum for

$$M = (A^H R^{-1} A)^{-1} A^H R^{-1}. \tag{9.56}$$

b) Show that the above choice of M ensures minimum error variance for each of the components of $\hat{X}$, that is $\| X_i - \hat{X}_i \|$ is minimum for M given by (9.56) ($i = 1, \ldots, p$).

c) We assume now that $V = [V_1^T \ V_2^T]^T$, where V_1 and V_2 are uncorrelated random vectors with the same size and respective covariance matrices denoted by Q_1 and Q_2. Splitting Y accordingly, show that

$$\hat{X}_2 = \hat{X}_1 + K(Z_2 - R_2\hat{X}_1), \tag{9.57}$$

where $\hat{X}_1 = X/Y_1$ and $\hat{X}_2 = X/ \, span\{Y_1, Y_2\}$, and calculate K.
(Hint: consider the discussion in the study of Kalman filtering.)

9.4 We consider the following state space model

$$
\begin{cases}
X_{n+1} = F_n X_n + V_n \\
Y_n \;\;\;= H_n X_n + W_n .
\end{cases}
\tag{9.58}
$$

a) Show that the process Y can also be described by means of a state space model without observation noise:

$$
\begin{cases}
X'_{n+1} = F'_n X'_n + V'_n \\
Y_n \;\;\;= H'_n X'_n .
\end{cases}
\tag{9.59}
$$

b) Give the recursion formula for the calculation of $V'_n / Y_{0:n}$.

9.5 (Kalman filter as a special case of filtering equations) In the particular case where the system (9.31) is linear and V and W are Gaussian processes, find the equations of the Kalman filter from the general filtering equations of state space models.

9.6 (Kalman smoothing filter) Let Y denote a stochastic process described by a stationary state space model. We are looking for the distribution of the successive states of the model $X_1, \ldots, X_n$, from the knowledge of the observed sequence $Y_1, \ldots, Y_n$. We assume that the model is of the form

$$
\begin{cases}
X_{k+1} = F X_k + V_{k+1} \\
Y_k = H X_k + W_k ,
\end{cases}
\tag{9.60}
$$

where V and W are independent white Gaussian noises with respective auto-covariance matrices Q_V and Q_W. In what follows, we use the same notations as in the study of the Kalman filter.

We are looking for the distribution of $X_{1:n}$ conditional to $Y_{1:n} = y_{1:n}$ and we denote by $p(x_{1:n}|y_{1:n})$ as the corresponding probability density function.

a) Show that $p(x_{1:n}|y_{1:n})$ may be expressed as a function of $p(x_k|y_{1:n}, x_{k+1:n})$, for $k = 1, \ldots, n-1$, and of $p(x_n|y_{1:n})$.

b) Show that $p(x_k|y_{1:n}, x_{k+1:n}) = p(x_k|y_{1:n}, x_{k+1})$.

c) Show that $p(x_k|y_{1,k}) \sim \mathcal{N}(m'_k, R'_k)$ and express the recurrence equation upon (m'_k, R'_k) for $k \leq n$. We shall note $(m_n, R_n) = (m'_n, R'_n)$.

d) For $k = n-1, n-2, \ldots, 1$, the parameters of the Gaussian density $p(x_k|y_{1:k}, x_{k+1})$ are calculated by means of the relation

$$
p(x_k|y_{1,n}, x_{k+1,n}) = p(x_{k+1}|x_k) p(x_k|y_{1:k}) / p(x_{k+1}|y_{1:k}).
\tag{9.61}
$$

Using the same notations as in Section 9.2, prove that $x_k|y_{1:n}, x_{k+1} \sim \mathcal{N}(m_k, R_k)$, where

$$m_k = \hat{X}_{k,k} + G_k(m_{k+1} - F_{k+1}\hat{X}_{k,k})$$

$$R_k = C_k - G_k B_{k+1} G_k^H \qquad (9.62)$$

$$G_k = C_k F_k^H B_{k+1}^{-1}.$$

(Hint: use the matrix inversion lemma 16.2.)

9.7 (Matched filtering and symbol transmission in the presence of noise) The signal $X_t = A\mathbb{1}_{[0,T]}(t)$ is transmitted to a receiver. A is a Bernoulli random variable with $P(A = 1) = p$ and $P(A = -1) = 1 - p$. The signal $Y_t = X_t + V_t$, where $V = (V_t)_{t \in \mathbb{R}}$ is a white Gaussian process independent of A, is observed at the receiver side.

a) At time T, we want to decide whether the symbol $+1$ or -1 has been transmitted. To do this, we apply a filter at the receiver side that outputs the maximum Signal to Noise Ratio (SNR) at time T. Give the expression of the impulse response of the filter and of its output.

b) Let Z_T denote the output of the filter at time T. A threshold at level S enables a decision to be made in the following way: if $Z_T > S$, then we decide that $A = 1$, otherwise we decide that $A = -1$. Calculate the probability P_e that a wrong decision is made. What is the value of S that achieves a minimum value for P_e? Consider the particular cases $p = 1/2$ and $p = 0$.

9.8 (Estimation of an unknown deterministic signal in the presence of noise) We consider a process $Y_t = g(t) + V_t$, where $g(t)$ is an unknown deterministic continuous function and $V = (V_t)_{t \in \mathbb{R}}$ a white noise with variance σ_V^2. Y_t is observed for $t \in [0, T]$. On this time interval, $g(t)$ is approximated by its truncated Fourier expansion $g_N(t) = \sum_{k=-N,N} \hat{g}_k e^{2i\pi kt/T}$.

a) Show that using a bank of matched filters to estimate the coefficients $(\hat{g}_k)_{k=-N,N}$ yields the following estimator of $g_N(t)$:

$$\tilde{g}_N(t) = \sum_{k=-n,N} \left(T^{-1} \int_{[0,T]} e^{-2i\pi kt/T} Y_t dt \right) e^{2i\pi kt/T}. \qquad (9.63)$$

b) Show that the average mean square error on $[0, T]$, defined by $T^{-1} \int_{[0,T]} \| g(t) - \tilde{g}_N(t) \|^2 dt$, is given by

$$T^{-1} \int_{[0,T]} \| g(t) - \tilde{g}_N(t) \|^2 dt = \left(\sum_{k>N} |\hat{g}_k|^2 \right) + \frac{2N+1}{T} \sigma_V^2. \qquad (9.64)$$

What would be a good choice for N?

9.9 We consider a WSS process $Y_n = X_n + V_n$, where V is a white noise uncorrelated with X, and $B(z) = \sum_{k=0,p} b_k z^{-k}$ is a transfer function. We are searching for $b = [b_0, \ldots, b_q]^T$ such that the Signal to Noise Ratio (SNR) of $[b(z)]Y_n$ defined by

$$\rho_b = \| [b(z)]X_n \|^2 \| [b(z)]V_n \|^{-2} \tag{9.65}$$

is maximum.

a) Show that this is achieved when b is the largest eigenvector of the matrix $T_{X,q}$, of size $(q+1) \times (q+1)$ and of general term $[T_{X,q}] = R_X(i-j)$ and give the corresponding value of ρ_b.

b) Solve the same problem when V is a correlated noise with covariance matrix $T_{V,q}$.

9.10 Let $X = (X_n)_{n \in \mathbb{Z}}$ be a WSS process observed through two sensors in the presence of additive noise, yielding the observations $Y_{1,n} = X_n + V_{1,n}$ and $Y_{2,n} = X_n + V_{2,n}$. We assume that X, V_1 and V_2 are zero mean uncorrelated processes with respective spectral densities $S_X(f)$, $S_{V_1}(f)$ and $S_{V_2}(f)$. We want to recover X from the observation of Y_1 and of Y_2 by means of two filters with transfer functions $h_1(z)$ and $h_2(z)$ as follows: we let

$$Z_n = [h_1(z)]Y_{1,n} + [h_2(z)]Y_{2,n}, \tag{9.66}$$

and we want $\| X_n - Z_n \|$ to be minimum. Show that the frequency response of these filters must be

$$h_1(e^{2i\pi f}) = \frac{S_X(f)S_{V_2}(f)}{(S_{V_1}(f) + S_{V_2}(f))S_X(f) + S_{V_1}(f)S_{V_2}(f)}, \tag{9.67}$$

and

$$h_2(e^{2i\pi f}) = \frac{S_X(f)S_{V_1}(f)}{(S_{V_1}(f) + S_{V_2}(f))S_X(f) + S_{V_1}(f)S_{V_2}(f)}. \tag{9.68}$$

10. Rational Spectral Densities

Purpose Processes with rational Power Spectrum Density (PSD) play a very important role for modelling second order properties of processes. In particular, piecewise continuous PSDs can be approximated with an arbitrary precision by rational PSDs (and even polynomial PSDs, see Appendix F). Moreover, these models involve few parameters, since in many situations it is sufficient to use rational PSDs of low degree. Finally, the identification or the estimation of the parameters of a rational PSD is relatively simple, as will be seen in the following chapters. We shall therefore indicate here some important results about these processes.

10.1 Difference Equations and Rational Spectral Densities

We consider the difference equation

$$X_n + \sum_{k=1,p} a_k X_{n-k} = \sum_{l=0,q} b_l U_{n-k}, \tag{10.1}$$

where $U = (U_n)_{n \in \mathbb{Z}}$ is a fixed WSS process, with an absolutely continuous spectrum. Moreover, we assume that the PSD $S_U(f)$ of U verifies $0 < m \le S_U(f) \le M < \infty$. If there is a stationary process X solution to Equation (10.1), then

$$a(e^{2i\pi f})d\hat{X}(f) = b(e^{2i\pi f})d\hat{U}(f), \tag{10.2}$$

with

$$a(e^{2i\pi f}) = 1 + \sum_{k=1,p} a_k e^{-2i\pi k f},$$

$$\text{and } b(e^{2i\pi f}) = \sum_{l=0,q} b_l e^{-2i\pi l f}. \tag{10.3}$$

We notice that if $a(z)$ is equal to zero at some points of the unit circle, then $b(e^{2i\pi f})a^{-1}(e^{2i\pi f})$ does not belong to $L^2(\mathcal{I}, B(\mathcal{I}), S_U(f)df)$, and the problem has no solution. We shall assume, therefore, that $a(e^{2i\pi f}) \ne 0$, $\forall f \in \mathcal{I}$. Then,

$$X_n = \int_{\mathcal{I}} e^{2i\pi n f} \frac{b(e^{2i\pi f})}{a(e^{2i\pi f})} d\hat{U}(f) \tag{10.4}$$

is the only solution to the problem.

We can also write $X_n = \sum_{k \in \mathbb{Z}} h_k U_{n-k}$, where the coefficients h_k are those of Laurent's series expansion of $b(z)a^{-1}(z)$ in the neighbourhood of the unit circle. We notice that $\sum_{k \in \mathbb{Z}} |h_k| < \infty$.

If, moreover, $b(z)a^{-1}(z)$ is the transfer function of a causal filter, that is, if $h_k = 0$ for $k < 0$, then the series $\sum_{k \in \mathbb{N}} h_k z^{-k}$ converges in a domain $|z| > 1 - \varepsilon$ ($\varepsilon > 0$). $b(z)a^{-1}(z)$ is therefore holomorphic in this domain and $a(z) \neq 0$ for $|z| \geq 1$. The converse is straightforward and the following theorem sums up these results.

Theorem 10.1 *If U is a regular WSS process whose PSD verifies*

$$0 < m \leq S_U(f) \leq M < \infty, \tag{10.5}$$

the difference equation

$$X_n + \sum_{k=1,p} a_k X_{n-k} = \sum_{l=0,q} b_l U_{n-l} \tag{10.6}$$

has a solution if and only if $a(z) = 1 + \sum_{k=1,p} a_k z^{-k} \neq 0$ for $|z| = 1$. We then have

$$X_n = \int_{\mathcal{I}} e^{2i\pi n f} \frac{b(e^{2i\pi f})}{a(e^{2i\pi f})} d\hat{U}(f), \tag{10.7}$$

with $b(z) = \sum_{k=0,q} b_k z^{-k}$. Moreover, the transfer function filter $b(z)a^{-1}(z)$ is causal if and only if $a(z) \neq 0$ for $|z| \geq 1$.

The PSD of X is given by

$$S_X(f) = \left| \frac{b(e^{2i\pi f})}{a(e^{2i\pi f})} \right|^2 S_U(f). \tag{10.8}$$

If U has a PSD that is a rational function of $e^{-2i\pi f}$, then this is also true for X. This is particularly the case when U is a white noise process since $S_U(f)$ is then constant. We shall assume henceforth that U is a white noise process. In this case, X is called an Auto Regressive Moving Average (ARMA) process. When $a(z) = 1$, X is called an MA process, and when $b(z)$ is constant, X is called an AR process. An ARMA model for which the degree of $a(z)$ and of $b(z)$ is respectively p and q will be referred to as an ARMA(p, q) model.

We now indicate an interesting property of the covariances of ARMA processes

Theorem 10.2 *The autocovariance coefficients of ARMA processes decrease exponentially.*

Proof It is clear that for $n > q$

$$[a(z)]R_X(n) = \mathbb{E}[(X_{m+n} + \sum_{k=1,p} a_k X_{m+n-k})X_m^*]$$

$$= 0,$$

$$(10.9)$$

for $X_{m+n} + \sum_{k=1,p} a_k X_{m+n-k} \in H_{U,m+n-q,m+n}$, $X_m \in H_{U,m}$, and U is a white noise process.

We now write $a(z)$ in the form

$$a(z) = \prod_{k=1,l} (1 - \alpha_k z^{-k})^{r_k},$$

$$(10.10)$$

with $r_1 + \ldots + r_l = p$. It is known that the sequences $(x_n)_{n \in \mathbb{N}}$, which satisfy the recurrence relation $x_n = \sum_{k=1,p} a_k x_{n-k}$, for $n \geq p$, constitute a vector space of dimension p whose elements are of the form

$$x_n = \sum_{k=1,l} (\sum_{s=0,r_k-1} c_{k,s} n^s) \alpha_k^n.$$

$$(10.11)$$

Then the result stems from the fact that the zeroes α_k of $a(z)$ lie inside the unit disk. $\square$

10.2 Spectral Factorisation of Rational Spectra

In the case of rational spectral processes, the minimum-phase causal factorisation theorem for regular processes, presented in the context of linear prediction theory, takes the following form

Theorem 10.3 *If X is an ARMA process, $S_X(f)$ can be factorised in the form*

$$S_X(f) = G(e^{2i\pi f}) = \left| \frac{b(e^{2i\pi f})}{a(e^{2i\pi f})} \right|^2,$$

$$(10.12)$$

where $a(z)$ and $b(z)$ are polynomials with no common zeroes, and $a(z) \neq 0$, for $|z| = 1$. In particular, there exists a single factorisation (up to a modulus 1 factor) for which $b(z) \neq 0$ for $|z| > 1$, $a(z) \neq 0$ for $|z| \geq 1$, and the numerator and denominator of $G(z)$ have degrees that are twice that of $b(z)$ and of $a(z)$ respectively. This factorisation coincides with the minimum-phase causal factorisation of $S_X(f)$.

Proof See Appendix H.

An ARMA process X may therefore be represented as the output of a filter with a white noise input process and having a rational transfer function

$b(z)a^{-1}(z)$. We then call it an ARMA representation of X. The ARMA representation associated with minimum-phase spectral factorisation defines a stable causal filter with a transfer function $b(z)a^{-1}(z) = \sum_{k \geq 0} h_k z^{-k}$, which gives the corresponding representation of X as a function of its normalised innovation ν: $X_n = \sum_{k \geq 0} h_k \nu_{n-k}$ and

$$X_n + \sum_{k=1,p} a_k X_{n-k} = \sum_{l=0,q} b_l \nu_{n-k}. \tag{10.13}$$

Conversely, as $H_{X,n} = H_{\nu,n}$, the transfer function filter $a(z)b^{-1}(z)$ gives the representation filter of ν as a function of X. We notice that the fact that $b(z)$ can be zero at some points of the unit circle is not a problem here (at least in theory), insofar as we get $a(e^{2i\pi f})b^{-1}(e^{2i\pi f}) \in L^2(\mathcal{I}, \mathcal{B}(\mathcal{I}), S_X(f)df)$ in any case.

We note that when we speak of the ARMA representation of a rational process, we generally mean the ARMA representation associated with minimum-phase causal factorisation.

10.3 State Space Representation of ARMA Models

ARMA models can be written in the form of a linear stationary state space model. We consider an ARMA model defined by the equation

$$Y_n + \sum_{k=1,p} a_k Y_{n-k} = \sum_{l=0,q} b_l U_{n-k} \tag{10.14}$$

and we begin by assuming that $p = q + 1$. We now consider the state space model

$$\begin{cases} \mathbf{X}_{n+1} = A\mathbf{X}_n + BU_{n+1}, \\ Z_n = C\mathbf{X}_n \end{cases} \tag{10.15}$$

with $\mathbf{X}_n = [X_n, \ldots, X_{n-p+1}]^T$, $B = [1, 0, \ldots, 0]^T$, $C = [b_0, \ldots, b_q]$, and

$$A = \begin{bmatrix} -a_1 & \ldots & & -a_p \\ 1 & 0 & . & . & 0 \\ 0 & 1 & 0 & . & 0 \\ & . & . & . & . \\ 0 & . & . & 1 & 0 \end{bmatrix}. \tag{10.16}$$

We show that $Z = Y$:

$$Z_n = \sum_{l=0,q} b_l X_{n-l},$$

$$\sum_{k=0,p} a_k Z_{n-k} = \sum_{l=0,q} b_l \left(\sum_{k=0,p} a_k X_{n-l-k} \right) \tag{10.17}$$

$$= \sum_{l=0,q} b_l U_{n-l}.$$

Therefore, $Z = Y$.

If $p \neq q + 1$, we can simply proceed in the same way by completing the shortest of the sequences $(a_k)_{k=1,p}$ and $(b_l)_{l=0,q}$ by coefficients equal to 0, until we obtain two sequences of the same length.

Conversely, we consider a linear stationary state space model of the form

$$\begin{cases} \mathbf{X}_{n+1} = A\mathbf{X}_n + BU_{n+1} \\ Y_n = C\mathbf{X}_n. \end{cases} \tag{10.18}$$

For such a model, $[I_p - Az^{-1}]\mathbf{X}_n = BU_n$, and, therefore,

$$Y_n = [C(I - Az^{-1})^{-1}B]U_n. \tag{10.19}$$

It is clear that $C(I - Az^{-1})^{-1}B$ is a rational transfer function, because the coefficients of the matrix $(I - Az^{-1})^{-1}$ are the ratio of the cofactors of $I - Az^{-1}$ and of the determinant of $I - Az^{-1}$. Therefore, if U is a white noise process, Y is an ARMA process if the zeroes of $|I - Az^{-1}|$, that is, the eigenvalues of A, are in the unit disk. Let us remark that we have $|I - Az^{-1}| = a(z)$, with $a(z) = \sum_{k=0,p} a_k z^{-k}$, which, in passing, shows that the calculation of the roots of a polynomial $a(z)$ amounts to that of the eigenvalues of the matrix A associated with it by relation (10.16), and which is called the *companion matrix* of $a(z)$.

Exercises

In the following exercises, $X = (X_n)_{n \in \mathbb{Z}}$ will denote an ARMA(p,q) process satisfying the recurrence equation

$$X_n + \sum_{k=1,p} a_k X_{n-k} = \sum_{l=0,q} b_l V_{n-l}, \tag{10.20}$$

where $V = (V_n)_{n \in \mathbb{Z}set}$ is a white noise with variance σ_V^2.

10.1 Find the minimum-phase factorisation of the following PSD:

$$S(f) = \frac{5 - 4\cos(2\pi f)}{25 - 9\cos^2(2\pi f)}. \tag{10.21}$$

10.2 We assume that X is an AR(1) process. Check that $\sigma_X^2 = \sigma_V^2(1 - |a|)^2$.

10.3 We consider the transfer function $h(z) = (1 + a_1 z^{-1} + a_2 z^{-2})^{-1}$. For which values of (a_1, a_2) in the plane $\mathbb{R}^2$ does $h(e^{2i\pi f})$ represent the minimum-phase factorisation of $S(f) = |h(e^{2i\pi f})|^2$?

10.4 (Difference equation with initial condition) Let $Y = (Y_n)_{n \in \mathbb{N}}$ be a process such that $Y_0 = 0$ and $Y_n = aY_{n-1} + V_n$ for $n > 0$, where $|a| < 1$ and the random variables $(V_n)_{n \in \mathbb{N}^*}$ are uncorrelated zero mean random variables with variance σ_V^2. Calculate $\text{cov}(Y_{m+n}, Y_m)$ and study its behaviour when m tends to $+\infty$.

10.5 (Causal expansion of an AR model) Let $X = (X_n)_{n \in \mathbb{Z}}$ denote an AR(p) process and denote by $(z_k)_{k=1,p}$ the roots of $z^n + \sum_{k=1,p} a_k z^{n-k} = 0$, with $|z_k| < 1$ for $k = 1, \ldots, p$. Show that the minimum-phase factorisation of $S_X(f)$ can also be written as

$$h(e^{2i\pi f}) = \sum_{n=0,\infty} \left(\frac{z_k^n}{\prod_{j \neq k}(1 - z_j z_k)} \right) e^{-2i\pi n f}. \tag{10.22}$$

10.6 We consider $Y_n = X_n + \sum_{k=1,K} \xi_k e^{2i\pi n f_k}$, where $X = (X_n)_{n \in \mathbb{Z}}$ is an ARMA(p, q) process. We assume that the random variables ξ_k are uncorrelated random variables, and are uncorrelated with X.

a) Express Y_n as the output of a state space model.

b) Parallelling the discussion in Section 10.3, explain how $(f_k)_{k=1,p}$ can be recovered from knowledge of $(R_Y(n))_{n=0,M}$, with $M > 2(K + p)$.

10.7 (Innovation of an ARMA model) Let X be an ARMA(2,2) process.

a) Write the state space model associated with this process.

b) We assume that V is Gaussian. Calculate $\mathbb{E}[V_n | Y_{0:n}]$ recursively by means of a Kalman filter.

10.8 (The Box-Jenkins forecasting method) We assume that the random variables $(V_n)_{n \in \mathbb{Z}}$ are independent. We are looking for an iterative technique to calculate $\tilde{X}_{n,n-m} = \mathbb{E}[X_n | H_{X,n-m}]$.

a) Check that $\tilde{X}_{n,n-1} + \sum_{k=1,p} a_k X_{n-k} = \sum_{l=1,q} b_l V_{n-l}$.

b) Show that $V_{n-k} = X_{n-k} - \tilde{X}_{n-k,n-k-1}$ for $k \geq 0$.

c) Assuming, for instance, that $p \geq q$, show that $\tilde{X}_{n,n-1}$ can be computed from $\tilde{X}_{n-1,n-2}, \ldots, \tilde{X}_{n-q,n-q-1}$ and $X_n, \ldots, X_{n-p+1}$ by means of the relation

$$\sum_{l=0,q} b_l \tilde{X}_{n-l,n-l-1} = \sum_{k=1,q} (b_k - a_k) X_{n-k+1} - \sum_{k=q+1,p} a_k X_{n-k+1}. \tag{10.23}$$

d) Generalise this result for the iterative calculation of $\tilde{X}_{n,n-m}$ for $m > 1$.

10.9 (Time continuous AR(1) model) We consider an electronic device consisting of an inductance L followed by a resistor R. The voltage across the circuit is modelled as a white noise with variance σ^2, supplied by a noise generator.

a) Check that the intensity in the circuit is of the form

$$X_t = \int_{\mathbb{R}} e^{2i\pi f t} \frac{1}{R + 2i\pi L f} d\hat{W}(f), \tag{10.24}$$

where $\| \hat{W}([a, b]) \|^2 = \sigma^2 (b - a)$.

b) Calculate the spectrum and the covariance function of X.

10.10 (Time continuous ARMA processes) We are looking for a process $X = (X_t)_{t \in \mathbb{R}}$ such that

$$\sum_{k=1,p} a_k X_t^{(k)} = \sum_{l=0,q} b_l V_t^{(l)}, \tag{10.25}$$

where $.^{(k)}$ denotes the kth derivative, and $V = (V_t)_{t \in \mathbb{R}}$ is a white noise.

a) Show that $V_t^{(l)}$ can be defined in the sense of generalised processes.

b) Find a condition that ensures the existence of a mean square continuous WSS solution X to Equation (10.25).

c) Check that the solution can be written in the form

$$X_t = \int_{\mathbb{R}} e^{2i\pi ft} \frac{\sum_{l=0,q} b_l (2i\pi f)^l}{1 + \sum_{k=1,p} a_k (2i\pi f)^k} d\hat{V}(f). \tag{10.26}$$

10.11 (Multivariate ARMA processes) Let $X = (X_n)_{n \in \mathbb{Z}}$ denote a multivariate ARMA(p,q) process of order d defined by the relation

$$X_n + \sum_{k=1,p} A_k X_{n-k} = \sum_{l=0,q} B_l V_{n-l}, \tag{10.27}$$

where the coefficients A_k and B_l are matrices of size $d \times d$. We assume that $V = (V_n)_{n \in \mathbb{Z}}$ is a zero mean multivariate white noise of order d with covariance matrix $\mathbb{E}[V_n V_n^H] = \Sigma^2$. Calculate the PSD matrix of X.

11. Spectral Identification of WSS Processes

Purpose In general, the spectrum of a process is not directly available and we only have knowledge of its first autocovariance coefficients, or of an estimation of them. From this partial knowledge, we recall how the spectra of ARMA processes can be identified. More generally, we explain how the set of spectra whose first autocovariance coefficients are given can be characterised. Important further results related to this problem are also presented.

11.1 Spectral Identification of ARMA Processes

We wish to identify the minimum-phase causal factorisation of an ARMA process from knowledge of its first autocovariance coefficients $(R_X(k))_{k=0,N}$. X is defined by

$$X_n + \sum_{k=1,p} a_k X_{n-k} = \sum_{l=0,q} b_l U_{n-l}, \tag{11.1}$$

where U is a white noise. We therefore wish to identify the coefficients $(a_k)_{k=1,p}$ and $(b_l)_{l=0,q}$.

11.1.1 Identification of the AR Part

To identify the coefficients $(a_k)_{k=1,p}$, we notice that

$$X_n = -\sum_{k=1,p} a_k X_{n-k} + \sum_{l=0,q} b_l U_{n-k}$$

$$= \sum_{k=0,\infty} h_k U_{n-k}. \tag{11.2}$$

As U is a white noise, $X_{n-l} \in H_{U,n-l}$ and $X_n + \sum_{k=1,p} a_k X_{n-k} \in H_{U,n-q,n}$. Therefore,

$$\mathbb{E}[(X_n + \sum_{k=1,p} a_k X_{n-k}) X_{n-l}^*] = R_X(l) + \sum_{k=1,p} a_k R_X(l-k)$$

$$= 0, \qquad \text{for } l > q. \tag{11.3}$$

The resolution of the linear system thus obtained yields the coefficients $(a_k)_{k=1,p}$.

11.1.2 Identification of the MA Part

To identify the coefficients $(b_l)_{l=0,q}$, we notice that $S_X(f)|a(e^{2i\pi f})|^2 = |b(e^{2i\pi f})|^2$. It is then clear that we can compute the coefficients $(b_l)_{l=0,q}$ by identifying the coefficients of polynomial equality

$$(\sum_{n=-(p+q)}^{p+q} R_X(n)e^{-2i\pi n f})|a(e^{2i\pi f})|^2 = |b(e^{2i\pi f})|^2. \tag{11.4}$$

We thus obtain non-linear relations whose optimum can be found by numerical techniques.

11.1.3 Identification of the State Space Representation

We saw in Section 10.3 that we may represent an ARMA process Y by means of a linear stationary state space model of the form

$$\begin{cases} \mathbf{X}_{n+1} = A\mathbf{X}_n + BU_{n+1} \\ Y_n \quad = C\mathbf{X}_n, \end{cases} \tag{11.5}$$

where U is a white noise. The process Y is therefore parameterised by the matrices A, B, and C.

The autocovariance coefficients of Y can then be expressed simply as a function of these matrices: for $k \geq 0$,

$$R_Y(k) = \mathbb{E}[Y_{n+k}Y_n^*]$$

$$= C\mathbb{E}[\mathbf{X}_{n+k}\mathbf{X}_n^*]C^H$$

$$= C\mathbb{E}[(A^k\mathbf{X}_n + \sum_{l=0,k-1} A^l BU_{n+1+k-l})\mathbf{X}_n^*]C^H \tag{11.6}$$

$$= CA^k PC^H,$$

where P represents the covariance matrix of $\mathbf{X}_n$. We notice that this expression clearly shows the exponential decrease of the coefficients $R_Y(k)$, already established in the previous chapter. Indeed, the eigenvalues of A have a modulus smaller than 1, and by noting $U\Lambda U^{-1}$ the eigenvalue decomposition of A, it results that $R_Y(k) = (CU)\Lambda^k(U^{-1}PC^H)$.

We shall see that the identification of the parameters (A, B, C) can be realised simply by considering the matrix

$$\mathcal{H}_K = \begin{bmatrix} R_Y(0) & R_Y(1) & . & R_Y(K) \\ R_Y(1) & . & . & . \\ . & . & . & . \\ R_Y(K) & . & . & R_Y(2K) \end{bmatrix}, \tag{11.7}$$

called a *Hankel matrix*, that is, its parallels to the second diagonal are made up of identical terms: $[\mathcal{H}_K]_{ij} = R_Y(i + j - 2)$. We shall assume that $K \geq r$, where $r = \max(p, q)$ represents the maximum of the degrees of the numerator and of the denominator of the spectral factorisation of the ARMA process Y.

Beforehand, we note that if M is an invertible matrix, we do not modify the process Y if in model (11.5) (A, B, C) is replaced by $(A', B', C') = (MAM^{-1}, MB, CM^{-1})$. The state vector then becomes $X'_n = MX_n$. It is therefore clear that the identification of (A, B, C), from knowledge of Y or of its autocovariances, can only be achieved up to a change of basis of the vectors X_n, defined by an invertible matrix M.

As $R_Y(k) = CA^k PC^H$, $\mathcal{H}_K$ can be expressed in the following way:

$$\mathcal{H}_K = \begin{bmatrix} C \\ CA \\ \vdots \\ CA^K \end{bmatrix} \times [PC, APC, \dots, A^K PC] = \mathcal{O}\mathcal{C}. \tag{11.8}$$

We notice that $(\mathcal{O}, \mathcal{C})$ is not unique and is in fact defined up to a change of basis defined by a matrix M: $\mathcal{H}_K = \mathcal{O}'\mathcal{C}'$, where $(\mathcal{O}', \mathcal{C}') = (\mathcal{O}M^{-1}, M\mathcal{C})$. This kind of factorisation may, for example, be obtained by performing the singular value decomposition of $\mathcal{H}_K$.

Let us consider any factorisation $(\mathcal{O}', \mathcal{C}')$ of $\mathcal{H}_K$. We denote by $\mathcal{O}'^{\uparrow}$ and $\mathcal{O}'^{\downarrow}$ the matrices obtained by deleting the last and the first line of $\mathcal{O}'$ respectively. It is clear that

$$\mathcal{O}'^{\uparrow} A' = \mathcal{O}'^{\downarrow}, \tag{11.9}$$

that is,

$$A' = [(\mathcal{O}'^{\uparrow})^H \mathcal{O}'^{\uparrow}]^{-1}(\mathcal{O}'^{\uparrow})^H \mathcal{O}'^{\downarrow}. \tag{11.10}$$

Moreover, the first line of $\mathcal{O}'$ gives vector C'. As for vector B', as for the calculation of the coefficient of $b(z)$ in the previous section, we may obtain it by solving a system of non-linear second degree equations. Indeed, it is clear that

$$Y_n = [C(I - Az^{-1})^{-1}B]U_n$$
$$= [a(z)^{-1}CN(z)B]U_n, \tag{11.11}$$

where $a(z) = |I - Az^{-1}|^{-1}$ and $N(z)$ is the polynomial matrix of the variable z^{-1} defined by the relation $(I - Az^{-1})^{-1} = a(z)^{-1}N(z)$. Therefore, $[a(z)]Y_n = [CN(z)B]U_n$, and the coefficients of B can be computed by solving the system of second degree equations, obtained by identifying the coefficients of the equality

$$|a(e^{-2i\pi f})|^2 \Big(\sum_{n=-(p+q)}^{p+q} R_Y(n)e^{-2i\pi nf} \Big) = |CN(e^{2i\pi f})B|^2. \qquad (11.12)$$

Remark We have assumed here that the matrix $(\mathcal{O}'^{\dagger})^H \mathcal{O}'^{\dagger}$ has an inverse. In order for this hypothesis to be justified, it is sufficient to check that the matrix $\mathcal{O}'$, of size $(K+1) \times r$ is of rank r $(K > r)$. As $\mathcal{O}' = \mathcal{O}M$, where M is an invertible matrix, it is sufficient to show that $\mathcal{O}$ is of rank r. But $\mathcal{O} = [C^T, (CA)^T, \dots, (CA^K)^T]^T$ and it is sufficient to show that the matrix $[C^T, (CA)^T, \dots, (CA^{r-1})^T]^T$ is of rank r in order for $\mathcal{O}$ to be so. To show that this matrix is full rank, we consider a linear combination $\sum_{k=0,r-1} \alpha_k CA^k$ of its lines. If $C(\sum_{k=0,r-1} \alpha_k A^k) = 0$,

$$(\textstyle\sum_{k=0,r-1} \alpha_k CA^k) A^l PC^H = \sum_{k=0,r-1} \alpha_k R_Y(k+l)$$
$$= 0, \qquad (11.13)$$

for any value of l. By denoting by $\alpha = [\alpha_0, \dots, \alpha_{r-1}]$ and T_{r-1} the matrix of size $r \times r$ and of general term $[T_{r-1}]_{ij} = R(i-j)$, it is then clear that $\alpha T_{r-1} \alpha^H = 0$. Therefore, if we had $\alpha \neq 0$, the matrix T_{r-1} would be singular. But from Caratheodory's theorem (Theorem 11.2) presented in the following paragraph, the spectrum of Y would then be carried by $r-1$ points at most, which is contradictory to the fact that Y is an ARMA process. The matrix $(\mathcal{O}'^{\dagger})^H \mathcal{O}'^{\dagger}$ is therefore invertible.

11.2 The Trigonometric Moment Problem

We wish here to characterise the set of all spectral measures whose first Fourier coefficients $(R(k))_{k=0,N}$ are given, a problem that is known as the *trigonometric moment problem*.

11.2.1 Condition of Existence of Solutions

We begin by recalling a necessary and sufficient condition in order for a set of coefficients to represent the first autocovariance coefficients of a certain process.

Theorem 11.1 *A sequence of coefficients $(R(k))_{k=0,N}$ represents the $N+1$ first autocovariance coefficients of a certain WSS process if and only if the Toeplitz matrix T_N of size $N+1$ and of general term $[T_N]_{i,j} = R(i-j)$ is positive.*

Proof See Appendix I.2.

When the matrix T_N is positive and singular we show, moreover, that the spectrum of the process is discrete and defined uniquely:

Theorem 11.2 (Caratheodory) *The matrix T_N is positive singular of rank $p < N+1$ if and only if there exists a unique positive measure μ, carried by p points, whose coefficients $(R(k))_{k=0,N}$ are the first Fourier coefficients.*

Proof See Appendix I.2 (see also, for example, [43]).

We notice that when T_N is singular, the corresponding discrete measure μ being denoted by $d\mu(f) = \sum_{k=1,p} \rho_k \delta_{f_k}$, T_N has a unique decomposition of the form

$$T_N = \sum_{k=1,p} \rho_k d(f_k) d(f_k)^H, \tag{11.14}$$

where $d(f_k) = [1, e^{2i\pi f_k}, \dots, e^{2i\pi N f_k}]^T$. This property is often used in order to estimate the frequencies of a set of sinusoids in the presence of noise, as will be seen below.

In order to be able to give a full description of the set of the positive measures μ such that

$$R(n) = \int_{\mathcal{I}} e^{2i\pi n f} d\mu(f), \quad n = 0, N, \tag{11.15}$$

when $T_N > 0$, we shall begin by presenting some results concerning orthogonal polynomials on the unit circle, and about certain classes of holomorphic functions.

11.2.2 Orthogonal Polynomials on the Unit Circle

Let $(R(n))_{n=0,N}$ be a sequence such that $T_N \geq 0$. Such a sequence is said to be positive. We then define a scalar product on the set of complex polynomials of order smaller than or equal to N by the relations

$$< z^m, z^n > = R(m - n), \quad (m, n) = 0, N. \tag{11.16}$$

In fact, for a polynomial $P(z)$, the norm property $\| P(z) \| = 0 \Rightarrow P = 0$ is only satisfied if $T_N > 0$. We shall examine this situation, since the case where T_N is positive and singular has already been considered in the Caratheodory theorem (Theorem 11.2).

Orthogonal Szegö Polynomials of the First Kind Orthogonal Szegö polynomials of the first kind are defined by the relations

$$\begin{cases} Q_0(z) = 1 \\ Q_n(z) = \sum_{k=0,n} q_{k,n} z^{n-k}, \text{ with } q_{0,n} = 1 \text{ and } n \leq N \\ < Q_m(z), Q_n(z) > = 0, \text{ for } m \neq n. \end{cases} \tag{11.17}$$

The polynomials $(Q_n(z))_{n=0,N}$ present a direct link with finite past linear prediction. More precisely, we can check that if the coefficients $R(k)$ match the autocovariance coefficients $R_X(k)$ of a process X, the coefficients $a_{k,n}$, which minimise the criterion $\| X_l - \sum_{k=1,n} a_{k,n} X_{l-k} \|^2$, are given by the relations $a_{k,n} = -q_{k,n}$. Indeed, the corresponding orthogonal polynomials satisfy the relations $< Q_n(z), z^l > = 0$, for $l = 0, n - 1$, which can be re-written as

$$R_X(n - l) + \sum_{k=1,n} q_{k,n} R_X(n - l - k) = 0, \quad \text{for } n - l = 1, n, \qquad (11.18)$$

and here we again find the Yule-Walker equations (8.15), with $a_{k,n} = -q_{k,n}$. The prediction error filter $a_n(z) = 1 - \sum_{k=1,n} a_{k,n} z^{-k}$ is therefore equal to $z^{-n} Q_n(z)$. Moreover, by denoting by σ_n^2 the order n prediction error variance, it is clear that $\sigma_n^2 = < Q_n(z), Q_n(z) >$.

Theorem 11.3 (Levinson's algorithm) *The polynomials $Q_n(z)$ can be obtained by means of the following recurrence relations:*

$$Q_0(z) \qquad\qquad = 1,$$

$$\sigma_0^2 \qquad\qquad = R(0),$$

for $n = 0, N - 1$,

$$Q_{n+1}(z) \qquad\qquad = z Q_n(z) - k_{n+1} \tilde{Q}_n(z), \qquad\qquad (11.19)$$

$$\tilde{Q}_{n+1}(z) \qquad\qquad = \tilde{Q}_n(z) - k_{n+1}^* z Q_n(z),$$

$$k_{n+1} \qquad\qquad = \sigma_n^{-2} \left(\sum_{k=0,n} q_{k,n} R(n + 1 - k) \right),$$

$$\sigma_{n+1}^2 \qquad\qquad = \sigma_n^2 (1 - |k_{n+1}|^2),$$

where $\tilde{Q}_n(z) = z^n Q_n^*(z^{-1}) = \sum_{k=0,n} q_{k,n}^* z^k$ (the exponent * of Q_n^* here represents the conjugation of the coefficients of the polynomial).

Proof See Appendix J for a geometrical proof of this result, and Exercise 11.2 for an algebraic proof.

The coefficients k_n associated with the sequence $(R(n))_{n=0,N}$ are called *reflection coefficients, partial correlation coefficients,* or *Schur coefficients,* depending on the specific terminology of each field of application where they are met.

The knowledge of $(k_n)_{n=1,N}$ and of $R(0)$ is equivalent to that of the coefficients $(R(n))_{n=0,N}$. Indeed, the coefficients k_n are obtained from the coefficients $R(n)$ by the Levinson algorithm, and conversely the coefficients $R(n)$ can be calculated iteratively from the coefficients k_n and from $R(0)$ by the relations

$$R(n+1) = \sigma_n^2 k_{n+1} - \sum_{k=1,n} q_{k,n} R(n+1-k), \tag{11.20}$$

for $n = 0, N-1$, and by noticing that the reflection coefficients and $R(0)$ completely define the polynomials $Q_n(z)$.

Levinson's algorithm clearly shows that $|k_n| \leq 1$, since the prediction error $\sigma_n^2 = \sigma_{n-1}^2(1 - |k_n|^2)$ is positive or equal to zero. In practice, the inequality $|k_n| \leq 1$ is of particular interest for signal compression techniques when we wish to encode the second order statistical properties of a signal. Their use is also interesting for parameterising the transfer functions of filters whose stability we wish to control. We deduce from the property $|k_n| \leq 1$ (for $n \geq 1$) that all the zeroes of $Q_n(z)$ are located in the unit disk:

Theorem 11.4 *If the coefficients $k_1, \ldots, k_n$ have a modulus strictly smaller than 1, then the zeroes of $Q_n(z)$ lie strictly inside the unit circle. If the coefficients $k_1, \ldots, k_{n-1}$ have a modulus strictly smaller than 1, and $|k_n| = 1$, then the zeroes of $Q_n(z)$ are on the unit circle and the only spectrum then corresponding to the coefficients $(R(k))_{k=0,n}$ is carried by the n points f_k such that $Q_n(e^{2i\pi f_k}) = 0$.*

Proof We proceed by induction. The property is true for $Q_0(z) = 1$ and $Q_1(z) = z - k_1$. We notice that the relation $Q_{p+1}(z) = 0$ is equivalent to $B_p(z) = zQ_p(z)\tilde{Q}_p^{-1}(z) = k_{p+1}$. We easily check that $B_p(z)$ has its modulus equal to 1 on the unit circle. Since, moreover, from the recurrence hypothesis, $\tilde{Q}_p(z)$ has no zero in the unit disk, $B_p(z)$ is holomorphic in the unit disk. The maximum principle (see Appendix K) then allows us to conclude that $B_p(z)$ has a modulus smaller than 1 in the unit disk, and therefore higher than 1 outside the unit disk, since we easily check that

$$[B_p(z^{-1})]^* = [B_p(z^*)]^{-1}. \tag{11.21}$$

Therefore, all the zeroes of equation $Q_{p+1}(z) = 0$ lie inside the unit disk.

Now if $|k_n| = 1$, $B_{n-1}(z)$ takes the value k_n at n points of the unit circle since, when z goes round the unit circle, $B_{n-1}(z)$ goes n times round the unit circle.

In this case, moreover, we have $\sigma_n^2 = \sigma_{n-1}^2(1 - |k_n|^2) = 0 = |T_n||T_{n-1}|^{-1}$. Therefore T_n is singular, whereas T_{n-1} is full rank. From Caratheodory's theorem, this shows that the corresponding spectrum is carried by n points. As $\sigma_n^2 = 0$, and

$$\sigma_n^2 = \int_{\mathcal{I}} |Q_n(e^{2i\pi f})|^2 d\mu(f), \tag{11.22}$$

it is clear that the n mass points of μ are the points for which $Q_n(e^{2i\pi f}) = 0$.

$\square$

One Step Extension of an Autocovariance Sequence Let $(R(n))_{n=0,N}$ be a sequence of autocovariances, that is, that the corresponding matrix T_N is positive. We then have the following result:

Theorem 11.5 *The set of coefficients $R(N+1)$ such that $(R(n))_{n=0,N+1}$ is a sequence of covariances is the closed disk $\mathbb{D}(C_N, \sigma_N^2)$, with centre*

$$C_N = -\sum_{k=1,N} q_{k,N} R(N+1-k) \tag{11.23}$$

and radius σ_N^2.

Proof See Appendix L.

Orthogonal Szegö Polynomials of the Second Kind In addition to the orthogonal polynomials of the first kind presented above, we need to define orthogonal Szegö polynomials of the second kind $(P_n(z))_{n=0,N}$ by the relations

$$P_0(z) = R(0),$$

$$P_n(z) = \int_{\mathcal{I}} \frac{e^{2i\pi f} + z}{e^{2i\pi f} - z}[Q_n(e^{2i\pi f}) - Q_n(z)]d\mu(f), \quad n = 1, N, \tag{11.24}$$

where μ is any positive measure whose first Fourier coefficients are the coefficients $(R(n))_{n=0,N}$.

In particular, the following result establishes the independence of $P_n(z)$ relative to a particular choice of μ among the set of positive measures whose first Fourier coefficients are the coefficients $(R(n))_{n=0,N}$.

Theorem 11.6

$$P_n(z) = [(R(0) + 2R(1)z^{-1} + ... + 2R(n)z^{-n})Q_n(z)]_+, \quad n = 0, N, \tag{11.25}$$

where $[.]_+$ represents the polynomial part for the variable z. Moreover, the polynomials $P_n(z)$ satisfy the following recurrence relations:

$$P_0(z) = R(0),$$

$$\text{and } P_{n+1}(z) = zP_n(z) + k_{n+1}\tilde{P}_n(z), \quad n = 0, N-1. \tag{11.26}$$

Proof See Appendix M.

11.2.3 Particular Classes of Holomorphic Functions

We denote the open unit disk by $\mathbb{D}$. In order to parameterise the set of spectra whose first autocovariance coefficients are the coefficients $(R(n))_{n=0,N}$, we shall recall the definition of Caratheodory functions and Schur functions.

Definition 11.1 *Let $f(z)$ be a holomorphic function in $\mathbb{D}$. If $\forall z \in \mathbb{D}$ $\mathcal{R}e[f(z)] \geq 0$, it is said to be a Caratheodory (or real positive) function. If $\forall z \in \mathbb{D}$ $|f(z)| \leq 1$, we say that it is a Schur function.*

We now envisage some of the properties of Caratheodory functions. The following theorem establishes a correspondence between the set of positive measures on $\mathcal{I}$ and the set of Caratheodory functions.

Theorem 11.7 *A function $F(z)$ is a Caratheodory function if and only if there exists a positive measure μ such that*

$$F(z) = i\beta + \int_{\mathcal{I}} \frac{e^{2i\pi f} + z}{e^{2i\pi f} - z} d\mu(f), \tag{11.27}$$

where β is a real coefficient.

Proof For a detailed justification of this result, refer to Chapter 11 of [37].

Moreover, it can be shown that the measure μ can be obtained from F by using the relation

$$\mu(B) = \lim_{r \to 1^-} \int_B \mathcal{R}e[F(re^{2i\pi f})] df, \tag{11.28}$$

satisfied for any Borel set B of $\mathcal{I}$. In particular, if $F(z)$ has a continuous extension on the unit circle, we have $d\mu(f) = \mathcal{R}e[F(e^{2i\pi f})] df$.

11.2.4 General Solution to the Problem

We have seen that any Caratheodory function can be written in the form

$$F(z) = i.\mathcal{I}m[F(0)] + \int_{\mathcal{I}} \frac{e^{2i\pi f} + z}{e^{2i\pi f} - z} d\mu(f), \tag{11.29}$$

where μ is a positive measure. From Lebesgue's dominated convergence theorem, it is clear that for $|z| < 1$

$$F(z) = i.\mathcal{I}m[F(0)] + \mu_0 + 2\mu_{-1}z + 2\mu_{-2}z^2 + \ldots \tag{11.30}$$

where $\mu_k = \int_{\mathcal{I}} e^{2i\pi k f} d\mu(f)$.

The problem of characterising the positive measures μ whose first Fourier coefficients are the coefficients $(R(n))_{n=0,N}$ is therefore equivalent to that of characterising Caratheodory functions whose series expansion is of the form

$$F(z) = R(0) + 2R(-1)z + \ldots + 2R(-N)z^N + O(z^{N+1}). \tag{11.31}$$

Characterising this set is given by the following result:

Theorem 11.8 *The set of positive measures μ whose first Fourier coefficients are the coefficients $(R(n))_{n=0,N}$ are those that correspond to the Caratheodory functions which can be written in the form*

$$F(z) = \int_{\mathcal{I}} \frac{e^{2i\pi f} + z}{e^{2i\pi f} - z} d\mu(f) = \frac{\tilde{P}_N(z) + S(z)z P_N(z)}{\tilde{Q}_N(z) - S(z)z Q_N(z)}, \tag{11.32}$$

where $S(z)$ is any Schur function. The Schur functions therefore parameterise the solutions of the trigonometric moment problem. When μ is an absolutely continuous measure with respect to Lebesgue's measure, its density $g(f)$ can be written almost everywhere in the form

$$g(f) = \frac{\sigma_N^2(1 - |S(e^{2i\pi f})|^2)}{|\tilde{Q}_N(e^{2i\pi f}) - S(e^{2i\pi f})e^{2i\pi f}Q_N(e^{2i\pi f})|^2}. \tag{11.33}$$

Proof See Appendix N.

11.2.5 Maximum Entropy Spectrum

When $S(z)$ is the null function, it is clear that relation (11.33) leads to the spectral density

$$g(f) = \frac{\sigma_N^2}{|\tilde{Q}_N(e^{2i\pi f})|^2}$$

$$= \frac{\sigma_N^2}{|Q_N(e^{2i\pi f})|^2}. \tag{11.34}$$

We check that this spectrum corresponds to the autoregressive maximum entropy model:

Theorem 11.9 *The spectrum with PSD $g(f) = \sigma_N^2|Q_N(e^{2i\pi f})|^{-2}$ represents the maximum entropy spectrum, that is, the PSD solution to the following constrained optimisation problem:*

$$\begin{cases} \max_g \int_{\mathcal{I}} \log[g(f)]df \\ R(n) = \int_{\mathcal{I}} e^{2i\pi nf}g(f)df, \quad n = -N, N. \end{cases} \tag{11.35}$$

Proof Let Y be an AR process defined by a difference equation of the form $[Q_N(z)]Y_n = V_{N+n}$, where V is a white noise process, and $Q_N(z)$ the orthogonal polynomial of degree N associated with the sequence $(R(n))_{n=0,N}$. It is clear that Y has an absolutely continuous spectrum, with PSD $\sigma_N^2 \times |Q_N(e^{2i\pi f})|^{-2}$.

From the study of the trigonometric moment problem, it is clear that the first $N + 1$ autocovariance coefficients of Y satisfy the relations

$$R_Y(n) = \int_{\mathcal{I}} e^{2i\pi nf}g(f)df, \quad n = 0, N. \tag{11.36}$$

Moreover, Y is a regular process whose innovation I_Y satisfies

$$\| I_{Y,n} \|^2 = \| Y_n - Y_n/H_{Y,n-1} \|$$

$$= \| Y_n - Y_n/H_{Y,n-N,n-1} \|$$

$$= \sigma_N^2 \tag{11.37}$$

$$= \exp(\int_{\mathcal{I}} \log g(f)df).$$

For any other process Z satisfying the autocovariance constraints, the innovation satisfies

$$\| I_{Z,n} \|^2 = \exp(\int_{\mathcal{I}} \log S_Z(f)df)$$

$$\leq \| Z_n - Z_n/H_{Z,n-N,n-1} \|^2. \tag{11.38}$$

But, $\| Z_n - Z_n/H_{Z,n-N,n-1} \|^2 = \sigma_N^2 = \exp(\int_{\mathcal{I}} \log g(f)df)$, for σ_N^2 depends only on $(R(n))_{n=0,N}$. Finally,

$$\int_{\mathcal{I}} \log S_Z(f)df \leq \int_{\mathcal{I}} \log g(f)df, \tag{11.39}$$

which indeed shows that the PSD of Y represents a global optimum of the problem (11.35). Moreover, this optimum is unique (in $L^1(\mathcal{I}, \mathcal{B}(\mathcal{I}), df)$) for the problem under consideration is that of maximising a strictly concave function under linear constraints. $\square$

We notice that the maximum entropy spectrum yields the largest innovation variance among spectra whose first $N+1$ autocovariance coefficients are $(R(n))_{n=0,N}$, for a WSS process $X \| I_{X,n} \|^2 = \exp(\int_{\mathcal{I}} \log S_X(f)df)$. In other words, the maximum entropy model is the "least predictable" one among the solutions of the trigonometric moment problem.

11.3 Line Spectra

We remark that from the Caratheodory theorem (Theorem 11.2), if the sequence $(R(n))_{n=0,N}$ is of the positive type and such that the corresponding Toeplitz matrix T_N has rank $k \leq N$, there exists a unique positive measure solution to the trigonometric moment problem, and it is carried by k mass points. Now, if T_N is full rank, it is possible to define a one step extension $R(N+1)$ of the sequence $(R(n))_{n=0,N}$ such that the corresponding Toeplitz $(N+2) \times (N+2)$ matrix, denoted by T_{N+1}, has rank $N+1$. From Section 11.2.2, this clearly amounts to choosing $R(N+1)$ on the circle $\partial D(C_N, \sigma_N^2)$, with

$$C_N = - \sum_{k=1,N} q_{k,N} R(N+1-k). \tag{11.40}$$

Equivalently, it appears from Levinson's algorithm that this amounts to choosing the $N + 1^{\text{th}}$ reflection coefficient, say k_{N+1}, on the unit circle. From this discussion and Theorem 11.4, we get the following result

Theorem 11.10 *If the sequence $(R(n))_{n=0,N}$ is such that the Toeplitz matrix T_N is positive definite, then for any point k_{N+1} on the unit circle there exists a solution to the trigonometric moment problem involving a discrete measure carried by $N + 1$ points given by the solutions f of the equation $B_N(e^{2i\pi f}) = k_{N+1}$.*

The mass points of the spectra obtained for $k_{N+1} = 1$ and for $k_{N+1} = -1$ are called the Line Spectrum Pairs (LSP). They represent interesting alternative coefficients to linear prediction coefficients or reflection coefficients in speech coding.

We now turn to the problem of identifying a process made up of a sum of harmonic components. In many applications, we observe such a process corrupted by an additive white noise. The process X observed is then of the form

$$X_n = \sum_{k=1,p} \xi_k e^{2i\pi n f_k} + B_n, \tag{11.41}$$

where B is a white noise process, uncorrelated with the $(\xi_k)_{k=1,p}$ and with variance σ_B^2. In this case, knowledge of the first autocovariance coefficients $(R_X(n))_{n=0,N}$ $(N \geq p)$ of X, makes it possible to identify the spectrum $d\mu_X(f) = \sum_{k=1,p} \| \xi_k \|^2 \delta_{f_k} + \sigma_B^2$, as the following result shows:

Theorem 11.11 *Let X be a WSS process of the form $X_n = \sum_{k=1,p} \xi_k e^{2i\pi n f_k} + B_n$, where B is a white noise process with variance σ_B^2. We denote by T_N the covariance matrix of X of size $N + 1$ $(N \geq p)$ and $U\Lambda U^H$ the eigenvalue decomposition of T_N, with*

$$U = [u_1, \ldots, u_{N+1}], \ \Lambda = diag(\lambda_1, \ldots, \lambda_{N+1}), \ and \ \lambda_1 \geq \ldots \geq \lambda_{N+1}. \tag{11.42}$$

Moreover, we note $d(f) = [1, e^{2i\pi f}, \ldots, e^{2i\pi N f}]^T$. Then,

$$\lambda_1 \geq \ldots \geq \lambda_p > \lambda_{p+1} = \ldots = \lambda_{N+1} = \sigma_B^2, \tag{11.43}$$

and

$$span\{u_1, \ldots, u_p\} = span\{d(f_1), \ldots, d(f_p)\}. \tag{11.44}$$

The frequencies $(f_k)_{k=1,p}$ $(f_k \in \mathcal{I})$ are the solutions to the system of equations

$$u_k^H d(f) = 0, \quad k = p+1, N+1. \tag{11.45}$$

To show this theorem, we shall require the following result:

Theorem 11.12 *For p distinct values $f_1, \ldots, f_p$ of $\mathcal{I}$ ($p \leq N + 1$), the vectors $d(f_1), \ldots, d(f_p)$ are independent.*

Proof (of Theorem 11.12) See Appendix I.2.

Proof (of Theorem 11.11) It is clear that T_N has the same eigenvectors as $T_N - \sigma_B^2 I_{N+1}$. $T_N - \sigma_B^2 I_{N+1}$ is the covariance matrix of the process Y defined by $Y_n = \sum_{k=1,p} \xi_k e^{2i\pi n f_k}$, and is therefore of rank p, from Caratheodory's theorem. The $N + 1 - p$ smallest eigenvalues of $T_N - \sigma_B^2 I_{N+1}$ are therefore equal to zero, and the space spanned by this matrix is generated by the vectors $(u_k)_{k=1,p}$. But we also have

$$T_N - \sigma_B^2 I_{N+1} = \sum_{k=1,p} \| \xi_k \|^2 \, d(f_k) d(f_k)^H. \tag{11.46}$$

As the vectors $(d(f_k))_{k=1,p}$ form an independent family (from Theorem 11.12),

$$span\{u_1, \ldots, u_p\} = span\{d(f_1), \ldots, d(f_p)\}. \tag{11.47}$$

The eigenvalues of T_N are deduced from those of $T_N - \sigma_B^2 I_{N+1}$ by adding σ_B^2. Moreover, as T_N is a Hermitian matrix, the eigenvectors of T_N form an orthonormal basis. Thus, from equality (11.47),

$$span\{d(f_1), \ldots, d(f_p)\} = [span\{u_{p+1}, \ldots, u_{N+1}\}]^{\perp}. \tag{11.48}$$

Therefore, the vectors $(d(f_k)_{k=1,p})$ satisfy equations (11.45). A value of f distinct from $f_1, \ldots, f_p$ cannot have the same property, otherwise $\{d(f), d(f_1), \ldots, d(f_p)\}$ would belong to a space of dimension p, which is impossible from Theorem 11.12. $\square$

Sometimes, we wish to visualise the values $(f_k)_{k=1,p}$ by tracing the function

$$\phi(f) = \frac{1}{\sum_{k=p+1,N+1} |u_k^H d(f)|^2}. \tag{11.49}$$

The desired frequencies correspond to the vertical asymptotes of the function.

In order to calculate the frequencies $(f_k)_{k=1,p}$, we solve the polynomial equation $z^{N+1} \sum_{k=p+1,N+1} [u_k^H e(z)][u_k^H e(1/z^*)]^* = 0$, where $e(z) = [1, z, \ldots, z^N]^T$, and consider the arguments of the roots.

If, in practice, model (11.41) is not quite coherent with the coefficients $(R(n))_{n=0,N}$, which are then only estimated values of the autocovariance coefficients, the $N - p$ smallest eigenvalues of T_N are not identical and we often choose to identify the frequencies $(f_k)_{k=1,p}$ by considering the p largest maxima of $\phi(f)$.

The techniques presented here above are very important in particular for array signal processing [47].

11.4 Lattice Filters

We shall end this chapter by mentioning an application of reflection coefficients, which somewhat extends the theme of spectral identification. Here, we are interested in the implementation of a Prediction Error Filter (PEF), or conversely, in the generation of a process from knowledge of its order p prediction error. These two operations are important in particular for applications involving data compression and decompression.

A technique that makes it possible to ensure that the zeroes of the order p PEF, denoted by $a_p(z)$, are in the unit disk involves parameterising $a_p(z)$ by the corresponding reflection coefficients $(k_i)_{i=1,p}$ and ensuring that the constraints $|k_i| \leq 1$ are satisfied. Indeed, it is much simpler to master these constraints than the constraints on the coefficients of $a_p(z)$, which ensure that its zeroes lie inside the unit disk. As $a_k(z) = z^{-k}Q_k(z)$, Levinson's algorithm leads to the relations

$$a_0(z) \quad = 1,$$

$$a_{l+1}(z) = a_l(z) - k_{l+1}z^{-1}\tilde{a}_l(z), \tag{11.50}$$

$$\tilde{a}_{l+1}(z) = z^{-1}\tilde{a}_l(z) - k_{l+1}^* a_l(z).$$

Consequently, defining by $E_{l,n}^d = [a_l(z)]X_n$ and $E_{l,n}^r = [\tilde{a}_l(z)]X_n$ the direct and the backward prediction errors of the process X, it results that

$$E_{l+1,n}^d = E_{l,n}^d - k_{l+1}E_{l,n-1}^r,$$

$$E_{l+1,n}^r = E_{l,n-1}^r - k_{l+1}^* E_{l,n}^d. \tag{11.51}$$

The vector $E_{l+1,n} = [E_{l+1,n}^d \ E_{l+1,n}^r]^T$ is then easily deduced from $E_{l,n}$ by matrix filtering:

$$E_{l+1,n} = \begin{bmatrix} 1 & -k_{l+1}[z^{-1}] \\ -k_{l+1}^* & [z^{-1}] \end{bmatrix} E_{l,n} \tag{11.52}$$

$$= [K_{l+1}(z)]E_{l,n}.$$

Letting $E_{0,n} = [X_n \ X_n]^T$, it thus results that the order p prediction error of X is given by

$$E_{p,n} = [K_p(z)] \times \ldots \times [K_1(z)]E_{0,n}. \tag{11.53}$$

From the point of view of the practical implementation of the filtering $E_{p,n}^d = [a_p(z)]X_n$, Equation (11.53) is characterised by a structure called a *lattice filter*, which is made up of cascaded filter cells $K_l(z)$ with two inputs and two outputs, as is clearly shown in Figure 11.1.a.

Conversely, if we wish to recover X from $E_{p,n}^d$ and from the coefficients $(k_i)_{i=1,p}$, we are led to a fairly similar implementation scheme, which simply expresses (see Figure 11.1.b) the re-writing of the first recurrence equation in (11.51) in the form

$$E_{l,n}^d = E_{l+1,n}^d + k_{l+1} E_{l,n-1}^r. \tag{11.54}$$

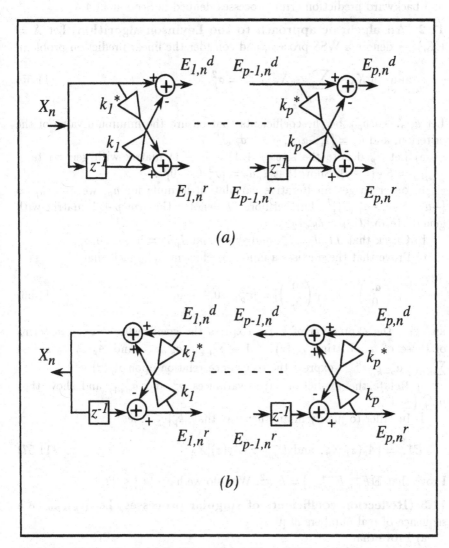

Fig. 11.1. Trellis filters: (a) Prediction Error Filter (PEF) of X, (b) Synthesis filter of X from the prediction error

Exercises

In the following exercises, we assume that the autocovariance sequences $(R(n))_{n=-N,N}$ under consideration define strictly positive Toeplitz matrices of size $N + 1$ with term (i, j) equal to $R(i - j)$, unless singularity of the matrix is specified.

11.1 Search for the orthogonality relationships that exist among the forward and backward prediction error processes defined in Section 11.4.

11.2 (An algebraic approach to the Levinson algorithm) Let $X = (X_n)_{n \in \mathbb{Z}}$ denote a WSS process, and consider the linear prediction problem

$$\min_{\alpha_1, \cdots, \alpha_p} \| X_n - \sum_{k=1,p} \alpha_k X_{n-k} \|^2 = \sigma_p^2. \tag{11.55}$$

Let $a_{1,p}, \cdots, a_{p,p}$ be the coefficients that ensure the minimum value of the criterion, and $a_p = [1, -a_{1,p}, \cdots, -a_{p,p}]^T$.

a) Let T_p denote the size $(p + 1) \times (p + 1)$ matrix with general term $[T_p]_{i,j} = R_X(i - j)$. Show that $T_p a_p = [\sigma_p^2, 0 \cdots 0]^T$.

In order to get an iterative calculation formula for a_p, we note $\tilde{a}_p = [-a_{p,p}^*, \cdots, -a_{1,p}^*, 1]^T$. Furthermore, J denotes the size $p + 1$ matrix with general term $[J]_{a,b} = \delta_{b,p+2-a}$.

b) Check that $J T_p J = T_p^*$, and show that $T_p \tilde{a}_p = [0, \cdots, 0, \sigma_p^2]^T$.

c) Prove that there exists a unique coefficient k_{p+1} such that

$$T_{p+1}[\begin{pmatrix} a_p \\ 0 \end{pmatrix} - k_{p+1} \begin{pmatrix} 0 \\ \tilde{a}_p \end{pmatrix}] = [\sigma_{p+1}^2, 0, \cdots, 0]^T, \tag{11.56}$$

and give the expression of k_{p+1}. Express the coefficients of a_{p+1} in terms of those of a_p. Letting $A_p(z) = 1 - \sum_{k=1,p} a_{k,p} z^{-k}$ and $\tilde{A}_p(z) = z^{-p} - \sum_{k=1,p} a_{p-k,p}^* z^{k-p}$, express the recurrence relation upon $A_p(z)$.

c) Relate the prediction error variances σ_p^2 and σ_{p+1}^2, and show that $|k_{p+1}| \leq 1$.

d) In order to justify in another way that $|k_p| \leq 1$, let

$$E_{p,n}^+ = [A_p(z)] X_n, \quad \text{and} \quad E_{p,n}^- = [\tilde{A}_p(z)] X_n. \tag{11.57}$$

Prove that $\mathbb{E}[E_{p,n}^+ E_{p,n-1}^{-*}] = k_p \sigma_p^2$. Why do we have $|k_p| \leq 1$?

11.3 (Reflection coefficients of singular processes) Let $(c_k)_{k>0}$ be a sequence of real numbers in $[0, 1]$.

a) Prove that

$$\sum_{n>0} c_n < \infty \Rightarrow \lim_{n \to \infty} \prod_{k=1,n} (1 - c_k) > 0. \tag{11.58}$$

To show this, check first that

$$\prod_{k=1,n} (1 - c_k) \geq 1 - \sum_{k=1,n} c_k, \tag{11.59}$$

and that

$$\forall \varepsilon, \ \exists N, \ \forall n > N, \ \prod_{k=1,n} (1 - c_k) \geq (1 - \varepsilon) \prod_{k=1,N-1} (1 - c_k). \tag{11.60}$$

b) Show that

$$\sum_{n>0} c_n = +\infty \Rightarrow \lim_{n \to \infty} \prod_{k=1,n} (1 - c_k) = 0. \tag{11.61}$$

c) Let X be a WSS process and denote by $(k_n)_{n \in \mathbb{N}}$ its reflection coefficients. Using a) and b), show that X is singular if and only if $\sum_{n>0} |k_n| = +\infty$.
(Hint: $\| I_{X,n} \| = R_X(0) \prod_{n>0} (1 - |k_n|^2)$.)

11.4 (Location of the zeroes of $Q_n(z)$) Let us recall the following result, known as Rouche's theorem (see, for instance, [33]):

Theorem 11.13 (Rouche) *We consider g and h, two functions holomorphic on an open domain, denoted by E, and continuous on $\overline{E}$. If for all $z \in \partial E$ $|g(z)| < |f(z)|$, then $f(z)$ and $f(z) + g(z)$ have the same number of roots in E.*

a) Prove Rouche's theorem. To do this, first prove Cauchy's theorem that states that the variation of the argument of $f(z)$ when z travels along ∂E is equal to 2π times the number of roots of $f(z)$ in E . Then, consider the argument of the function $f(z) + g(z) = f(z)(1 + f^{-1}(z)g(z))$.
(Hint: to prove Cauchy's theorem, apply the theorem of residues to the function $f^{-1}(z)f'(z)$.)

b) Let X denote a WSS process, and denote by $Q_n(z)$ the corresponding degree n first order Szegö polynomial. Using Rouche's theorem, show that all the roots of $Q_n(z)$ lie in the open unit disk $\mathbb{D}$, provided that the reflection coefficients $(k_i)_{i=1,n}$ lie in $\mathbb{D}$.

11.5 (Mass point spectra) Let $(R(n))_{n=-N,N}$ denote an autocovariance sequence and f_0 a point of $\mathcal{I}$. Show that there exists a unique positive measure, denoted by μ, carried by N points of $\mathcal{I}$, among which f_0, with first $N + 1$ Fourier coefficients matching $(R(n))_{n=0,N}$.

11.6 (Band-limited spectra) Let $X = (X_t)_{t \in \mathbb{R}}$ denote a WSS process, the spectrum of which is carried by a sub-interval of $\mathbb{R}$, denoted by $[-B, B]$. Show that if the autocovariance function $R_X(t)$ of X is known on a finite interval of the form $[-T, T]$, where T is any strictly positive number, then $R_X(t)$ is uniquely defined on $\mathbb{R}$.

11.7 (Constant reflection coefficients sequences) Let $(R(n))_{n \in \mathbb{Z}}$ denote an autocovariance sequence. We want to calculate the corresponding PSD, denoted by $S(f)$, given that $R(0) = 1$ and that the sequence $(k_n)_{n \in \mathbb{N}^*}$ of associated reflection coefficients is constant: $k_n = \alpha$ for all $n \in \mathbb{N}^*$, with $|\alpha| < 1$. In order to calculate $S(f)$, let us consider the following Caratheodory function:

$$F(z) = R(0) + 2(\sum_{n>0} R(n)z^n). \tag{11.62}$$

Denote by $S_{n+1}(z)$ the Schur function of the representation of $F(z)$ in terms of degree n orthogonal Szegö polynomials:

$$F(z) = \frac{zP_n(z) + S_{n+1}(z)\tilde{P}_n(z)}{zQ_n(z) - S_{n+1}(z)\tilde{Q}_n(z)}. \tag{11.63}$$

a) Show that $S_n(z)$ does not depend on n.
b) Express $S_{n+1}(z)$ in terms of $S_n(z)$ and calculate $S_n(z)$.
c) Calculate $F(z)$.
d) Calculate the PSD $S(f)$.

11.8 (Uniqueness of the maximum entropy spectrum) Let us denote by $(R(n))_{n=-N,N}$ an autocovariance sequence.

a) Calculating the maximum entropy spectrum estimator associated with $(R(n))_{n=-N,N}$ as the solution of a constrained maximisation problem by means of Lagrange multipliers, show that this yields a PSD of the form $S(f) = (\sum_{k=-N,N} c_k e^{-2i\pi k f})^{-1}$.

b) Show that there exists a unique spectrum density of the form $(\sum_{k=-N,N} c_k e^{-2i\pi k f})^{-1}$ that matches the autocovariance constraints.

11.9 (The Christoffel-Darboux formula) Show that the first order orthogonal polynomials satisfy the relations

$$\sum_{k=0,n} \frac{1}{\sigma_k^2} Q_k(z)[Q_k(z')]^* = \frac{\tilde{Q}_n(z)[\tilde{Q}_n(z')]^* - z\bar{z}'Q_n(z)[Q_n(z')]^*}{\sigma_n^2(1 - z\bar{z}')}$$

$$= \frac{\tilde{Q}_{n+1}(z)[\tilde{Q}_{n+1}(z')]^* - Q_{n+1}(z)[Q_{n+1}(z')]^*}{\sigma_{n+1}^2(1 - z\bar{z}')}, \tag{11.64}$$

for $z, z' \in D$.
(Hint: prove this by induction.)

11.10 (Mass values of mass point spectra) Let $X = (X_n)_{n \in \mathbb{Z}}$ denote a WSS process.

a) Let $f_0 \in \mathcal{I}$ and denote by μ_{X, f_0} the unique mass point measure carried by n distinct points that has mass at point f_0 (see Exercise 11.4). Show that the mass points of μ_{X,f_0} are the solutions to an equation of the form $B_n(e^{2i\pi f}) = e^{i\alpha}$, where $B_n(e^{2i\pi f}) = e^{2i\pi f} Q_n(e^{2i\pi f})/\tilde{Q}_n(e^{2i\pi f})$, and give the expression of $e^{i\alpha}$.

b) Let us denote by $d\mu_{X,f_0}(f) = \sum_{k=0,n-1} \rho_k \delta_{f_k}(f)$ and $Q_{n+1,\mu_{X,f_0}}(z)$ the first order Szegö polynomial of degree $n+1$ associated with the measure μ_{X,f_0}. Show that

$$\rho_k = \int_I \frac{Q_{n+1,\mu_{X,f_0}}(e^{2i\pi f})}{(e^{2i\pi f} - e^{2i\pi f_k})[\frac{d}{dz}Q_{n+1,\mu_{X,f_0}}(z)]_{z=e^{2i\pi f_k}}} d\mu_{X,f_0}(f). \qquad (11.65)$$

c) Expressing $Q_{n+1,\mu_{X,f_0}}$ in terms of $Q_n(z)$ and using the Christoffel-Darboux formula (see the exercise above), prove that

$$\rho_k = \frac{1}{\sum_{k=0,n} \sigma_k^{-2}|Q_k(e^{2i\pi f})|}. \qquad (11.66)$$

11.11 (Inverse Levinson's algorithm) We consider a covariance sequence $(R(n))_{n=0,N}$ and the corresponding reflection coefficients and first order Szegö polynomials denoted by $(k_n)_{n=1,N}$ and $(Q_n)_{n=0,N}$ respectively.

a) Calculate $(R(n))_{n=0,N}$ iteratively from $(k_n)_{n=1,N}$ and σ_N^2.

b) Check that the sequence $(k_n)_{n=1,N}$ can be recovered from $Q_N(z)$ and σ_N^2 by means of the following recursion formula: for $n = N-1, \ldots, 0$,

$$k_{n+1} = -Q_{n+1}(0),$$
$$Q_n(z) = \frac{z^{-1}}{1 - |k_{n+1}|^2}[Q_{n+1}(z) + k_{n+1}\tilde{Q}_{n+1}(z)]. \qquad (11.67)$$

11.12 (The multivariate Levinson's algorithm)[20][12] Let X be a size d multivariate WSS process and consider the forward and backward linear prediction problems:

$$\min_{L_1,\ldots,L_p} \| X_n + \sum_{k=1,p} L_k X_{n-k} \|^2,$$
$$\min_{M_1,\ldots,M_p} \| X_n + \sum_{k=1,p} M_k X_{n+k} \|^2. \qquad (11.68)$$

Let T_p and T_p^* represent block Toeplitz matrices with general block terms $[T_p]_{ij} = R_X(i-j) = \mathbb{E}[X_{n+i}X_{n+j}^H]$ and $[T^*_p]_{ij} = R_X(j-i)$.

a) Check that the solutions to the forward and backward linear prediction problems, denoted by $(A_k)_{k=1,p}$ and $(B_k)_{k=1,p}$ respectively, where the A_k and B_k are $d \times d$ matrices satisfy the following linear systems of equations:

$$[I_d, A_{p,1}^H, \ldots, A_{p,p}^H]T_p = [\Gamma_{p-1}^+, 0, \ldots, 0],$$
$$\text{and } [I_d, B_{p,1}^H, \ldots, B_{p,p}^H]T_p^* = [\Gamma_{p-1}^-, 0, \ldots, 0]. \qquad (11.69)$$

Check that Γ_{p-1}^+ and Γ_{p-1}^- represent the prediction error matrices of the linear forward and backward prediction problems respectively.

b) We note $A_0(z) = B_0(z) = I_d$. $A_k(z) = z^k + \sum_{m=1,k} A_{k,m} z^{m-k}$ and $B_k(z) = z^k + \sum_{m=1,k} B_{k,m} z^{m-k}$ denote the order k linear predictors. In addition, let $\Gamma_0^+ = \Gamma_0^- = R_X(0)$. Show that for $k = 1, \ldots, p$ $A_k(z)$ and $B_k(z)$ can be calculated recursively using the following multivariate Levinson's algorithm:

$$C_k = \sum_{l=0,k-1} R_X(-k-l+1) A_{k-1,l},$$

$$A_k(z) = z A_{k-1}(z) - \tilde{B}_{k-1}(z)(\Gamma_{k-1}^-)^{-1} C_k,$$

$$B_k(z) = z B_{k-1}(z) - \tilde{A}_{k-1}(z)(\Gamma_{k-1}^+)^{-1} C_k^H, \qquad (11.70)$$

$$\Gamma_k^+ = \Gamma_{k-1}^+ - C_k^H (\Gamma_{k-1}^-)^{-1} C_k,$$

$$\Gamma_k^- = \Gamma_{k-1}^- - C_k (\Gamma_{k-1}^+)^{-1} C_k^H.$$

12. Non-parametric Spectral Estimation

Purpose In practice, the methods for identifying the spectrum of a process considered in the previous chapter come up against the fact that its autocovariance coefficients are not known exactly but are only estimated. Here, we address the problem of non-parametric estimation of these coefficients, hence that of non-parametric spectral estimation by means of the periodogram estimator.

12.1 Definitions

To begin with, we recall some basics about statistical estimation. We consider a probability space $(\Omega, \mathcal{A}, dP)$ where P is unknown. In the case of non-parametric estimation, the information on P is not very precise (for instance, P is absolutely continuous). In the case of parametric estimation, we assume that P belongs to a set $\{P_\theta; \theta \in \Theta\}$, where θ is a vector of parameters that completely characterises P_θ. We generally arrange it so that there is a one to one correspondence between Θ and $\{P_\theta; \theta \in \Theta\}$.

We define a statistical model as a triplet $(\Omega, \mathcal{A}, \mathcal{P})$, where $(\Omega, \mathcal{A})$ is a measurable space and $\mathcal{P}$ a family of probability distributions defined on $\mathcal{A}$. In the case of parametric estimation, we will have $\mathcal{P} = \{P_\theta; \theta \in \Theta\}$, and a model of the form $(\Omega, \mathcal{A}, \{P_\theta; \theta \in \Theta\})$.

Now we shall recall the following basic definitions:

Observation The observation x consists of a realisation of a random variable X, or of a sequence $(X_1, ..., X_n)$ of random variables. If the random variables $(X_i)_{i=1,n}$ are independent and have the same distribution, we call it a *sample* of size n.

Estimator The aim of parameter estimation is to evaluate some information characterised by the probability P and denoted by $g(P)$, or more simply g, from the observation x. In the case of parametric estimation, the probability measure depends on the value θ of the parameter. g is then a function of θ. We estimate g by a function $f(x)$ of the observation. The random variable $f(X)$ is called the estimator of g.

Quality of an estimator In order to evaluate the quality of an estimator, we define a cost function $C(g, f(X))$, which quantises the dissimilarity between g and its estimator $f(X)$. Cost functions are also called *loss functions*.

We often look for an estimator f for which $f_1 \leq f$ when $f_1 \neq f$, where the inequality $f_1 \leq f$ is defined by

$$f_1 \leq f \text{ if } \forall P \in \mathcal{P}, \ \mathbb{E}_P\left[C(g, f_1(X))\right] \leq \mathbb{E}_P\left[C(g, f_2(X))\right], \tag{12.1}$$

where $\mathbb{E}_P$ represents the mathematical expectation for the probability distribution P. $\mathbb{E}[C(g, f(X))]$ is called the *mean cost*, or *risk*, of the estimator $f(X)$. Unfortunately, for this inequality, two estimators cannot always be compared. We are thus often led to restrict the set estimators to one subfamily of the form $\{f(X); f \in F\}$ inside which we can define a best estimator, that is, find an estimator $\hat{f}(X)$ such that

$$\forall P \in \mathcal{P}, \ \hat{f} = \arg\min_{f \in F} \mathbb{E}_P\left[C(g, f(X))\right]. \tag{12.2}$$

The mean square error criterion defined by

$$\| f(X) - g \|^2 = \operatorname{var}[f(X)] + |\mathbb{E}[f(X)] - g|^2, \tag{12.3}$$

is a risk function often used in signal processing. $\mathbb{E}[f(X)] - g$ is called the *bias* of $f(X)$. We often restrict the class of estimators to unbiased estimators, in which case the existence of a best estimator $\hat{f}$ is ensured. Criterion (12.3) then presents the advantage of often leading to a solution $\hat{f}$ that is easy to compute. Furthermore, this criterion has a nice interpretation since it represents the mean power of the estimation error.

Convergence A sequence of estimators $f_n(X_1, \ldots, X_n)$, which converges in probability towards g when the size n of the observation tends towards $+\infty$, is said to be weakly consistent and, when the convergence is almost sure, we speak of a strongly consistent estimator. Moreover, we shall say that a sequence of estimators of g is asymptotically unbiased if the sequence of their means converges towards g.

12.2 Elements of Non-parametric Estimation

12.2.1 Independent Data Sequences

For a sample of size n, $\mathbf{X}_n = [X_1, \ldots, X_n]^T$, we define the empirical probability distribution and the empirical distribution function by

$$\hat{P}_n \quad = \frac{1}{n} \sum_{i=1,n} \delta_{X_i},$$

$$\tag{12.4}$$

$$\text{and } \hat{F}_n(y) = \frac{1}{n} \sum_{i=1,n} \mathbb{1}_{X_i \leq y},$$

respectively. The empirical estimator of $g = f(F)$, where F is the distribution function associated with P, is defined by $f(\hat{F}_n)$. For many functions f, the use of this estimator is justified by the following results (see, for example, [22], Chapter 1).

Theorem 12.1 *At any point y, $\lim_{n \to \infty} \hat{F}_n(y) \overset{a.s.}{=} F_X(y)$, where $F_X(y)$ is the distribution function of X.*

Proof $\mathbb{E}[\mathbb{1}_{X \leq y}] = P(X \leq y) = F_X(y)$, and from the strong law of large numbers we obtain $\lim_{n \to \infty} \hat{F}_n(y) \overset{a.s.}{=} F_X(y)$. $\square$

Theorem 12.2 *Let G be a function of the form*

$$G(F) = h\left(\int_{\mathbb{R}} g(x)dF(x)\right),\tag{12.5}$$

where h is continuous at point $\int_{\mathbb{R}} g(x)dF_X(x)$. Then, $\lim_{n \to \infty} G(\hat{F}_n) \overset{a.s.}{=} G(F_X)$.

Proof Consider

$$\begin{aligned}S_n &= \int_{\mathbb{R}} g(x)d\hat{F}_n(x)\\&= \frac{1}{n}\sum_{i=1,n} g(X_i).\end{aligned}\tag{12.6}$$

From the strong law of large numbers, S_n converges almost surely towards

$$\mathbb{E}[g(X)] = \int_{\mathbb{R}} g(x)dF_X(x),\tag{12.7}$$

when $\mathbb{E}[g(X)]$ exists and is finite. Therefore,

$$P\left(\{\omega; \lim_{n \to \infty} S_n(\omega) = \mathbb{E}[g(X)]\}\right) = 1,\tag{12.8}$$

and from the continuity of h in $\mathbb{E}[g(X)]$,

$$P(\{\omega; \lim_{n \to \infty} h(S_n(\omega)) = h(\mathbb{E}[g(X)])\}) = 1\tag{12.9}$$

for

$$\{\omega; \lim_{n \to \infty} S_n(\omega) = \mathbb{E}[g(X)]\} \subset \{\omega; \lim_{n \to \infty} h(S_n(\omega)) = h(\mathbb{E}[g(X)])\}.\tag{12.10}$$

Therefore,

$$\lim_{n \to \infty} G(\hat{F}_n) \overset{a.s.}{=} G(F_X).\square\tag{12.11}$$

In particular, this result ensures almost sure convergence of the empirical moments: noting $\hat{m}_{X^k,n} = n^{-1}\sum_{i=1,n} X_i^k$,

$$\begin{aligned}\lim_{n \to \infty} \hat{m}_{X^k,n} &\overset{a.s.}{=} \mathbb{E}[X^k],\\\lim_{n \to \infty} \frac{1}{n}\sum_{i=1,n}(X_i - \hat{m}_{X,n})^k &\overset{a.s.}{=} \mathbb{E}(X - \mathbb{E}[X])^k].\end{aligned}\tag{12.12}$$

Unfortunately, when the random variables $(X_n)_{n \in \mathbb{Z}}$ is a random process, we can no longer apply the above-mentioned results directly since, except in particular situations, the random variables X_n are not independent.

12.2.2 Ergodic Processes

We recall that for a stationary and strict sense ergodic process X,

$$\lim_{n \to \infty} \frac{1}{n} \sum_{i=1,n} f(X_{i+1}, \ldots, X_{i+k}) \overset{\text{a.s.}}{=} \mathbb{E}[f(X_1, \ldots, X_k)]. \qquad (12.13)$$

In particular, we have

$$\lim_{n \to \infty} \hat{m}_{X,n} = \lim_{n \to \infty} \frac{1}{n} \sum_{i=1,n} X_i$$

$$\overset{\text{a.s.}}{=} m_X,$$

$$(12.14)$$

$$\text{and } \lim_{n \to \infty} \hat{R}_{X,n}(k) = \lim_{n \to \infty} \frac{1}{n} \sum_{i=1,n} X_i X_{i+k} - \hat{m}_{X,n}^2$$

$$\overset{\text{a.s.}}{=} R_X(k).$$

When the process is only WSS and wide sense ergodic, the properties in (12.14) are still satisfied if we replace the almost sure convergence by the mean square convergence.

In the following section, we shall present conditions that enable us to guarantee the almost sure convergence of mean and autocovariance functions, for a large class of processes.

12.3 Empirical Mean and Autocovariances

Since in practice we generally have to work from a sampled version of a realisation of the process of interest, we shall only address the discrete case in what follows. It is clear that sampling a stationary or ergodic process at regular time intervals provides a discrete process that is still stationary or ergodic, respectively. We shall now give some results concerning empirical estimators of the mean and autocovariance functions of WSS discrete time processes.

First, however, we shall introduce the notion of WSS linear process and make some remarks about the Fourier series expansion of the PSD of a process.

12.3.1 Linear Processes

In what follows, we shall often have to consider linear processes for the empirical estimators, of which it is fairly simple to establish a certain number of results.

Definition 12.1 *A process X defined by*

$$X_n = m_X + \sum_{k \in \mathbb{Z}} h_k V_{n-k}, \tag{12.15}$$

where V is a zero mean white noise process with finite variance σ^2 and $\sum_{k \in \mathbb{Z}} |h_k| < \infty$, is called a linear process.

In particular, the class of linear processes contains ARMA processes.

For a linear process X, Lebesgue's dominated convergence theorem yields the relations

$$
\begin{aligned}
R_X(k) \quad &= \mathbb{E}[X_{n+k} X_n^*] \\
&= \sigma^2 \sum_{l \in \mathbb{Z}} h_{k+l} h_l^*, \\
\sum_{k \in \mathbb{Z}} |R_X(k)| &\le \sigma^2 (\sum_{k \in \mathbb{Z}} |h_k|)^2, \\
\sum_{k \in \mathbb{Z}} R_X(k) &= \sigma^2 |\sum_{k \in \mathbb{Z}} h_k|^2.
\end{aligned}
\tag{12.16}
$$

We note that the above sums are finite. Furthermore, absolute summability of the autocovariances implies the continuity of the PSD:

Theorem 12.3 *For a second order stationary process X, the hypothesis*

$$\sum_{k \in \mathbb{Z}} |R_X(k)| < \infty \tag{12.17}$$

implies the continuity of $S_X(f)$.

Proof Let first us note that for fixed $n \in \mathbb{N}$

$$
\begin{aligned}
|S_X(f_1) - S_X(f_2)| \\
&\le \sum_{k \in \mathbb{Z}} |R_X(k)||1 - e^{-2i\pi k(f_1 - f_2)}| \\
&\le 2\pi |f_1 - f_2| \sum_{|k| \le n} |k||R_X(k)| + 2 \sum_{|k| > n} |R_X(k)|.
\end{aligned}
\tag{12.18}
$$

This latter inequality holds because

$$
\begin{aligned}
|1 - e^{-2i\pi k(f_1 - f_2)}| &\le 2|sin(\pi k(f_1 - f_2))| \\
&\le 2\pi |k||f_1 - f_2|.
\end{aligned}
\tag{12.19}
$$

Moreover, since

$$\forall \varepsilon > 0, \exists n \in \mathbb{N}, \quad \sum_{k > n} |R_X(k)| < \frac{\varepsilon}{4}, \tag{12.20}$$

it thus results that for $|f_1 - f_2| < (\varepsilon/2)(2\pi \sum_{|k|\leq n} |k| |R_X(k)|)^{-1}$,

$$|S_X(f_1) - S_X(f_2)| < \varepsilon, \tag{12.21}$$

which completes the proof. $\square$

Conversely, we note that the hypothesis $\sum_{k\in\mathbb{Z}} |R_X(k)| < \infty$ often made in this book and verified for linear processes is also verified for processes whose spectral measure is absolutely continuous with respect to Lebesgue's measure and whose PSD is continuous, piecewise derivable and with bounded derivatives (see Appendix F).

Finally, we recall that for a process X whose PSD is of class $\mathcal{C}^1$ in the neighbourhood of f,

$$\sum_{k\in\mathbb{Z}} e^{-2i\pi k f} R_X(k) = S_X(f), \tag{12.22}$$

from Dirichlet's criterion (see Appendix F).

12.3.2 Empirical Mean

Let $X = (X_n)_{n\in\mathbb{Z}}$ denote a wide sense stationary process. Its empirical mean, defined by $\hat{m}_{X,n} = n^{-1} \sum_{k=1,n} X_k$, is an unbiased estimator:

$$\mathbb{E}[\hat{m}_{X,n}] = \mathbb{E}[X_n]$$

$$= m_X. \tag{12.23}$$

The variance of $\hat{m}_{X,n}$ is clearly given by

$$\| \hat{m}_{X,n} - m_X \|^2 = \frac{1}{n} \sum_{|j|<n} (1 - \frac{|j|}{n}) R_X(j). \tag{12.24}$$

Therefore,

$$\| \hat{m}_{X,n} - m_X \|^2 \leq \frac{1}{n} \sum_{|j|<n} |R_X(j)|, \tag{12.25}$$

and if $\lim_{n\to\infty} R_X(n) = 0$, $\hat{m}_{X,n}$ mean square converges towards m_X. In fact, the convergence of the right hand side of (12.25) towards 0 is equivalent to the first order ergodicity property (see Slutsky's theorem, Theorem 2.6).

If, moreover, $\sum_{n\in\mathbb{Z}} |R_X(j)| < \infty$, it results from (12.24) and Lebesgue's dominated convergence theorem that

$$\lim_{n\to\infty} n \| \hat{m}_{X,n} - m_X \|^2 = \sum_{j\in\mathbb{Z}} R_X(j). \tag{12.26}$$

In addition, if $S_X(f)$ is of class $\mathcal{C}^1$ in the neighbourhood of 0, $\sum_{n\in\mathbb{Z}} R_X(j) = S_X(0)$ from Dirichlet's criterion. The fact that the variance of $\hat{m}_{X,n}$ is asymptotically equivalent to $n^{-1} S_X(0)$ thus suggests the possible existence of a central limit theorem for $\hat{m}_{X,n}$. The following result formalises this idea:

Theorem 12.4 *If X is a linear process with*

$$X_n = m_X + \sum_{k \in \mathbb{Z}} h_k V_{n-k}, \tag{12.27}$$

where $\sum_{k \in \mathbb{Z}} h_k \neq 0$, then

$$\lim_{n \to \infty} \sqrt{n}[\hat{m}_{X,n} - m_X] \sim \mathcal{N}(0, \sigma^2 | \sum_{k \in \mathbb{Z}} h_k |^2). \tag{12.28}$$

Proof See Appendix O.

We note that the central limit theorem ensures almost sure convergence of a sequence of variables towards the mean value of the limit. This stems directly from Borel-Cantelli's lemma. Hence, under the hypotheses of Theorem 12.4, $\lim_{n \to \infty} \hat{m}_{X,n} \overset{\text{a.s.}}{=} m_X$.

More general results about extensions of the central limit theorem to sequences of non-independent random variables can be found in [3].

12.3.3 Empirical Autocovariance Coefficients

We generally use the following empirical autocovariance estimators, defined for $k \geq 0$ by

$$\tilde{R}_{X,n}(k) = \frac{1}{n-k} \sum_{j=1,n-k} (X_{j+k} - \hat{m}_{X,n})(X_j - \hat{m}_{X,n})^*, \tag{12.29}$$

and $\hat{R}_{X,n}(k) = \frac{1}{n} \sum_{j=1,n-k} (X_{j+k} - \hat{m}_{X,n})(X_j - \hat{m}_{X,n})^*,$

and $\tilde{R}_{X,n}(k) = \tilde{R}_{X,n}^*(-k)$, and $\hat{R}_{X,n}(k) = \hat{R}_{X,n}^*(-k)$, for $k < 0$.

These estimators are biased, and

$$\mathbb{E}[\hat{R}_{X,n}(k)] = \frac{n-k}{n} \mathbb{E}[\tilde{R}_{X,n}(k)]$$

$$= \frac{1}{n} \sum_{j=1,n} \mathbb{E}[(X_{j+k} - \hat{m}_{X,n})(X_j - \hat{m}_{X,n})^*] + O(\tfrac{k}{n})$$

$$= \frac{1}{n} \sum_{j=1,n} \mathbb{E}[(X_{j+k} - m_X)(X_j - m_X)^*]$$

$$- \| \hat{m}_{X,n} - m_X \|^2 + O(\tfrac{k}{n})$$

$$= R_X(k) - \frac{1}{n} \sum_{j=-n,n} (1 - \tfrac{|j|}{n}) R_X(j) + O(\tfrac{k}{n}). \tag{12.30}$$

It is clear that if $\lim_{n \to \infty} R_X(n) = 0$, the estimators $\tilde{R}_{X,n}(k)$ and $\hat{R}_{X,n}(k)$ are asymptotically unbiased.

We notice that if the mean m_X of the process X is known, we can re-place $\hat{m}_{X,n}$ by m_X in the definitions of $\tilde{R}_{X,n}(k)$ and of $\hat{R}_{X,n}(k)$. In this case, $\tilde{R}_{X,n}(k)$ becomes an unbiased estimator. However, we often prefer to use the biased estimator $\hat{R}_{X,n}(k)$ for, unlike $(\tilde{R}_{X,n}(k))_{k=0,n-1}$, the sequence $(\hat{R}_{X,n}(k))_{k=0,n-1}$ is of the positive type, and therefore there exists a spectral measure whose coefficients $(\hat{R}_{X,n}(k))_{k=0,n-1}$ are the first Fourier coefficients. Thus, in what follows, we shall always work with the estimator $\hat{R}_{X,n}(k)$. For the sake of simplicity, we shall assume that the process X under consideration is zero mean.

Theorem 12.5 *Let X denote a linear process of the form $X_n = \sum_{k \in \mathbb{Z}} h_k \times V_{n-k}$, where $\mathbb{E}[|V_n|^4] = \nu \sigma^4 < \infty$. Then, if X is real-valued,*

$$\lim_{n \to \infty} n.\mathrm{cov}[\hat{R}_{X,n}(k), \hat{R}_{X,n}(l)]$$

$$= (\nu - 3) R_X(k) R_X(l) \tag{12.31}$$

$$+ \sum_{p \in \mathbb{Z}} [R_X(p) R_X(p - k + l) + R_X(p + l) R_X(k - p)].$$

If X is a second order circular complex process, that is, if $\| \mathcal{R}e[X_n] \| = \| \mathcal{I}m[X_n] \|$, and $\mathbb{E}[X_n^2] = 0$,

$$\lim_{n \to \infty} n.\mathrm{cov}[\hat{R}_{X,n}(k), \hat{R}_{X,n}(l)]$$

$$\tag{12.32}$$

$$= (\nu - 2) R_X(k) R_X(l) + \sum_{p \in \mathbb{Z}} R_X(p + l) R_X(k - p).$$

Proof See Appendix P.

Remark Infinite sums such as $\sum_{p \in \mathbb{Z}} R_X(p + l) R_X(k - p)$ in the above ex-pressions can be estimated in practice. Indeed, we have as an example

$$\sum_{p \in \mathbb{Z}} R_X(p + l) R_X(k - p) = \int_{\mathcal{I}} S_X^2(f) e^{2i\pi(k+l)f} \, df. \tag{12.33}$$

Denoting by $\hat{S}_{X,n}(f)$ the spectral estimator of the periodogram studied be-low, we shall see at the end of this chapter that $\int_{\mathcal{I}} \hat{S}_{X,n}^2(f) e^{2i\pi(k+l)f} \, df$ is an example of a consistent estimator of the right-hand term of (12.33).

With hypotheses similar to those of Theorem 12.5, we can also establish that estimated autocovariances have Gaussian asymptotic distributions:

Theorem 12.6 *If X is a linear process, with $X_n = \sum_{k \in \mathbb{Z}} h_k V_{n-k}$, $\mathbb{E}[|V_n|^2] = \sigma^2$, $\sum_{k \in \mathbb{Z}} h_k \neq 0$, and $\mathbb{E}[|V_n|^4] = \nu \sigma^4 < \infty$,*

$$\lim_{n \to \infty} \sqrt{n} \left(\begin{bmatrix} \hat{R}_{X,n}(0) \\ \cdot \\ \cdot \\ \hat{R}_{X,n}(N) \end{bmatrix} - \begin{bmatrix} R_X(0) \\ \cdot \\ \cdot \\ R_X(N) \end{bmatrix} \right) \sim \mathcal{N}(0, \Gamma_{X,N}), \tag{12.34}$$

where

$$[\Gamma_{X,N}]_{kl} = (\nu - 3)R_X(k)R_X(l)$$

$$+ \sum_{p \in \mathbb{Z}}[R_X(p)R_X(p - k + l) + R_X(p + l)R_X(k - p)],$$
(12.35)

if X is a real-valued process, and if X is a complex circular process,

$$[\Gamma_{X,N}]_{kl} = (\nu - 2)R_X(k)R_X(l) + \sum_{p \in \mathbb{Z}} R_X(p + l)R_X(k - p).$$
(12.36)

Proof See Appendix Q.

12.4 Empirical PSD: the Periodogram

The periodogram estimator of the PSD of a zero mean WSS process X is defined by

$$\hat{S}_{X,n}(f) = \frac{1}{n} | \sum_{k=1,n} X_k e^{-2i\pi kf} |^2.$$
(12.37)

It is easy to check that $\hat{S}_{X,n}(f)$ is also given by

$$\hat{S}_{X,n}(f) = \sum_{k=-n+1}^{n-1} \hat{R}_{X,n}(k)e^{-2i\pi kf}.$$
(12.38)

Comparing this expression with the relation $S_X(f) = \sum_{k \in \mathbb{Z}} R_X(k)e^{-2i\pi kf}$, which gives the expression of the PSD when $\sum_{k \in \mathbb{Z}} |R_X(k)| < \infty$, suggests that the periodogram would be a convenient estimator of the PSD.

Unfortunately, we shall see below that the periodogram is not a consistent estimator! We shall, however, show that it is possible to obtain a consistent estimator by smoothing the periodogram by means of a filter.

From (12.38), it results that

$$\mathbb{E}[\hat{S}_{X,n}(f)] = \sum_{k=-n+1}^{n-1} (1 - \frac{|k|}{n})R_X(k)e^{-2i\pi kf},$$
(12.39)

and when $\sum_{k \in \mathbb{Z}} |R_X(k)| < \infty$, Lebesgue's dominated convergence theorem yields

$$\lim_{n \to \infty} \mathbb{E}[\hat{S}_{X,n}(f)] = \sum_{k \in \mathbb{Z}} R_X(k)e^{-2i\pi kf}$$

$$= S_X(f).$$
(12.40)

We remark that the mean convergence is then uniform in f for

$$|\mathbb{E}[\hat{S}_{X,n}(f)] - S_X(f)| \leq \frac{1}{\sqrt{n}} \sum_{|k| \leq \sqrt{n}} |R_X(k)| + \sum_{|k| > \sqrt{n}} |R_X(k)|, \quad (12.41)$$

and the two right-hand terms tend towards 0 when n tends towards $+\infty$.

We are now going to study the asymptotic properties of the periodogram in terms of variance. As the general case is not simple, we shall restrict ourselves, as we did when establishing properties of empirical mean estimators and autocovariance coefficients, to the large class of linear processes.

We shall begin by considering the case where X is a white noise process. To be concise, we shall mainly restrict ourselves to the case of real-valued processes. The circular complex case can be studied by a similar approach, and leads to similar conclusions. The difference in results comes from the difference that we obtain for the expression of the fourth order moments $\mathbb{E}[X_a X_b^* X_c X_d^*]$. Additional results can be found in [18] or [10], in particular.

12.4.1 The White Noise Case

We first remark that for a white noise process X, $\hat{S}_{X,n}(f)$ is an unbiased estimator.

We now note

$$c_n(f) = \frac{1}{\sqrt{n}}[\cos(2\pi f), \ldots, \cos(2\pi nf)]^T,$$

$$s_n(f) = \frac{1}{\sqrt{n}}[\sin(2\pi f), \ldots, \sin(2\pi nf)]^T, \quad (12.42)$$

and $\mathbf{X}_n = [X_1, \ldots, X_n]^T.$

Hence,

$$\hat{S}_{X,n}(f) = \frac{1}{n}|\textstyle\sum_{k=1,n} X_k e^{-2i\pi kf}|^2$$

$$= |c_n(f)^T \mathbf{X}_n|^2 + |s_n(f)^T \mathbf{X}_n|^2. \quad (12.43)$$

We show that for $f \in \mathcal{I} - \{0, 1/2\}$, $\hat{S}_{X,n}$ converges towards an exponential distribution, and that for distinct values of f the corresponding random variables $\hat{S}_{X,n}(f)$ are independent:

Theorem 12.7 *If X is a white noise process with variance σ^2, $\hat{S}_{X,n}(f)$ converges in distribution to $\mathcal{E}(\sigma^{-2})$ if $f \in \mathcal{I} - \{0, 1/2\}$ and to a $\chi^2(1)$ distribution with mean σ^2 if $f \in \{0, 1/2\}$.*
Moreover, for $0 < f_1 < f_2 < 1/2$, the vector

$$[c_n(f_1)^T \mathbf{X}_n, s_n(f_1)^T \mathbf{X}_n, c_n(f_2)^T \mathbf{X}_n, s_n(f_2)^T \mathbf{X}_n]^T \quad (12.44)$$

converges in distribution towards a Gaussian random vector with covariance matrix $(\sigma^2/2)I_4$, and $\hat{S}_{X,n}(f_1)$ and $\hat{S}_{X,n}(f_2)$ are asymptotically independent.

Proof See Appendix R.

The evolution of the covariance of the vector $[\hat{S}_{X,n}(f_1), \hat{S}_{X,n}(f_2)]^T$ is given by the following result, which shows that, unfortunately, the variance of the periodogram does not decrease towards 0 when n tends towards $+\infty$.

Theorem 12.8 *If* $\mathbb{E}[X_n^4] = \nu\sigma^4 < \infty$, *then*

$$\text{var}[\hat{S}_{X,n}(f)] \qquad = 2\sigma^4 + n^{-1}(\nu - 3)\sigma^4 + o(n^{-1}) \; \text{if} \; f = 0, 1/2,$$

$$\text{var}[\hat{S}_{X,n}(f)] \qquad = \sigma^4 + n^{-1}(\nu - 3)\sigma^4 + o(n^{-1}) \; \text{if} \; f \neq 0, 1/2,$$

$$\text{cov}[\hat{S}_{X,n}(f_1), \hat{S}_{X,n}(f_2)] = n^{-1}(\nu - 3)\sigma^4 + o(n^{-1}) \; \text{if} \; 0 < f_1 < f_2 < 1/2.$$
$$(12.45)$$

Proof

$$\mathbb{E}[\hat{S}_{X,n}(f_1)\hat{S}_{X,n}(f_2)]$$

$$= \frac{1}{n^2} \sum_{(a,b,c,d)=1,n} \mathbb{E}[X_a X_b X_c X_d] e^{-2i\pi(a-b)f_1} e^{-2i\pi(c-d)f_2}$$

$$= \frac{1}{n}\mathbb{E}[X_a^4] + \frac{n-1}{n} \parallel X_a^2 \parallel^2$$

$$+ \frac{1}{n^2}(|\sum_{k=1,n} e^{-2i\pi k(f_1+f_2)}|^2 - n) \parallel X_a^2 \parallel^2$$

$$+ \frac{1}{n^2}(|\sum_{k=1,n} e^{-2i\pi k(f_1-f_2)}|^2 - n) \parallel X_a^2 \parallel^2$$

$$= \sigma^4 + \frac{\sigma^4}{n}(\nu - 3) + \frac{\sigma^4}{n^2}(|\frac{\sin(\pi n(f_1 + f_2))}{\sin(\pi(f_1 + f_2))}|^2 + |\frac{\sin(\pi n(f_1 - f_2))}{\sin(\pi(f_1 - f_2))}|^2),$$
$$(12.46)$$

where the four terms are obtained for $a = b = c = d$, $a = b \neq c = d$, $a = c \neq b = d$, and $a = d \neq b = c$ respectively. The different cases considered in the theorem are then simply obtained from expression (12.46) and from the fact that $\mathbb{E}[\hat{S}_{X,n}(f)] = \sigma^2$. $\square$

12.4.2 The Periodogram of Linear Processes

We now assume that X is a linear process defined by $X_n = \sum_{k\in\mathbb{Z}} h_k V_{n-k}$, where the variance of V is equal to 1. The PSD of X is therefore

$$S_X(f) = |\sum_{k\in\mathbb{Z}} h_k e^{-2i\pi kf}|^2 S_V(f)$$
$$(12.47)$$
$$= |h(e^{2i\pi f})|^2.$$

As $\sum_{k\in\mathbb{Z}} |R_X(k)| \leq (\sum_{k\in\mathbb{Z}} |h_k|)^2 < \infty$, relation (12.41) shows that $\mathbb{E}[S_{X,n}(f)]$ converges towards $S_X(f)$, uniformly in f.

In order to take advantage of the results of the previous paragraph to calculate the variance of $\hat{S}_{X,n}(f)$, we now point out the link that exists between $\hat{S}_{X,n}(f)$ and $\hat{S}_{V,n}(f)$.

Theorem 12.9

$$\hat{S}_{X,n}(f) = |h(e^{2i\pi f})|^2 \hat{S}_{V,n}(f) + R_n(f)$$

$$= S_X(f)\hat{S}_{V,n}(f) + R_n(f), \tag{12.48}$$

with

$$\lim_{n\to\infty} \left(\sup_{f\in\mathcal{I}} \mathbb{E}[|R_n(f)|]\right) = 0. \tag{12.49}$$

Proof See Appendix S.

Then we are able to express the asymptotic behaviour of $\hat{S}_{X,n}(f)$.

Theorem 12.10 *For a linear process X with non-zero power spectral density $S_X(f)$, $\hat{S}_{X,n}(f)$ converges in distribution towards an $\mathcal{E}(S_X(f)^{-1})$ random variable if $0 < f < 1/2$, and to a $\chi^2(1)$ random variable with mean $S_X(f)$ when $f \in \{0, 1/2\}$. Moreover, if $\mathbb{E}[|V_n|^4] = \nu < \infty$ $(\sigma^2 = 1)$ and $\sum_{k\in\mathbb{Z}} \sqrt{|k|}|h_k| < \infty$, we obtain the following expressions for the asymptotic variance:*

$$\mathrm{var}[\hat{S}_{X,n}(f)] \qquad = 2S_X^2(f) + O(n^{-1/2}) \quad \textit{if } f \in \{0, 1/2\},$$

$$\mathrm{var}[\hat{S}_{X,n}(f)] \qquad = S_X^2(f) + O(n^{-1/2}) \quad \textit{if } 0 < f < 1/2,$$

$$\mathrm{cov}[\hat{S}_{X,n}(f_1), \hat{S}_{X,n}(f_2)] = O(n^{-1/2}) \qquad\qquad \textit{if } 0 < f_1 < f_2 < 1/2. \tag{12.50}$$

The terms in $O(n^{-1/2})$ decrease towards 0 uniformly in f on any compact set of $]0, 1/2[\times]0, 1/2[-\{(f, f); f \in]0, 1/2[\}$.

Remark We can, in fact, demonstrate the more precise result (see, for example, [10] p.348)

$$\mathrm{cov}[\hat{S}_{X,n}(f_1), \hat{S}_{X,n}(f_2)] = O(n^{-1}) \text{ if } 0 < f_1 < f_2 < 1/2. \tag{12.51}$$

Proof See Appendix T.

12.4.3 The Case of Line Spectra

To complete the study of the periodogram, we consider the case of a process X defined by $X_n = \sum_{k=1,p} \xi_k e^{2i\pi n f_k}$. For this, we begin by recalling the following result:

Theorem 12.11 *The Fejer kernel of order n defined by*

$$K_n(f) = \frac{1}{n}\left(\frac{\sin(\pi n f)}{\sin \pi f}\right)^2, \tag{12.52}$$

satisfies the following equalities

$$\int_{\mathcal{I}} K_n(f)df = 1,$$

$$\tag{12.53}$$

and $\forall \varepsilon \in]0, 1/2[$, $\lim_{n\to\infty} \int_{-\varepsilon}^{\varepsilon} K_n(f)df = 1$.

Proof We note that

$$K_n(f) = \frac{1}{n}|\sum_{k=1,n} e^{2i\pi kf}|^2$$

$$\tag{12.54}$$

$$= \frac{1}{n}\sum_{k=-n,n}(n - |k|)e^{2i\pi kf}.$$

The first equality is a straightforward consequence of relation (12.54). For $\varepsilon \in]0, 1/2[$ and $|f| \in]\varepsilon, 1/2[$,

$$K_n(f) < \frac{1}{n\sin^2(\pi\varepsilon)}. \tag{12.55}$$

It is therefore clear that

$$\lim_{n\to\infty}\int_{-\varepsilon}^{\varepsilon} K_n(f)df = \lim_{n\to\infty}\int_{\mathcal{I}} K_n(f)df$$

$$= 1._{\square} \tag{12.56}$$

Denoting by $[x]$ the integer part of x, we now remark that

$$\hat{S}_{X,n}(f) = \frac{1}{n}|\sum_{k=1,n}(\sum_{l=1,p}\xi_l e^{2i\pi kf_l})e^{-2i\pi kf}|^2$$

$$= \sum_{l=1,p}|\xi_l|^2 K_n(f_l - f)$$

$$+ \sum_{l\neq m}\xi_l\xi_m^* e^{i\pi[n/2](f_l-f_m)}K_{[n/2]}^{1/2}(f_l - f)K_{[n/2]}^{1/2}(f_m - f)$$

$$+ O(1/\sqrt{n}). \tag{12.57}$$

From Theorem 12.11, it is clear that for any continuous function g,

$$\lim_{n \to \infty} \int_{\mathcal{I}} g(f) \hat{S}_{X,n}(f) df = \sum_{l=1,p} |\xi_l|^2 g(f_l). \qquad (12.58)$$

It therefore appears that $\hat{S}_{X,n}(f)$, as a distribution, converges towards the stochastic measure $\sum_{l=1,p} |\xi_l|^2 \delta_{f_l}$.

12.5 Smoothed Periodogram

The periodogram spectral estimator is not very satisfactory. Indeed, it can be seen from Theorem 12.10 that it converges in the mean but that it does not converge in the mean square sense. Moreover, the periodogram is asymptotically uncorrelated at different frequencies, which causes an irregular shape of the periodogram, independently of the regularity of the true PSD $S_X(f)$.

12.5.1 Integrated Periodogram

However, when the spectrum of X appears in integral relations of the form

$$\int_{\mathcal{I}} g(f) S_X(f) df, \qquad (12.59)$$

where g is a continuous function, the use of the periodogram estimator can often be envisaged in a satisfactory way for a linear process X defined by $X_n = \sum_{k \in \mathbb{Z}} h_k V_{n-k}$, which satisfies

$$\mathbb{E}[|V_n|^4] \quad < \infty,$$

$$\text{and } \sum_{k \in \mathbb{Z}} \sqrt{|k|} |h_k| < \infty. \qquad (12.60)$$

Indeed, we have the following result:

Theorem 12.12 *For a process X defined by $X_n = \sum_{k \in \mathbb{Z}} h_k V_{n-k}$, where V is a white noise that satisfies (12.60), and a continuous function g on $\mathcal{I}$,*

$$\lim_{n \to \infty} \int_{\mathcal{I}} g(f) \hat{S}_{X,n}(f) df \overset{m.s.}{=} \int_{\mathcal{I}} g(f) S_X(f) df. \qquad (12.61)$$

Proof

$$\| \int_{\mathcal{I}} g(f) \hat{S}_{X,n}(f) df - \int_{\mathcal{I}} g(f) S_X(f) df \|^2$$

$$= \int_{\mathcal{I} \times \mathcal{I}} g(f_1) g(f_2) \text{cov}[\hat{S}_{X,n}(f_1), \hat{S}_{X,n}(f_2)] df_1 df_2. \qquad (12.62)$$

From Theorem 12.10, $\text{cov}[\hat{S}_{X,n}(f_1), \hat{S}_{X,n}(f_2)]$ is uniformly bounded on $\mathcal{I} \times \mathcal{I}$ and converges uniformly towards 0 on any compact set of $\mathcal{I} \times \mathcal{I}$ whose intersection with the set of $\{(f,f); f \in \mathcal{I}\}$ is empty. Therefore, the right-hand term of (12.62) converges towards 0 when n tends towards $+\infty$, which completes the proof. $\square$

12.5.2 Smoothed Periodogram

The above result justifies that, in practice, we often replace the periodogram estimator by a smoothed version of it, for which we estimate the spectrum at frequency f by averaging the values of the periodogram over a window of frequencies centred on f. We denote the weighting function used by $W(f)$. Extending $W(f)$ outside $\mathcal{I}$ periodically, the new spectral estimator is given by

$$\hat{S}_{X,n}^W(f) = \int_{\mathcal{I}} W(f-u)\hat{S}_{X,n}(u)du. \tag{12.63}$$

In the case where X is a white noise process, in order for $\hat{S}_{X,n}^W(f)$ to be an asymptotically unbiased estimator of $S_X(f)$ it is clearly necessary that $\int_{\mathcal{I}} W(u)du = 1$. Moreover, to ensure the positivity of $\hat{S}_{X,n}^W(f)$, and the constraint $\hat{S}_{X,n}^W(f) = \hat{S}_{X,n}^W(-f)$ when X is a real-valued process, it is natural to take a positive function for W that satisfies the symmetry of property $W(-f) = W(f)$. In fact, to account for all these constraints, we consider a sequence $(W_n)_{n\in\mathbb{N}^*}$ of weighting functions such that

$$W_n(f) \geq 0$$

$$W_n(-f) = W_n(f)$$

$$\int_{\mathcal{I}} W_n(f)df = 1 \tag{12.64}$$

the support $[-f_n, f_n]$ of W_n verifies $\lim_{n\to\infty} f_n = 0$.

We show that if $S_X(f)$ is continuous, then $\hat{S}_{X,n}^{W_n}(f)$ is asymptotically unbiased. From the constraint $\int_{\mathcal{I}} W_n(f)df = 1$,

$$|\mathbb{E}[\hat{S}_{X,n}^{W_n}(f)] - S_X(f)| \leq \int_{\mathcal{I}} W_n(f-u)|\mathbb{E}[\hat{S}_{X,n}(u)] - S_X(f)|du$$

$$= \int_{\mathcal{I}} W_n(f-u)|S_X(u) - S_X(f)|du$$

$$+ \int_{\mathcal{I}} W_n(f-u)|\mathbb{E}[\hat{S}_{X,n}(u)] - S_X(u)|du \tag{12.65}$$

$$\leq \max_{u\in[f-f_n,f+f_n]} |S_X(u) - S_X(f)|$$

$$+ \int_{\mathcal{I}} W_n(f-u)|\mathbb{E}[\hat{S}_{X,n}(u)] - S_X(u)|du.$$

Since $\lim_{n\to\infty} f_n = 0$, $S_X(u)$ is continuous in f and $\mathbb{E}[\hat{S}_{X,n}(u)] - S_X(u)$ converges to 0 uniformly in u, it is clear that $\mathbb{E}[\hat{S}_{X,n}^{W_n}(f)] - S_X(f)$ tends to 0 when n tends towards $+\infty$, which shows that $\hat{S}_{X,n}^{W_n}(f)$ is asymptotically unbiased.

We now remark that

$$\text{cov}[\hat{S}_{X,n}^{W_n}(f_1), \hat{S}_{X,n}^{W_n}(f_2)]$$

$$= \int_{\mathcal{I}\times\mathcal{I}} W_n(f_1 - u_1)W_n(f_2 - u_2)\text{cov}[\hat{S}_{X,n}(u_1), \hat{S}_{X,n}(u_2)]du_1 du_2.$$
(12.66)

Therefore, for $f_1 \neq f_2$ Theorem 12.10 easily leads to the inequality

$$|\text{cov}[\hat{S}_{X,n}^{W_n}(f_1), \hat{S}_{X,n}^{W_n}(f_2)]| \leq \frac{c}{n}\left(\int_{\mathcal{I}} W_n(f)df\right)^2,$$
(12.67)

where c is a constant. Similarly,

$$\text{var}[\hat{S}_{X,n}^{W_n}(f),] \leq c_1 S_X(f)^2 \left(\int_{\mathcal{I}} W_n(f)df\right)^2 + \frac{c_2}{n}\left(\int_{\mathcal{I}} W_n(f)df\right)^2,$$
(12.68)

with $c_1 = 2$ or 1, depending on whether f belongs to $\{0, 1/2\}$ or not, and where c_2 is a constant.

We should like to obtain a smoothed estimator with variance equal to zero. In order to do this, we often consider a modified version of the periodogram, defined, for example, by

$$\tilde{S}_{X,n}(f) = \hat{S}_{X,n}(f_{n,k}) \text{ for } f \in \mathcal{I}_k,$$
(12.69)

with $f_{n,k} = k/n$ and

$$\begin{cases} \mathcal{I}_k = [f_{n,k}, f_{n,k} + \frac{1}{n}[, \frac{-n}{2} \leq k \leq \frac{n}{2} - 1, \text{ for } n \text{ even} \\ \\ \mathcal{I}_k = [f_{n,k} - \frac{1}{2n}, f_{n,k} + \frac{1}{2n}[, \frac{1-n}{2} \leq k \leq \frac{n-1}{2}, \text{ for } n \text{ odd.} \end{cases}$$
(12.70)

Then, letting

$$W_n = \sum_{|k|\leq m_n} W_{n,k}\delta_{f_{n,k}},$$
(12.71)

the smoothed version of $\tilde{S}_{X,n}(f)$ is given by

$$\tilde{S}_{X,n}^{W_n}(f) = \sum_{|l|\leq m_n} \tilde{S}_{X,n}(f - f_{n,l})W_{n,l}$$

$$= \sum_{|l|\leq m_n} \hat{S}_{X,n}(f_{n,k} - f_{n,l})W_{n,l},$$
(12.72)

for $f \in \mathcal{I}_k$. Here, the conditions (12.64) are expressed by

$$W_{n,k} \geq 0$$

$$W_{n,-k} = W_{n,k}$$

$$\sum_{|k| \leq m_n} W_{n,k} = 1 \tag{12.73}$$

$$\lim_{n \to \infty} m_n/n = 0.$$

Theorem 12.10 then makes it possible to establish the following result:

Theorem 12.13 *Let there be a sequence of smoothing functions* $W_n = \sum_{|k| \leq m_n} W_{n,k} \delta_{f_{n,k}}$, *which satisfies conditions (12.73) and a linear process* X *defined by* $X_n = \sum_{k \in \mathbb{Z}} h_k V_{n-k}$, *with* $\mathbb{E}[|V_n|^4] < \infty$ *and* $\sum_{k \in \mathbb{Z}} \sqrt{|k|}|h_k| < \infty$. *Thus, the smoothed periodogram is asymptotically unbiased and its asymptotic covariance satisfies the relations*

$$\lim_{n \to \infty} (\sum_{|k| \leq m_n} W_{n,k}^2)^{-1} \mathrm{var}[\tilde{S}_{X,n}^{W_n}(f)] = 2S_X^2(f) \; \textit{if } f \in \{0, 1/2\},$$

$$\lim_{n \to \infty} (\sum_{|k| \leq m_n} W_{n,k}^2)^{-1} \mathrm{var}[\tilde{S}_{X,n}^{W_n}(f)] = S_X^2(f) \;\; \textit{if } 0 < f < 1/2,$$

$$\lim_{n \to \infty} (\sum_{|k| \leq m_n} W_{n,k}^2)^{-1} \mathrm{cov}[\tilde{S}_{X,n}^{W_n}(f_1), \tilde{S}_{X,n}^{W_n}(f_2)] = 0,$$

$$\textit{if } 0 < f_1 < f_2 < 1/2. \tag{12.74}$$

Proof We consider Theorem 12.10 and relations similar to relations (12.65), (12.67) and (12.68), for $\tilde{S}_{X,n}^{W_n}(f)$ and for the window $(W_n)_{|n| \leq m_n}$. Thus, the result stems from the inequality

$$n^{-1}(\sum_{|k| \leq m_n} W_{n,k})^2 \leq \frac{2m_n + 1}{n} \sum_{|k| \leq m_n} W_{n,k}^2, \tag{12.75}$$

and from using the condition $\lim_{n \to \infty} m_n/n = 0$. $\square$

From the above result, it appears that $\tilde{S}_{X,n}^{W_n}(f)$, which is asymptotically unbiased, has a zero asymptotic variance if the additional condition

$$\lim_{n \to \infty} \sum_{|k| \leq m_n} W_{n,k}^2 = 0 \tag{12.76}$$

is satisfied. We note that this condition and the condition $\sum_{|k| \leq m_n} W_{n,k} = 1$ imply that $\lim_{n \to \infty} m_n = +\infty$.

Conditions (12.73) and (12.76) are satisfied when we take, for example, a rectangular smoothing window with amplitude $W_{n,k} = 1/(2m_n + 1)$, or a triangular smoothing window with amplitude $W_{n,k} = (m_n - |k|)/(m_n^2 - m_n)$, for $|k| \leq m_n$, and, for example, $m_n = \sqrt{n}$.

Remark The interest of sampling $\hat{S}_{X,n}(f)$ as shown in this section comes from the fact that it is possible to satisfy conditions $\forall n \in \mathbb{N}, \sum_{|k| \leq m_n} W_{n,k} = 1$ and $\lim_{n \to \infty} \sum_{|k| \leq m_n} W_{n,k}^2 = 0$ simultaneously, where Cauchy-Schwarz's inequality

$$\int_{\mathcal{I}} W(f) df \leq \left(\int_{\mathcal{I}} W(f)^2 df \right)^{1/2} \tag{12.77}$$

clearly shows that we cannot simultaneously have $\forall n \in \mathbb{N}$, $\int_{\mathcal{I}} W_n(f) df = 1$ and $\lim_{n \to \infty} \int_{\mathcal{I}} W_n^2(f) df = 0$, which makes it difficult to ensure zero asymptotic variance of the periodogram. Another justification for the approach adopted here is that, in practice, the periodogram is generally computed by means of a fast Fourier transform that provides a sampled version of $\hat{S}_{X,n}(f)$.

12.5.3 Averaged Periodogram

Another approach, often used jointly with the previous one, yields a reduced variance of the periodogram by averaging several realisations of it. In particular, in the case of independent realisations of the periodogram estimator, the variance is reduced by a factor K^{-1}, where K is the number of realisations of the periodogram that are averaged.

Exercises

In this section, $X = (X_n)_{n \in \mathbb{Z}}$ denotes a WSS process.

12.1 We consider an AR(1) process $X = (X_n)_{n \in \mathbb{Z}}$ with $X_n = aX_{n-1} + V_n$ and $|a| < 1$. Calculate the variance of the empirical mean $N^{-1} \sum_{n=1,N} X_n$ and compare it to the variance of the empirical mean of a white noise.

12.2 (Autocovariance estimator in the time continuous case) We consider a WSS process $Y = (Y_t)_{t \in \mathbb{R}}$ and $\hat{R}_{Y,T}(t) = T^{-1} \int_{[0,T-t]} Y_{t+u} Y_u^* du$ an estimator of its autocovariance function for $0 \leq t \leq T$.

a) Calculate $\mathbb{E}[\hat{R}_{Y,T}(t)]$.

b) If X is Gaussian, show that

$$\lim_{T \to \infty} T.cov(\hat{R}_{Y,T}(t), \hat{R}_{Y,T}(t'))$$

$$= \frac{1}{T} \int_{[0,T-t]} (e^{2i\pi f(t-t')} + e^{2i\pi f(t+t')}) S_Y(f) df. \tag{12.78}$$

c) Compare this result to the discrete case.

12.3 (Missing data) We consider a situation where there are missing or obvious gross error observations, yielding $Y_n = g_n X_n$, where $g_n = 1$ if X_n is recorded and $g_n = 0$ if X_n is missing. We define

$$R_g(n) = \lim_{N \to \infty} N^{-1} \sum_{k=1,N-n} g(k+n)g(k). \tag{12.79}$$

Show that $\hat{R}_{Y,N}(n) R_g^{-1}(n)$ is an asymptotically unbiased estimator of $R_X(n)$.

12.4 (Autocorrelation estimator) We consider the autocorrelation estimator of X defined by $\hat{\rho}_{X,n}(k) = \hat{R}_{X,n}(k)\hat{R}_{X,n}^{-1}(0)$.

   a) Check that $\mathbb{E}[\hat{\rho}_{X,n}(k)] = (1 - N^{-1}|k|)\rho_{X,n}(k)$.

   b) Calculate the asymptotic expression of $n.cov(\hat{\rho}_{X,n}(k), \hat{\rho}_{X,n}(l))$.

12.5 (Averaged periodogram) Express the averaged periodogram in terms of the empirical covariance matrix $\hat{T}_{X,p}^K = K^{-1} \sum_{k=1,p} \mathbf{X}_{kp}\mathbf{X}_{kp}^H$, where $\mathbf{X}_n = [X_n, \ldots, X_{n-p+1}]^T$, and study its behaviour as K tends to $+\infty$.

12.6 (Correlogram) A PSD estimator alternative to the periodogram of X is the correlogram PSD estimator, defined by

$$\tilde{S}_X(f) = \sum_{n=-N,N} \tilde{R}_X(n)e^{-2i\pi n f}, \tag{12.80}$$

where the coefficients $(\tilde{R}_X(n))_{n=-N,N}$ are given by (12.29).

   a) Express $\mathbb{E}[\hat{S}_X(f)]$ and $\mathbb{E}[\tilde{S}_X(f)]$ in terms of $S_X(f)$ and of the Dirichlet kernel defined by $D_n(f) = \sum_{k=-n,n} e^{-2i\pi k f}$.

   b) Show that $\tilde{S}_X(f)$ might not be a positive function on $\mathcal{I}$.

12.7 (Positive constraint matching via alternate projections)[11] We denote by $(\overline{R}_X(n))_{n=-N,N}$ any estimator of the autocovariance sequence $(R_X(n))_{n=-N,N}$, with $\overline{R}_X(-n) = \overline{R}_X^*(n)$ for $n = 1, \ldots, N$. We apply the following algorithm to the sequence $(\overline{R}_X(n))_{n=-N,N}$:

1. Extend the sequence $(\overline{R}_X(n))_{n=-N,N}$ to $(\overline{R}_X(n))_{n=-L,L}$ $(L \gg N)$ with zeroes: $\overline{R}_X(n) = 0$ for $|n| > |N|$. Initialise the algorithm with the sequence $(\overline{R}_X(n))_{n=-L,L}$,
2. calculate the Fast Fourier Transform (FFT) of the sequence,
3. set to 0 the negative components of the FFT,
4. perform the inverse FFT,
5. set to 0 the components of index n such that $|n| > |N|$,
6. go to step 2.

   a) Show that the above procedure converges.

   b) What is this algorithm useful for? Consider in particular the correlogram estimator defined in the previous exercise.

12.8 We suppose that X is a zero mean Gaussian circular process. What is the distribution of $[\hat{S}_{X,N}(f_1), \ldots, \hat{S}_{X,N}(f_p)]^T$?

12.9 We assume that X is periodic with period N and we define Y_n by $Y_n = \sum_{k=0,q} b_k X_{n-k}$. Denoting $B(f) = \sum_{k=0,q} b_k e^{-2i\pi kf}$, and $f_l = -(1/2) + (k/N)$ ($l = 0, \ldots, N-1$), show that

$$\hat{S}_{Y,N}(f_k) = |B(f_k)|^2 S_X(f_k). \tag{12.81}$$

12.10 (Frequency response of conventional windows) We note $t_n = \frac{1}{[N-1]}(n - [N-1]/2)$, where $[.]$ represents the integer part.

a) Give the frequency response of the following windows $(W_n)_{n=0,N-1}$,

- rectangle: $W_n = 1$;
- triangle (Bartlett window): $W_n = 1 - 2|t_n|$;
- squared cosine (Hann window): $W_n = \cos^2(\pi t_n) = (1 + \cos(2\pi t_n))/2$.

b) By means of a computer program, plot $(W_n)_{n=0,N-1}$ and its frequency response for the above windows as well as for the Hamming window defined by $W_n = 0.54 + 0.46\cos(2\pi t_n)$ (choose, for instance, $N = 101$). Check that it may be more interesting to choose the Hamming window rather than the closely defined Hann window since it shows lower side lobes.

12.11 (Welch periodogram) From the sample $(X_n)_{n=0,N}$ of X, the spectrum is estimated as follows: we consider a window function $(W_n)_{n=0,D}$ that is shifted by S samples ($S \leq D$) P times. $P = [(N - D)/S] + 1$ with $[.]$ representing the integer part. Then, we define the sequences $(X_n^{(p)})_{n=0,D-1}$, for $p = 1, \ldots, P$, by $X_n^{(p)} = W_n X_{n+(p-1)S}$. The Welch periodogram is defined by

$$S_{We}(f) = \frac{1}{\sum_{n=0,D-1} |W_n|^2} \frac{1}{P} \sum_{p=1,P} D^{-1}| \sum_{n=0,D-1} X_n^{(p)} e^{-2i\pi nf}|^2. \tag{12.82}$$

Calculate $\mathbb{E}[S_{We}(f)]$ in terms of $S_X(f)$ and $W(f) = \sum_{n=0,D-1} W_n e^{-2i\pi nf}$.

13. Parametric Spectral Estimation

Purpose The use of the periodogram empirical spectrum estimator or of the smoothed periodogram is not well-suited to some situations, in particular when only a small number of autocovariance coefficients of the process being studied are available from the observation. In such situations, parametric spectral estimation techniques are often preferred, and in particular techniques involving rational spectra models.

13.1 Introduction

As we saw in the context of non-parametric spectral estimation, spectral estimation is undertaken more or less directly from the knowledge of an estimator of the first autocovariance coefficients associated with the spectral measure.

In some situations, the data available only enables the estimation of a very small number of autocovariance coefficients of the process. In such cases, the periodogram estimator will have a very poor resolution (we may define resolution as the minimum frequency shift from which it is possible to distinguish two harmonic components, in the sense of a certain criterion to be specified).

To illustrate this problem, we consider the following example: in array signal processing, we often consider a process $X = (X_z)_{z \in \mathbb{R}}$, indexed by the spatial position z, and regularly sampled at points $1, \ldots, n$ by means of n sensors, yielding the random vector $\mathbf{X} = [X_1, \ldots, X_n]^T$. Observing $\mathbf{X}$ at different instants generally yields a large number of available realisations of $\mathbf{X}$. As the number of sensors is generally small, we first estimate the n first autocovariance coefficients of the process $(X_n)_{n \in \mathbb{Z}}$ or, equivalently, the covariance matrix of $\mathbf{X}$. Then, we estimate the spectral measure of this process by means of parametric techniques that offer a better resolution than a Fourier transform performed on n points.

Moreover, unlike non-parametric estimation, the parametric approach makes it possible to integrate constraints upon the autocovariance coefficients and upon the spectrum, relating to possible prior information about the process. Thus, for some processes, we have specific information about the absolutely continuous part and about the singular part of the spectrum, such as: the absolutely continuous part is constant, the singular part is carried by p points, the support of the spectral measure is known, and so on.

There are several parametric methods for estimating the first autocovariance coefficients of a stationary process, which can integrate various constraints, the most conventional probably being that of the positivity of the estimated covariance matrix of X_n.

We have already mentioned the interest of ARMA modelling in the chapter dedicated to rational spectrum processes. We can estimate the parameters of an ARMA model by optimising conventional statistical criteria like the maximum likelihood criterion.

Before dealing with parametric spectral estimation techniques, we shall recall a few results about parametric estimation whose justification will be found, for instance, in the books on statistics cited in the references. We shall also give some indications about the estimation of the autocovariance coefficients of wide sense stationary processes.

13.2 Elements of Parametric Spectral Estimation

13.2.1 Cramer-Rao Lower Bound (CRLB)

We assume that the distributions P_θ of the statistical model $(\Omega, \mathcal{A}, \{P_\theta; \theta \in \Theta\})$ studied are absolutely continuous (with respect to Lebesgue's measure), with densities p_θ, such that $p_\theta(x)$ is twice differentiable with respect to θ, almost everywhere in x. We assume, moreover, that Θ is an open set of $\mathbb{R}^p$, that for any element A of $\mathcal{A}$, $\int_A p_\theta(x)dx$ is twice differentiable with respect to θ, and that the operations of derivation and integration can be interchanged. Then, for any unbiased estimator $T(\mathbf{X})$ ($\mathbf{X} = [X_1, \ldots, X_n]^T$) of $g(\theta)$, the covariance matrix Σ_T of $T(\mathbf{X})$ is lower bounded by the so-called Cramer-Rao Lower Bound (CRLB) [26], [28]:

$$\Sigma_T \geq \frac{\partial g(\theta)}{\partial \theta} I_\theta^{-1} \frac{\partial g(\theta)^H}{\partial \theta}, \tag{13.1}$$

where I_θ denotes the Fisher information matrix. The general term of I_θ is given by

$$[I_\theta]_{i,j} = \mathbb{E}[(\frac{\partial}{\partial \theta_i} \log p_\theta)(\frac{\partial}{\partial \theta_j} \log p_\theta)] \tag{13.2}$$

$$= -\mathbb{E}[\frac{\partial^2}{\partial \theta_i \partial \theta_j} \log p_\theta],$$

and the general term of the matrix $\frac{\partial}{\partial \theta} g(\theta)$ is given by

$$[\frac{\partial g(\theta)}{\partial \theta}]_{i,j} = \frac{\partial g_i(\theta)}{\partial \theta_j}. \tag{13.3}$$

An unbiased estimator whose covariance matrix reaches the CRLB is said to be *efficient*. This bound indicates the optimal performance that we can hope to obtain in terms of variance for an unbiased estimator, but it cannot always be reached.

13.2.2 Maximum Likelihood Estimators

For the statistical model $(\Omega, \mathcal{A}, \{P_\theta; \theta \in \Theta\})$, the maximum likelihood estimator of θ is the estimator of θ that maximises $P_\theta(\mathbf{X})$.

Estimation in the maximum likelihood sense is a most appreciated technique in statistics, when it can be implemented. This comes from the fact that if for the statistical model $(\Omega, \mathcal{A}, \{P_\theta; \theta \in \Theta\})$ there exists an efficient estimator of θ, then it coincides with the maximum likelihood estimator of θ [26], [28]. In addition, it can be shown that the maximum likelihood estimator of $g(\theta)$ is $g(\hat{\theta})$, where $\hat{\theta}$ is the maximum likelihood estimator of θ.

13.2.3 Minimum Variance Linear Unbiased Estimators

In the absence of a precise statistical model, we sometimes assume that the observation vector $\mathbf{X}$ is described by a linear model of the form $\mathbf{X} = H\theta + W$, where W is a zero mean random vector with a known covariance matrix, denoted by C. We look for a linear unbiased estimator of θ of the form $\hat{\theta} = M\mathbf{X}$. Here, we consider the real case. We easily check that the minimum variance linear unbiased estimator of θ is the solution of the following constrained optimisation problem

$$\begin{cases} \min_M Tr(MCM^T) \\ MH = I. \end{cases} \tag{13.4}$$

Using Lagrange multipliers [86], we obtain the solution:

$$\hat{\theta} = (H^T C^{-1} H)^{-1} H^T C^{-1} \mathbf{X}. \tag{13.5}$$

The covariance matrix of $\hat{\theta}$ is then equal to $(H^T C^{-1} H)^{-1}$. This estimator is sometimes called the Best Linear Unbiased Estimator and referred to as BLUE. When W is Gaussian, it is clear that $\hat{\theta}$ also represents the maximum likelihood estimator of θ for the model $\mathbf{X} = H\theta + W$ (see Exercise 13.1).

13.2.4 Least Squares Estimators

For a linear model of the form $\mathbf{X} = H\theta + W$, where W is a random vector with unknown covariance matrix, the least squares estimator of θ is given by

$$\hat{\theta} = \arg\min_\theta \parallel \mathbf{X} - H\theta \parallel_K^2$$

$$= (H^T K H)^{-1} H^T K \mathbf{X},$$

(13.6)

where the norm $\parallel . \parallel_K$ is defined by $\parallel u \parallel_K^2 = u^T K u$. When K is equal to the inverse of the covariance matrix C of W, we again find the linear unbiased estimator with minimum variance.

If the model is not linear: $\mathbf{X} = g(\theta) + W$, the criterion $J(\theta) = \parallel \mathbf{X} - g(\theta) \parallel_K^2$ to be minimised is no longer a quadratic function of the parameter, and we must generally implement descent algorithms such as the gradient algorithm (see Chapter 16) to search for the optimum.

13.3 Estimation of the Autocovariance Coefficients

The autocovariance coefficients $(R_X(k))_{k=-N,N}$ of a wide sense stationary process have an important property: the covariance matrix T_N of size $N+1$ and with general terms $[T_N]_{ij} = R_X(i-j)$ is positive Hermitian and Toeplitz. We can account for this particular algebraic structure to estimate the coefficients of T_N.

We note $\mathbf{X}_k = [X_k, \ldots, X_{k+N}]^T$, and we consider the empirical estimator of T_N given by

$$\hat{T}_N = \frac{1}{n} \sum_{k=1,n} \mathbf{X}_k \mathbf{X}_k^H.$$

(13.7)

This matrix is Hermitian and positive, but does not have a Toeplitz structure in general. We can resolve this by projecting $\hat{T}_N$ on the vector space of Toeplitz matrices of size $N+1$. We easily verify that this amounts to averaging $\hat{T}_N$ along its parallels to the diagonal. Unfortunately, the Toeplitz matrix thus obtained may no longer be positive.

In this case, we can use an alternate projection technique, which involves projecting iteratively the estimator on the set of Toeplitz matrices, then on that of the positive Hermitian matrices (which can be done simply by setting the negative eigenvalues of the matrix to 0). After convergence of this procedure, we obtain a positive Hermitian Toeplitz matrix, but it may be singular, of rank $p \le N$. This can be a drawback for, as we saw in the chapter relating to spectral identification, this is one specific property of the processes made up of p harmonic components.

Another possible approach involves estimating T_N in the maximum likelihood sense, by using a parameterisation of T_N coherent with its positive Hermitian and Toeplitz structure. In the case where X is Gaussian, the maximum likelihood approach clearly leads to minimising

$$\log |S| + Tr(\hat{T}_N S^{-1}),$$

(13.8)

which is the opposite of the log-likelihood of $(\mathbf{X}_k)_{k=1,n}$, over the set of positive Hermitian Toeplitz matrices S. Criterion (13.8) can, in particular, be optimised by parameterising the autocovariance coefficients via their reflection coefficients [27], or via the parameters of some discrete solutions to the trigonometric moment problem [23].

Rather than using the non-convex criterion (13.8), we sometimes prefer to minimise the quadratic criterion

$$\| (S - \hat{T}_N)\hat{T}_N^{-1} \|^2, \tag{13.9}$$

which also yields an efficient estimator.

To end this section, we indicate here a fairly flexible and easy to implement parameterisation of positive Hermitian Toeplitz matrices. We can show that the cone

$$K_M^+ = \{S = \sum_{i=1,M} \alpha_i d(f_i) d(f_i)^H; \ \alpha_i \geq 0, \ f_i = M^{-1}(2i - M - 1)\} \tag{13.10}$$

for large enough M represents a good approximation of the set of positive Hermitian matrices. In this context, the optimisation of criterion (13.9) can clearly be reduced to a quadratic minimisation problem, with a positivity constraint upon the parameters:

$$\begin{cases} \min_{\alpha_1,\ldots,\alpha_M} \| (\sum_{i=1,M} \alpha_i d(f_i) d(f_i)^H - \hat{T}_N)\hat{T}_N^{-1}) \|^2 \\ \alpha_i \geq 0 \quad i = 1, M. \end{cases} \tag{13.11}$$

This problem is solved simply by means of conventional optimisation techniques (Uzawa-Arrow-Hurwicz's algorithm). With this approach, information such as the support of the spectrum of X can easily be taken into account by setting to zero the coefficients α_i for which f_i does not belong.

13.4 Spectrum Estimation of ARMA Models: Mean Square Criteria

13.4.1 Estimation of Rational Spectra

ARMA modelling of the spectrum of a process involves identifying this process as the output of a transfer function filter

$$h(z) = \frac{b(z)}{a(z)} = \frac{\sum_{l=0,q} b_l z^{-l}}{1 + \sum_{k=1,p} a_k z^{-k}}, \tag{13.12}$$

with unit variance white noise input. The spectrum is then estimated by

$$\tilde{S}_X(f) = \frac{|b(e^{2i\pi f})|^2}{|a(e^{2i\pi f})|^2}. \tag{13.13}$$

Estimation of the AR Part We shall denote here by $(X_k)_{k=1,N}$ the observed variables. We know that

$$R_X(n) + \sum_{k=1,p} a_k R_X(n-k) = 0, \quad n > q. \tag{13.14}$$

For $k = 0, N-1$, we estimate $R_X(k)$ by

$$\hat{R}_{X,N}(k) = \frac{1}{N} \sum_{n=1,N-|k|} X_{n+|k|} X_n^* \tag{13.15}$$

$$= \hat{R}_{X,N}^*(-k).$$

In general, we do not use all the values $\hat{R}_{X,N}(k)$ available, since the variances of the estimators of $R_X(k)$ obtained for k close to N are high. We note $(\hat{R}_{X,N}(k))_{k=-M,M}$, with $M < N$, the sequence used to perform spectrum estimation.

The estimated autocovariance coefficients satisfy the relations

$$\hat{R}_{X,N}(n) + \sum_{k=1,p} a_k \hat{R}_{X,N}(n-k) = \varepsilon_n, \quad n > q, \tag{13.16}$$

where ε_n is an error term related to the estimation of the autocovariance coefficients. The estimation of $a = [a_1, \ldots, a_p]^T$ in the least squares sense from $(\hat{R}_{X,N}(k))_{k=-M,M}$ is therefore given by

$$\hat{a} = \arg\min \sum_{l=q+1,M} |\hat{R}_{X,N}(n) + \sum_{k=1,p} a_k \hat{R}_{X,N}(n-k)|^2, \tag{13.17}$$

that is, if we assume that $M - q \geq p$,

$$\hat{a} = -(\hat{R}^H \hat{R})^{-1} \hat{R}^H [\hat{R}_{X,N}(q+1), \ldots, \hat{R}_{X,N}(M)]^T, \tag{13.18}$$

with

$$\hat{R} = \begin{bmatrix} \hat{R}_{X,N}(q) & .. & \hat{R}_{X,N}(q+1-p) \\ . & ... & \\ \hat{R}_{X,N}(M-1) & .. & \hat{R}_{X,N}(M-p) \end{bmatrix}. \tag{13.19}$$

Estimation of the MA Part (Durbin's method) We now write

$$Y_n = X_n + \sum_{k=1,p} \hat{a}_k X_{n-k}. \tag{13.20}$$

We shall try to identify Y_n with $\sum_{l=0,q} b_l \nu_{n-l}$, where ν is the normalised innovation of X. We may approximate the spectrum of Y by that of an AR model with high order, denoted by L. We thus obtain the estimator

$$\hat{S}_Y(f) = \frac{\sigma_L^2}{|c(e^{2i\pi f})|^2}$$

$$= \frac{\sigma_L^2}{|1 + \sum_{k=1,L} c_k e^{-2i\pi k f}|^2}. \tag{13.21}$$

The coefficients $(c_k)_{k=1,L}$ that minimise the prediction error of Y also minimise the quadratic criterion

$$\sum_{k,l=0}^{L} c_k c_l^* R_Y(l-k), \tag{13.22}$$

where $c_0 = 1$. The minimum of this criterion provides an estimator of the variance of the linear prediction error of Y. In what follows, we shall denote $c_k = 0$ for $k < 0$ or $k > L$, and $b_k = 0$ for $k < 0$ or $k > q$. We now remark that the autocovariance coefficients associated with a spectrum with PSD $|\sum_{l=0,q} b_l e^{-2i\pi l f}|^2$ are of the form $R(n) = \sum_{l \in \mathbb{Z}} b_l b_{n-l}^*$. By replacing $R_Y(l-k)$ by $R(l-k)$ in criterion (13.22), it can be re-written as

$$\sum_{k,l=0}^{L} c_k c_l^* \left(\sum_{n \in \mathbb{Z}} b_{n-l} b_{n-k}^*\right) = \sum_{n \in \mathbb{Z}} |\sum_{k=0,L} c_k b_{n-k}^*|^2$$

$$= \sum_{k,l=1,q} b_k b_l^* \left(\sum_{n \in \mathbb{Z}} c_{n-k}^* c_{n-l}\right)$$

$$+ 2\mathcal{R}e[b_0 \sum_{l=1,q} b_l^* \left(\sum_{n \in \mathbb{Z}} c_n^* c_{n-l}\right)]$$

$$+ |b_0|^2 \left(\sum_{n \in \mathbb{Z}} |c_n|^2\right). \tag{13.23}$$

As the variances of the innovations of the spectra with respective PSD $|\sum_{k=0,q} b_l e^{-2i\pi l f}|^2$ and $\sigma_L^2 |1 + \sum_{k=1,L} c_k e^{-2i\pi k f}|^{-2}$ are equal to $|b_0|^2$ and σ_L^2 respectively, we shall identify b_0 with $\hat{b}_0 = \sigma_L$. Consequently, the minimum of criterion (13.23) with respect to $b = [b_1, \ldots, b_q]^T$ is given by

$$\hat{b} = -\sigma_L T_c^{-1} r_c,$$

$$\text{with } [T_c]_{k,l} = \frac{1}{L+1} \sum_{n=\max(k,l)}^{L+\min(k,l)} c_{n-k} c_{n-l}^* \quad (k,l) = 1, q, \tag{13.24}$$

$$\text{and } [r_c]_k = \frac{1}{L+1} \sum_{n=k}^{L} c_n^* c_{n-k} \quad k = 1, q.$$

Remarks
1) We have identified b_0 with the order L prediction error variance of Y.

Therefore, for large L, $|b_0|^2$ can be seen as an estimator of the variance of the innovation of Y. Taking into account the results relating to the minimum-phase causal factorisation of the PSD of regular processes, it therefore appears that the factorisation of the MA part estimated by Durbin's method is the minimum-phase causal factorisation.

2) The previous method provides an estimator of the spectral factorisation of the PSD whose calculation is not necessary if only the PSD of X is required. Indeed, it then suffices to put

$$| \sum_{l=0,q} b_l e^{-2i\pi lf} |^2 = \sum_{k=-q,q} \beta_k e^{-2i\pi kf}, \tag{13.25}$$

and to notice that the Fourier coefficients of Y satisfy $R_Y(k) = \beta_k$ ($k = -q, q$). We then identify the estimators of the coefficients β_k by means of the relations

$$\hat{\beta}_k = \hat{R}_{Y,N}(k) = [1 \quad \hat{a}^H] \begin{bmatrix} \hat{R}_{X,N}(k) & . & \hat{R}_{X,N}(k-p) \\ . & & \\ \hat{R}_{X,N}(k+p) & . & \hat{R}_{X,N}(k) \end{bmatrix} \begin{bmatrix} 1 \\ \hat{a} \end{bmatrix}. \tag{13.26}$$

13.4.2 Rational Filter Synthesis

Along the same lines as the ARMA spectral estimation discussed above, where we were concerned with finding the coefficients of a rational PSD from its estimated autocovariance coefficients, we can try to estimate the rational transfer function that best corresponds, in the mean square sense, to a fixed impulse response. Although this problem does not directly fall within the context of the estimation of rational spectra, the approach proposed nevertheless presents similarities to the above study.

To realise a stable causal filter whose impulse response is $(h_{d,n})_{n\in\mathbb{N}}$, we wish to approach this filter by a stable causal filter with rational transfer function

$$h(z) = \frac{b(z)}{a(z)} = \frac{\sum_{l=0,q} b_l z^{-l}}{1 + \sum_{k=1,p} a_k z^{-k}}, \tag{13.27}$$

that we also note as $h(z) = \sum_{n\geq 0} h_n z^{-n}$. We can then look for the coefficients $(a_k)_{k=1,p}$ and $(b_l)_{l=0,q}$ that minimise the criterion

$$J = \sum_{n=0,N} |h_{d,n} - h_n|^2, \tag{13.28}$$

which takes into account the $N + 1$ first coefficients of the impulse response to be identified.

By considering, for example, the simple case where

$$h(z) = b(1 + az^{-1})^{-1}$$

$$= \sum_{k \in \mathbb{N}} b(-a)^k z^{-k},$$

(13.29)

with $|a| < 1$, it appears that J is a non-linear function of the parameters to be estimated. Therefore, to simplify the problem, we generally use Prony's method, which involves minimising the distance between the coefficients of the impulse responses of the transfer function filters $h(z)a(z)$ and $b(z)$.

Thus, letting $a_0 = 1$, and $b_n = 0$ for $n > q$, we are led to minimise

$$J(a, b) = \sum_{n=0,N} \left| \sum_{k=0,p} a_k h_{d,n-k} - b_n \right|^2.$$

(13.30)

Considering $J(a, b)$ as a function of the variable b parameterised by a, we classically compute the optimum $b = g(a)$ then we look for the optimum in a of $J(a, g(a))$. Thus, as

$$J(a, b) = \| H_0 [1 \ a^T]^T - b \|^2 + \| v_h + Ha \|^2,$$

(13.31)

with

$$H_0 = \begin{bmatrix} h_{d,0} & 0 & & . . & 0 \\ h_{d,1} & h_{d,0} & & 0 . 0 \\ . & . & & . . . \\ h_{d,q} & h_{d,q-1} & . . & h_{d,q-p} \end{bmatrix},$$

$$H = \begin{bmatrix} h_{d,q} & . . & h_{d,q-p+1} \\ . & . . . & \\ h_{d,N-1} & . . & h_{d,N-p} \end{bmatrix},$$

(13.32)

and $v_h = [h_{d,q+1}, \ldots, h_{d,N}]^T$,

assuming that $N - q > p$, we obtain

$$\hat{a} = -(H^H H)^{-1} H^H v_h,$$

(13.33)

and $\hat{b} = H_0 [1 \ \hat{a}^T]^T$.

13.5 Asymptotic Log-likelihood of Gaussian Processes

Before considering maximum likelihood spectral estimation of processes with rational PSD, in this section we present some important results. We shall see that for Gaussian processes the maximum likelihood estimator is the optimum of a criterion which, when the size of the observation vector tends towards $+\infty$, can be expressed as a function of the spectral density of the parametric model. We will thus obtain new estimation criteria that approximate the likelihood criterion and for which we shall point out the convergence properties of the corresponding estimators.

13.5.1 Gaussian Log-likelihood

We assume that X is a regular, real or circular complex, zero mean Gaussian process, parameterised by (θ, σ^2), where σ^2 represents the variance of the innovation of X. These parameters therefore characterise the second order moments of X, and consequently its spectral measure.

The parametric model of the PSD of X will be denoted by $\sigma^2 S_\theta(f)$. $S_\theta(f)$ therefore represents the PSD of a regular process whose innovation variance is equal to 1. The true PSD of X will be denoted by $S_X(f)$.

For this modelling, the maximum likelihood estimator of (θ, σ^2) associated with $\mathbf{X}_n = [X_1, \ldots, X_n]^T$ is the minimum of the function

$$L_n(\theta, \sigma^2) = \log \sigma^2 + \frac{1}{n} \log |T_{\theta,n}| + \frac{1}{n} \sigma^{-2} \mathbf{X}_n^H T_{\theta,n}^{-1} \mathbf{X}_n, \tag{13.34}$$

where $T_{\theta,n}$ represents the covariance matrix associated with the PSD $S_\theta(f)$. The optimum in σ^2 of $L_n(\theta, \sigma^2)$ is given by

$$\hat{\sigma}_n^2 = \frac{1}{n} \mathbf{X}_n^H T_{\theta,n}^{-1} \mathbf{X}_n. \tag{13.35}$$

Replacing σ^2 by $\hat{\sigma}_n^2$ in (13.34), we obtain the new expression

$$l_n(\theta) = \frac{1}{n} \log |T_{\theta,n}| + \log(\frac{1}{n} \mathbf{X}_n^H T_{\theta,n}^{-1} \mathbf{X}_n). \tag{13.36}$$

13.5.2 Asymptotic Behaviour of the Log-likelihood

We shall now show that we can obtain an asymptotic expression of $l_n(\theta)$ when n tends towards $+\infty$, which can be expressed directly as a function of the PSD $S_\theta(f)$.

From Theorems 8.2 and 8.5 in Chapter 8,

$$\lim_{n \to \infty} \frac{1}{n} \log |T_{\theta,n}| = \int_{\mathcal{I}} \log S_\theta(f) df. \tag{13.37}$$

As the innovation of a process with PSD $S_\theta(f)$ is equal to 1, $\int_{\mathcal{I}} \log S_\theta(f) df = \log 1 = 0$, and

$$\lim_{n \to \infty} \frac{1}{n} \log |T_{\theta,n}| = 0. \tag{13.38}$$

We now study the limit of the term $n^{-1} \mathbf{X}_n^H T_{\theta,n}^{-1} \mathbf{X}_n$ of (13.36). For this, we only consider cases for which for any θ in the set Θ of parameters, and any f in $\mathcal{I}$, $0 < m < S_\theta(f) < M < +\infty$. We then have the following result:

Theorem 13.1 *If a parametric model $S_\theta(f)$ of the PSD satisfies*

$$\forall \theta \in \Theta, \quad \forall f \in \mathcal{I}, \quad 0 < m < S_\theta(f) < M < +\infty, \tag{13.39}$$

and $S_\theta(f)$ is continuous for any value of θ, then almost surely

$$\lim_{n\to\infty} \frac{1}{n}\mathbf{X}_n^H T_{\theta,n}^{-1}\mathbf{X}_n = \int_{\mathcal{I}} \frac{S_X(f)}{S_\theta(f)}df. \tag{13.40}$$

The proof of this result is not straightforward, and we are going to break it down into several steps. We note that a different proof can be found in [10]. The approach followed here provides intermediate results of interest for some applications in signal processing.

Theorem 13.2 We denote by $Q_{\theta,k}(z)$ the orthogonal Szegö polynomial of the first kind and of degree k, and $\sigma_{\theta,k}^2$ the variance of the corresponding prediction error associated with the PSD $S_\theta(f)$. Then,

$$\frac{1}{n}\mathbf{X}_n^H T_{\theta,n}^{-1}\mathbf{X}_n = \frac{1}{n} \sum_{k=0,n-1} \sigma_{\theta,k}^{-2}|[Q_{\theta,k}(z)]X_1|^2. \tag{13.41}$$

Proof We consider the scalar product defined by

$$< z^k, z^l > = \int_{\mathcal{I}} e^{2i\pi(k-l)f} S_\theta(f)df$$

$$= R_\theta(k-l). \tag{13.42}$$

The orthogonality relations among the polynomials $Q_{\theta,k}(z) = \sum_{l=0,k} q_{k,l}z^{l-k}$, with $q_{k,0} = 1$, are expressed by $L_\theta T_{\theta,n} L_\theta^H = D_\theta$, where $T_{\theta,n}$ is the Toeplitz matrix of size $n \times n$ and of general term $[T_{\theta,n}]_{i,j} = R_\theta(i-j)$,

$$L_\theta = \begin{pmatrix} 1 & 0 & .. & 0 \\ q_{n-1,1} & 1 & .. & 0 \\ . & . & . & . \\ q_{n-1,n-1} & q_{n-2,n-2} & .. & 1 \end{pmatrix}, \tag{13.43}$$

and $D_\theta = diag(\sigma_{\theta,0}^2, \sigma_{\theta,1}^2, \ldots, \sigma_{\theta,n-1}^2)$. It is then clear that

$$T_{\theta,n}^{-1} = L_\theta^H D_\theta^{-1} L_\theta, \tag{13.44}$$

and from (13.44), it directly results that

$$\frac{1}{n}\mathbf{X}_n^H T_{\theta,n}^{-1}\mathbf{X}_n = \frac{1}{n}\sum_{k=0,n-1} \sigma_{\theta,k}^{-2}|\sum_{l=0,k} q_{k,l}X_{k+1-l}|^2$$

$$= \frac{1}{n}\sum_{k=0,n-1} \sigma_{\theta,k}^{-2}|[Q_{\theta,k}(z)]X_1|^2. \quad \Box \tag{13.45}$$

Remark In the above proof, we have established that the Levinson algorithm, which provides the linear prediction orthogonal polynomials $Q_{\theta,k}(z)$, also supplies us with a Cholevski factorisation (that is, a factorisation in the form MM^H, where M is a triangular matrix) of the inverse of a positive Toeplitz Hermitian matrix of size n with a computational cost of about n^2 operations.

In order to prove relation (13.40), we also have to refer to the following result:

Theorem 13.3 *If* $\forall f \in \mathcal{I}$, $0 < m < S_\theta(f) < M < +\infty$ *and* $S_\theta(f)$ *is continuous, then at any point* f

$$\lim_{n \to \infty} \sigma_{\theta,k}^{-2} |Q_{\theta,k}(e^{2i\pi f})|^2 = S_\theta^{-1}(f), \tag{13.46}$$

and the convergence is uniform in f.

Proof We write $\nu_n^{(k)} = \sigma_{\theta,k}^{-2}[z^{-k}Q_{\theta,k}(z)]Y_n$, where Y is a process with PSD $S_\theta(f)$. It is clear that $\nu_n^{(k)} \in H_{Y,n}$, and that $\nu_n^{(k)} \perp Y_{n-1}, \ldots, Y_{n-k}$. Moreover, $\nu_n^{(k)} - \nu_n^{(l)} \perp Y_n, Y_{n-1}, \ldots, Y_{n-\min\{k,l\}}$. Indeed,

$$\mathbb{E}[(\nu_n^{(k)} - \nu_n^{(l)})Y_n^*] = 0 \text{ since, for } \mathbb{E}[\nu_n^{(m)}Y_n^*] = \frac{\sigma_{\theta,m}^2}{\sigma_{\theta,m}^2} = 1, \quad \forall m \in \mathbb{N}. \tag{13.47}$$

Therefore, $\nu_n^{(k)} - \nu_n^{(l)} \in H_{Y,n-1-\min\{k,l\}}$ and $\lim_{k,l \to \infty}(\nu_n^{(k)} - \nu_n^{(l)}) \in H_{Y,-\infty}$. Since $H_{Y,-\infty} = \{0\}$, $\lim_{k,l \to \infty} \| \nu_n^{(k)} - \nu_n^{(l)} \| = 0$. $(\nu_n^{(k)})_{k \in \mathbb{N}}$ is therefore a Cauchy sequence of elements of $H_{Y,n}$. It is therefore convergent in $H_{Y,n}$ and we denote the limit of this sequence by ν_n. Since $\nu_n^{(k)} \in H_{Y,n} \ominus H_{Y,n-k,n-1}$, it is clear that $\nu_n \in H_{Y,n} \ominus H_{Y,n-1}$. But $H_{Y,n} \ominus H_{Y,n-1} = span\{I_{Y,n}\}$, therefore $\nu_n = \alpha Y_{Y,n}$, where $\alpha \in \mathbb{C}$. Since $\mathbb{E}[\nu_n^{(m)}Y_n^*] = 1$, it results from the continuity of the scalar product that $\mathbb{E}[\nu_n Y_n^*] = \alpha \mathbb{E}[I_{Y,n}Y_n^*] = \alpha \| I_{Y,n} \|^2 = \alpha = 1$, since the innovation of Y is equal to its normalised innovation ($\| I_{Y,n} \|^2 = \exp(\int_{\mathcal{I}} \log S_\theta(f)df) = 1$). Consequently, $\nu_n = [h^{-1}(z)]Y_n$, where $h(z)$ is the minimum-phase causal factorisation of $S_\theta(f)$ ($h^{-1}(z)$, is a causal transfer function).

We show now that $e^{-2i\pi kf}\sigma_{\theta,k}^{-2}Q_{\theta,k}(e^{2i\pi f})$ converges towards $h^{-1}(e^{2i\pi f})$ in $L^2(\mathcal{I}, \mathcal{B}(\mathcal{I}), df)$:

$$\lim_{k \to \infty} \int_{\mathcal{I}} |e^{-2i\pi kf}\sigma_{\theta,k}^{-2}Q_{\theta,k}(e^{2i\pi f}) - h^{-1}(e^{2i\pi f})|^2 df$$

$$\leq \frac{1}{m} \lim_{k \to \infty} \int_{\mathcal{I}} |e^{-2i\pi kf}\sigma_{\theta,k}^{-2}Q_{\theta,k}(e^{2i\pi f}) - h^{-1}(e^{2i\pi f})|^2 S_\theta(f)df$$

$$\leq \frac{1}{m} \lim_{k \to \infty} \| \nu_n^{(k)} - \nu_n \|^2$$

$$\leq 0, \tag{13.48}$$

which shows that $e^{-2i\pi kf}\sigma_{\theta,k}^{-2}Q_{\theta,k}(e^{2i\pi f})$ converges towards $h^{-1}(e^{2i\pi f})$ in $L^2(\mathcal{I}, \mathcal{B}(\mathcal{I}), df)$. Since the functions $e^{-2i\pi kf}\sigma_{\theta,k}^{-2}Q_{\theta,k}(e^{2i\pi f})$ and $h^{-1}(e^{2i\pi f})$ are continuous on $\mathbb{R}$ and periodic with period 1, it is easy to check that the convergence in $L^2(\mathcal{I}, \mathcal{B}(\mathcal{I}), df)$ leads to uniform convergence on $\mathcal{I}$. Moreover,

$\lim_{k\to\infty} \sigma_{\theta,k}^2 = 1$, and $\sigma_{\theta,k}^{-2}|Q_{\theta,k}(e^{2i\pi f})|^2$ converges uniformly on $\mathcal{I}$ towards $|h^{-1}(e^{2i\pi f})|^2 = S_\theta^{-1}(f)$. $\square$

Some additional calculations are necessary to prove that

$$\lim_{n\to\infty} |\sigma_{\theta,n}^{-1}[Q_{\theta,n}(z)]X_1|^2 = \int_{\mathcal{I}} \frac{S_X(f)}{S_\theta(f)} df. \tag{13.49}$$

We shall especially require the following theorem, which provides a particular version of the strong law of large numbers.

Theorem 13.4 *If a sequence* $(Y_n)_{n\in\mathbb{N}^*}$ *of independent random variables is such that*

$$\lim_{n\to\infty} \frac{1}{n} \sum_{k=1,n} \mathbb{E}[Y_k] = a$$

$$\lim_{n\to\infty} \frac{1}{k^2} \sum_{k=1,n} \text{var}[Y_k] < \infty, \tag{13.50}$$

then the sequence $n^{-1}\sum_{k=1,n} Y_k$ *converges almost surely towards* a.

Proof See Appendix V.

We are now able to present the completion of the proof of Theorem 13.1:

Proof (of Theorem 13.1) The random variables $(\sigma_{\theta,k}^{-2}[Q_{\theta,k}(z)]X_1)_{k\in\mathbb{N}}$ are uncorrelated, for $\sigma_{\theta,k}^{-2}[Q_{\theta,k}(z)]X_1 \perp X_l$ for $l = 1, k$. Furthermore, they are Gaussian, and thus independent. Consequently, the random variables $(|\sigma_{\theta,k}^{-2}[Q_{\theta,k}(z)]X_1|^2)_{n\in\mathbb{N}}$ are independent. We now remark that

$$\mathbb{E}[\sigma_{\theta,k}^{-2}|[Q_{\theta,k}(z)]X_1|^2] = \sum_{l,m=0}^k \sigma_{\theta,k}^{-2} q_{k,l} q_{k,m}^* R_X(m-l)$$

$$= \int_{\mathcal{I}} \sigma_{\theta,k}^{-2}|Q_{\theta,k}(e^{2i\pi f})|^2 S_X(f) df. \tag{13.51}$$

From Lebesgue's dominated convergence theorem and Theorem 13.3, we therefore have

$$\lim_{n\to\infty} \frac{1}{n} \sum_{k=0,n-1} \mathbb{E}[\sigma_{\theta,k}^{-2}|[Q_{\theta,k}(z)]X_1|^2]$$

$$= \int_{\mathcal{I}} \lim_{n\to\infty} (\frac{1}{n} \sum_{k=0,n-1} \sigma_{\theta,k}^{-2}|Q_{\theta,k}(z)|^2) S_X(f) df \tag{13.52}$$

$$= \int_{\mathcal{I}} \frac{S_X(f)}{S_\theta(f)} df.$$

In addition, since X is Gaussian,

$$\text{var}[\sigma_{\theta,k}^{-2}|[Q_{\theta,k}(e^{2i\pi f})]X_1|^2] = C(\mathbb{E}[\sigma_{\theta,k}^{-2}|[Q_{\theta,k}(z)]X_1|^2])^2$$

$$= C(\int_{\mathcal{I}} \sigma_{\theta,k}^{-2}|Q_{\theta,k}(e^{2i\pi f})|^2 S_X(f) df)^2, \tag{13.53}$$

with $C = 1$ in the circular complex case and $C = 2$ in the real case. Moreover,

$$\lim_{n \to \infty} \sum_{k=0,n-1} \frac{1}{k^2} \text{var}[\sigma_{\theta,k}^{-2} |[Q_{\theta,k}(z)] X_1|^2]$$

$$= \lim_{n \to \infty} \sum_{k=0,n-1} \frac{C}{k^2} (\int_{\mathcal{I}} \sigma_{\theta,k}^{-2} |Q_{\theta,k}(z)|^2 S_X(f) df)^2$$

$$\leq C[\sup_{k \in \mathbb{N}} (\int_{\mathcal{I}} \sigma_{\theta,k}^{-2} |Q_{\theta,k}(z)|^2 S_X(f) df)^2] (\sum_{k \in \mathbb{N}} \frac{1}{k^2})$$

$$< \infty.$$

(13.54)

Theorem 13.4 leads to the conclusion that almost surely

$$\lim_{n \to \infty} \frac{1}{n} \sum_{k=0,n-1} \sigma_{\theta,k}^{-2} |[Q_{\theta,k}(z)] X_1|^2$$

$$= \lim_{n \to \infty} \frac{1}{n} \sum_{k=0,n-1} \mathbb{E}[\sigma_{\theta,k}^{-2} |[Q_{\theta,k}(z)] X_1|^2]$$

$$= \int_{\mathcal{I}} \frac{S_X(f)}{S_\theta(f)} df.$$

(13.55)

Taking into account Theorem 13.2 makes it possible to complete the proof of Theorem 13.1: almost surely,

$$\lim_{n \to \infty} \frac{1}{n} \mathbf{X_n}^H \Sigma_\theta^{-1} \mathbf{X_n} c = \lim_{n \to \infty} \frac{1}{n} \sum_{k=0,n-1} \sigma_{\theta,k}^{-2} |[Q_{\theta,k}(z)] X_1|^2$$

$$= \int_{\mathcal{I}} \frac{S_X(f)}{S_\theta(f)} df.$$

(13.56)

$\square$

13.6 Approximate Maximum Likelihood Estimation

13.6.1 Principle of the Method

We notice that the asymptotic approximation

$$l(\theta) \sim \log \left(\int_{\mathcal{I}} \frac{S_X(f)}{S_\theta(f)} df \right)$$

(13.57)

cannot be used in practice since $S_X(f)$ is not known. However, under the hypotheses of Theorem 13.1, $S_\theta^{-1}(f)$ is continuous, and the results relating to the convergence of the integral functions of the periodogram allow us to conclude that almost surely

$$\lim_{n\to\infty} \int_{\mathcal{I}} \frac{\hat{S}_{X,n}(f)}{S_\theta(f)} df = \int_{\mathcal{I}} \frac{S_X(f)}{S_\theta(f)} df. \tag{13.58}$$

We can, therefore, look for the estimator of (θ, σ^2) given by

$$\hat{\theta} = \arg\min_\theta \int_{\mathcal{I}} \frac{\hat{S}_{X,n}(f)}{S_\theta(f)} df$$
$$\hat{\sigma}^2 = \frac{1}{n} \mathbf{X}_n^H T_{\hat{\theta}}^{-1} \mathbf{X}_n. \tag{13.59}$$

We can hope that the estimator thus obtained is close to the maximum likelihood estimator. We shall consider this point a little further on.

The optimum $\hat{\theta}$ is generally obtained by means of iterative optimisation techniques and there are often local optima of the criterion. But we can note that under the hypothesis that $S_X(f) = \sigma_0^2 S_{\theta_0}(f)$, the minimum of $\int_{\mathcal{I}} S_X(f) S_\theta^{-1}(f) df$ is equal to 1, and this minimum is only reached when θ reaches the true value θ_0 of the vector parameter. This remark makes it possible to detect the local optima of the criterion and to start the optimisation again with a different initialisation of θ until a minimum equal to 1, or in practice close to 1, is obtained.

Minimising an integral function is generally not very practical. Therefore, we sometimes prefer to modify the above-mentioned criterion in order to obtain a simpler criterion to optimise, which can be done by using the following result:

Theorem 13.5 *We note* $f_{n,k} = -1/2 + k/n$, $(k = 0, n-1)$. *Then, almost surely*

$$\lim_{n\to\infty} \frac{1}{n} \sum_{k=0,n-1} \frac{\hat{S}_{X,n}(f_{n,k})}{S_\theta(f_{n,k})} = \int_{\mathcal{I}} \frac{S_X(f)}{S_\theta(f)} df. \tag{13.60}$$

Proof Let there be a fixed m, with $m < n$, and let us note

$$\sigma_{\theta,m}^{-2} |Q_{\theta,m}(e^{2i\pi f})|^2 = B_m(e^{2i\pi f})$$
$$= \sum_{k=-m,m} b_{m,k} e^{-2i\pi k f}. \tag{13.61}$$

$$\frac{1}{n} \sum_{k=0,n-1} \sigma_{\theta,m}^{-2} |Q_{\theta,m}(e^{2i\pi f_{n,k}})|^2 \hat{S}_{X,n}(f_{n,k})$$
$$= \frac{1}{n} \sum_{k=0,n-1} \left(\sum_{l=-m,m} b_{m,l} e^{-2i\pi l f_{n,k}} \right) \hat{S}_{X,n}(f_{n,k})$$
$$= \sum_{l=-m,m} \sum_{t=-n+1,n-1} \hat{R}_{X,n}(t) b_{m,l}^* \frac{1}{n} \sum_{k=0,n-1} e^{2i\pi(l-t)f_{n,k}}$$
$$= \sum_{l=-m,m} \hat{R}_{X,n}(l) b_{m,l}^* + 2\mathrm{Re}[\sum_{l=1,m} \hat{R}_{X,n}(n-l) b_{m,l}^*]. \tag{13.62}$$

Almost surely, $\lim_{n\to\infty} \hat{R}_{X,n}(n-l) = 0$. Moreover, almost surely,

$$\lim_{n\to\infty} \sum_{l=-m,m} \hat{R}_{X,n}(l)b_{m,l}^*$$

$$= \sum_{l=-m,m} R_X(l)b_{m,l}^*$$

$$= \int_{\mathcal{I}} \sigma_{\theta,m}^{-2}|Q_{\theta,m}(e^{2i\pi f})|^2 \sum_{l=-m,m} R_X(l)e^{-2i\pi l f}\,df \qquad (13.63)$$

$$= displaystyle int_{\mathcal{I}}\, \sigma_{\theta,m}^{-2}|Q_{\theta,m}(e^{2i\pi f})|^2 S_X(f)df.$$

Therefore,

$$\lim_{n\to\infty} \frac{1}{n}\sum_{k=0,n-1} \sigma_{\theta,m}^{-2}|Q_{\theta,m}(e^{2i\pi f_{n,k}})|^2 \hat{S}_{X,n}(f_{n,k})$$

$$\qquad (13.64)$$

$$= \int_{\mathcal{I}} \sigma_{\theta,m}^{-2}|Q_{\theta,m}(e^{2i\pi f})|^2 S_X(f)df,$$

and the result can then be simply deduced from the uniform convergence of $\sigma_{\theta,m}^{-2}|Q_{\theta,m}(e^{2i\pi f})|^2$ towards $S_\theta^{-1}(f)$ when m tends towards $+\infty$. $\square$

We note that computation of the values $\hat{S}_{X,n}(f_k)$ can be carried out simply by means of a fast Fourier transform algorithm applied to $\mathbf{X}_n$.

13.6.2 Convergence of the Estimators

We now present without proofs convergence results that justify using the estimators associated with the approximate likelihood criteria obtained above. Letting $S_{(\theta,\sigma^2)}(f) = \sigma^2 S_\theta(f)$, we have the following theorem:

Theorem 13.6 *If $S_X(f) = S_{(\theta_0,\sigma_0^2)}(f)$, the sequence of estimators $\hat{\theta}_n$ that minimise one of the criteria*

$$l(\theta), \quad \frac{1}{n}\mathbf{X}_n T_{\theta,n}^{-1}\mathbf{X}_n^H, \quad or \quad \frac{1}{n}\sum_{k=0,n-1} \frac{\hat{S}_{X,n}(f_k)}{S_\theta(f_k)}, \qquad (13.65)$$

converges almost surely towards θ_0, and $\hat{\sigma}^2 = n^{-1}\mathbf{X}_n^H \Sigma_{\hat{\theta}}^{-1}\mathbf{X}_n$ converges almost surely towards σ_0^2. Moreover, $(\hat{\theta}, \hat{\sigma}^2)$ is an efficient estimator of (θ_0,σ_0^2), and $\sqrt{n}[(\hat{\theta}, \hat{\sigma}^2) - (\theta_0,\sigma_0^2)]$ converges in distribution towards a zero mean Gaussian random variable, with covariance matrix $I_{(\theta_0,\sigma_0^2)}^{-1}$, with

$$I_{(\theta,\sigma^2)} = \frac{1}{2}\int_{\mathcal{I}} [\nabla_{(\theta,\sigma^2)}\log S_{(\theta,\sigma^2)}(f)][\nabla_{(\theta,\sigma^2)}\log S_{(\theta,\sigma^2)}(f)]^H df, \qquad (13.66)$$

where $\nabla_{(\theta,\sigma^2)}$ represents the gradient with respect to (θ,σ^2).

Proof A proof of this theorem can be found in [10].

We notice that $n^{-1}I_\theta^{-1}$ represents the Cramer-Rao lower bound for the estimators of the parameter θ, and I_θ the Fisher information matrix. To intuitively justify the expression of (13.66), we consider the case of a real-valued Gaussian process X. By using relations (13.37) and (13.56), the expression of the likelihood can be approximated by

$$p_{(\theta,\sigma^2)}(\mathbf{X_n}) \sim \frac{\exp(-\frac{n}{2}\int_I \frac{\hat{S}_{X,n}(f)}{S_{(\theta,\sigma^2)}(f)}df)}{(2\pi)^{n/2}[\exp(n\int_I \log S_{(\theta,\sigma^2)}(f)df)]^{1/2}}. \tag{13.67}$$

We denote by L the dimension of the vector θ, and we denote by convention $\sigma^2 = \theta_{L+1}$. For $i,j \leq L+1$, the component (i,j) of the Fisher information matrix satisfies

$$[nI_{(\theta,\sigma^2)}]_{i,j}$$

$$= -\mathbb{E}[\frac{\partial^2}{\partial\theta_i\partial\theta_j}\log p_\theta(\mathbf{X_n})]$$

$$= \frac{n}{2}\frac{\partial^2}{\partial\theta_i\partial\theta_j}(\int_I [\log S_{(\theta,\sigma^2)}(f) + \frac{S_X(f)}{S_{(\theta,\sigma^2)}(f)}]df)$$

$$= \frac{n}{2}\int_I [\frac{-1}{S_{(\theta,\sigma^2)}^2(f)} + \frac{2S_X(f)}{S_{(\theta,\sigma^2)}^3(f)}][\frac{\partial}{\partial\theta_i}S_{(\theta,\sigma^2)}(f)][\frac{\partial}{\partial\theta_j}S_{(\theta,\sigma^2)}(f)]df$$

$$+ \frac{n}{2}\int_I [\frac{1}{S_{(\theta,\sigma^2)}(f)} - \frac{S_X(f)}{S_{(\theta,\sigma^2)}^2(f)}]\frac{\partial^2}{\partial\theta_i\partial\theta_j}S_{(\theta,\sigma^2)}(f)df. \tag{13.68}$$

For $S_X(f) = \sigma_0^2 S_{\theta_0}(f)$, and $(\theta,\sigma^2) = (\theta_0,\sigma_0^2)$, it thus results that

$$[I_{(\theta,\sigma^2)}]_{i,j} = \frac{1}{2}\int_I \frac{\partial^2}{\partial\theta_i\partial\theta_j}\log S_{(\theta,\sigma^2)}(f)df \tag{13.69}$$

$$= \frac{1}{2}\int_I [\frac{\partial}{\partial\theta_i}\log S_{(\theta,\sigma^2)}(f)][\frac{\partial}{\partial\theta_j}\log S_{(\theta,\sigma^2)}(f)]df.$$

We shall therefore have, in particular,

$$[I_{(\theta,\sigma^2)}]_{L+1,L+1} = -\mathbb{E}[\frac{\partial^2}{\partial\sigma^2\partial\sigma^2}\log p_\theta(\mathbf{X_n})] \tag{13.70}$$

$$= \frac{1}{2\sigma^4},$$

and, for $i \leq L$,

$$[I_{(\theta,\sigma^2)}]_{i,L+1} = \frac{1}{2\sigma^2} \int_\mathcal{I} \frac{\partial}{\partial \theta_i} \log S_{(\theta,\sigma^2)}(f) df$$

$$= \frac{1}{2\sigma^2} \frac{\partial}{\partial \theta_i} [\int_\mathcal{I} \log S_{(\theta,\sigma^2)}(f) df]$$

$$= \frac{1}{2\sigma^2} \frac{\partial \sigma^2}{\partial \theta_i}$$

$$= 0. \tag{13.71}$$

13.7 Maximum Likelihood Estimation of ARMA Models

We now assume that X is a Gaussian ARMA process and, to make things simpler, we consider that X is real-valued, the circular complex case hardly causing any more problems.

13.7.1 The General Case

X is characterised by a difference equation of the form

$$X_n + \sum_{k=1,p} a_k X_{n-k} = \sum_{l=0,q} c_l Y_{n-l}, \tag{13.72}$$

where Y is a white noise process. This equation may also be written as

$$X_n + \sum_{k=1,p} a_k X_{n-k} = I_n + \sum_{l=1,q} b_l I_{n-l}, \tag{13.73}$$

where I is the innovation process of X. We write $a = (a_1, \ldots, a_p)$, $b = (b_1, \ldots, b_q)$, $\theta = (a, b)$, and $\sigma^2 = \| I_n \|^2$. An approximate maximum likelihood estimator of (a, b, σ^2) can be given by

$$\hat{\theta} = \arg\min_\theta \int_\mathcal{I} \frac{\hat{S}_{X,n}(f)}{S_\theta(f)} df, \tag{13.74}$$

$$\hat{\sigma}^2 = n^{-1} \mathbf{X}_n^H \Sigma_{\hat{\theta}}^{-1} \mathbf{X}_n,$$

or variants like

$$\hat{\theta} = \arg\min_\theta (\frac{1}{n} \sum_{k=0,n-1} \frac{\hat{S}_{X,n}(f_{n,k})}{S_\theta(f_{n,k})}). \tag{13.75}$$

Theorem 13.7 *The Fisher information matrix for the parameter (a, b, σ^2) is given by*

$$I_{(a,b,\sigma^2)} = \frac{1}{\sigma^2} \begin{bmatrix} T_a & -T_{ab} & 0 \\ -T_{ab}^H & T_b & 0 \\ 0 & 0 & 1/(2\sigma^2) \end{bmatrix}, \tag{13.76}$$

where T_a, T_{ab} and T_b are matrices of sizes $p \times p$, $p \times q$, and $q \times q$ respectively, with general terms

$$[T_a]_{k,l} = \int_{\mathcal{I}} \frac{\sigma^2}{|a(e^{2i\pi f})|^2} e^{2i\pi(k-l)f} df,$$

$$[T_{ab}]_{k,l} = \int_{\mathcal{I}} \frac{\sigma^2}{a^*(e^{2i\pi f})b(e^{2i\pi f})} e^{2i\pi(k-l)f} df, \tag{13.77}$$

$$[T_b]_{k,l} = \int_{\mathcal{I}} \frac{\sigma^2}{|b(e^{2i\pi f})|^2} e^{2i\pi(k-l)f} df.$$

Proof We notice that

$$\int_{\mathcal{I}} \frac{\partial}{\partial a_k} \log S_{(\theta,\sigma^2)}(f) df = \int_{\mathcal{I}} \frac{\sigma^2 a(e^{2i\pi f}) e^{2i\pi k f} + \sigma^2 [a(e^{2i\pi f})]^* e^{-2i\pi k f}}{|a(e^{2i\pi f})|^2} df, \tag{13.78}$$

and

$$\frac{1}{2} \int_{\mathcal{I}} \frac{\partial^2}{\partial a_k \partial a_l} \log S_{(\theta,\sigma^2)}(f) df = \frac{\sigma^2}{2} \int_{\mathcal{I}} 2\mathcal{R}e\left[\frac{e^{2i\pi(k-l)f}}{|a(e^{2i\pi f})|^2}\right] df$$
$$- \frac{\sigma^2}{2} \int_{\mathcal{I}} 2\mathcal{R}e\left[\frac{e^{-2i\pi(k+l)f}}{(a(e^{2i\pi f}))^2}\right] df. \tag{13.79}$$

But, since $(a(e^{2i\pi f}))^{-2} \times e^{-2i\pi(k+l)f}$ is a strictly causal function of the form $\sum_{u=k+l,\infty} c_u e^{-2i\pi u f}$, the second integral of the right-hand side of (13.79) is equal to zero. Finally, as the function $g(f) = |a(e^{-2i\pi f})|^{-2} e^{2i\pi(k-l)f}$ satisfies $g(-f) = g^*(f)$,

$$[T_a]_{k,l} = \int_{\mathcal{I}} \frac{\sigma^2}{|a(e^{2i\pi f})|^2} e^{2i\pi(k-l)f} df. \tag{13.80}$$

Similar considerations provide the expressions of T_{ab} and of T_b. As for the terms involving derivations with respect to σ^2, their expression has already been calculated in the previous paragraph. □

We shall now consider more specifically the case of AR models for which it is possible to calculate an approximation of the maximum likelihood estimator iteratively in a simple way.

13.7.2 AR Models

In the case of an AR model of the form

$$S_{(\theta,\sigma^2)}(f) = \frac{\sigma^2}{|1 + \sum_{k=1,p} a_k e^{-2i\pi kf}|^2}, \tag{13.81}$$

the Fisher information matrix of (θ, σ^2) is simply given by the approximate formula

$$I_{(\theta,\sigma^2)} = \frac{1}{\sigma^2} \begin{bmatrix} T_a & 0 \\ 0 & (2\sigma^2)^{-2} \end{bmatrix}, \tag{13.82}$$

as a particular case of the Fisher information matrix of an ARMA model. Consequently, $\mathrm{var}[\hat{a}_k] \geq \dfrac{\sigma^2}{n}[T_a^{-1}]_{k,k}$, and $\mathrm{var}[\hat{\sigma}^2] \geq \dfrac{2\sigma^4}{n}$.

Theorem 13.8 *Given $(X_k)_{k=1,n}$, the approximate maximum likelihood estimator of an AR model, defined by*

$$\hat{a} = \arg\min_a \int_{\mathcal{I}} |a(e^{2i\pi f})|^2 \hat{S}_{X,n}(f) df,$$

$$\hat{\sigma}^2 = \int_{\mathcal{I}} |\hat{a}(e^{2i\pi f})|^2 \hat{S}_{X,n}(f) df, \tag{13.83}$$

is obtained by solving the Yule-Walker equations, where the autocovariances are estimated by

$$\hat{R}_{X,n}(k) = \frac{1}{n} \sum_{l=1,n-k} X_{l+k} X_l^*. \tag{13.84}$$

Proof It is easy to check that the optimality conditions obtained by setting to zero the partial derivatives of $\int_{\mathcal{I}} |a(e^{2i\pi f})|^2 \hat{S}_{X,n}(f) df$ are given by

$$\int_{\mathcal{I}} e^{2i\pi kf} a(e^{2i\pi f}) \hat{S}_{X,n}(f) df = 0, \quad k = 1, p, \tag{13.85}$$

that is, for $k = 1, p$,

$$\int_{\mathcal{I}} (e^{2i\pi kf} + \sum_{l=1,p} a_l e^{2i\pi(k-l)f}) \hat{S}_{X,n}(f) df$$

$$= \hat{R}_{X,n}(k) + \sum_{l=1,p} a_l \hat{R}_{X,n}(k-l) \tag{13.86}$$

$$= 0,$$

which proves the result. $\square$

We note that the AR model thus estimated is causal and stable, since the sequence $(\hat{R}_{X,n}(k))_{k=0,p}$ is of the positive type.

13.7.3 AR Models Parameterised by Reflection Coefficients

As we have seen in the context of spectral identification, the reflection coefficients that appear in Levinson's algorithm represent a set of coefficients
equivalent to the linear prediction coefficients and have many interesting
properties. We may, therefore, try to estimate an AR model by parameterising it directly by means of its reflection coefficients. Here, we present Burg's
method that enables the reflection coefficients to be estimated in such a way
that the numerical values obtained for the estimated reflection coefficients are
guaranteed to have a modulus smaller than or equal to 1. This makes it possible to ensure the stability of the AR model thus estimated. We also present
an approximate maximum likelihood estimation of the reflection coefficients.

The set of parameters considered here is $\theta = [k_0, k_1, \ldots, k_p]^T$, where
$k_0 = R_X(0)$.

Burg's method We can compute the coefficients k_i iteratively by minimising the variance of the direct prediction error defined by

$$
\begin{aligned}
E_{l,k}^d &= [Q_{\theta,l}(z)]X_{k-l} \\
&= [zQ_{\theta,l-1}(z)]X_{k-l} - k_l[\tilde{Q}_{\theta,l-1}(z)]X_{k-l},
\end{aligned}
\tag{13.87}
$$

and noting that $Q_{\theta,l-1}(z)$ depends only on $(k_i)_{i=1,l-1}$. The variance of $E_{l,k}^d$
can be estimated by

$$
\hat{\sigma}_l^{2,d} = \frac{1}{n-l} \sum_{k=l+1,n} |E_{l,k}^d|^2.
\tag{13.88}
$$

Instead, in order to ensure that, numerically, the estimator $\hat{k}_l$ of k_l has a
modulus smaller than or equal to 1, we minimise

$$
\rho_l = \frac{1}{2}(\hat{\sigma}_l^{2,d} + \hat{\sigma}_l^{2,r}),
\tag{13.89}
$$

where $\hat{\sigma}_l^{2,r}$ represents the estimator $(n-l)^{-1}\sum_{k=l+1,n}|E_{l,k}^r|^2$ of the backward
prediction error

$$
\begin{aligned}
E_{l,k}^r &= [\tilde{Q}_{\theta,l}(z)]X_{k-l} \\
&= [\tilde{Q}_{\theta,l-1}(z)]X_{k-l} - k_l^*[zQ_{\theta,l-1}(z)]X_{k-l}.
\end{aligned}
\tag{13.90}
$$

We notice that the direct and backward prediction errors satisfy the recurrence equations

$$
\begin{aligned}
E_{l,k}^d &= E_{l-1,k}^d - k_l E_{l-1,k-1}^r \\
E_{l,k}^r &= E_{l-1,k-1}^r - k_l^* E_{l-1,k}^d.
\end{aligned}
\tag{13.91}
$$

It results, therefore, that

$$\rho_l = \frac{1}{2(n-l)} \sum_{k=l+1,n} [|E^d_{l-1,k} - k_l E^r_{l-1,k-1}|^2 + |E^r_{l-1,k-1} - k^*_l E^d_{l-1,k}|^2].$$

(13.92)

The minimum of this expression provides Burg's estimator of k_l:

$$\hat{k}_l = \frac{-2\sum_{k=l+1,n}(E^d_{l-1,k})^* E^r_{l-1,k-1}}{\sum_{k=l+1,n}(|E^d_{l-1,k}|^2 + |E^r_{l-1,k-1}|^2)},$$

(13.93)

which clearly has a modulus smaller than or equal to 1. The coefficients k_l are thus computed iteratively for increasing values of l.

Approximate Maximum Likelihood In the Gaussian case, an approximate estimator of the maximum likelihood of the reflection coefficients k_i associated with an AR model can be calculated iteratively in a simple way.

For this, we begin by calculating the exact maximum likelihood estimator associated with a first order AR model, which provides an estimator of k_1. We then use a second order AR model for which the value of k_1 is equal to that obtained above, which enables k_2 to be estimated simply. We iterate this procedure until the estimation of all the required reflection coefficients has been performed.

This method leads to the exact maximum likelihood estimator only in the case of a first order autoregressive model, since otherwise at each iteration we carry out the estimation of a single reflection coefficient conditional to knowledge of the previous ones.

Exercises

In this section, $X = (X_n)_{n\in\mathbb{Z}}$ denotes a WSS process.

13.1 (Estimation of the amplitude of a noisy sinusoid) Let X denote a WSS process of the form

$$X_n = \xi e^{2i\pi n f_0} + V_n,$$

(13.94)

where ξ is a zero mean complex random variable and $V = (V_n)_{n\in\mathbb{Z}}$ is a circular white noise, uncorrelated with ξ. f_0 is assumed to be known, as well as the first autocovariance coefficients of X, denoted by $(R_X(k))_{k=0,N}$. We are looking for the value of ξ.

a) ξ is estimated as a linear transform of $X_n, \ldots, X_{n+N}$:

$$\hat{\xi} = \sum_{k=0,N} h^*_k X_{n-k}.$$

(13.95)

We search for the vector $h = [h_1, \ldots, h_N]^T$ as the solution to the following constrained minimisation problem:

$$\begin{cases} \min_h \| h^H \mathbf{X_n} \|^2 \\ h^H d(f_0) = 1, \end{cases} \qquad (13.96)$$

where $\mathbf{X_n} = [X_n, \ldots, X_{n+N}]^T$ and $d(f) = e^{2i\pi n f}[1, e^{2i\pi f}, \ldots, e^{2i\pi N f}]^T$. Give an interpretation of this criterion. Let $T_{X,N}$ denote the covariance matrix of $\mathbf{X_n}$. Show that the optimum of (13.96) is given by

$$\hat{h} = \frac{T_{X,N}^{-1} d(f_0)}{d(f_0)^H T_{X,N}^{-1} d(f_0)}. \qquad (13.97)$$

b) Let $T_{V,N}$ denote the covariance matrix of $\mathbf{V_n} = [V_n, \ldots, V_{n+N}]^T$ and $T_{S,N} = T_{X,N} - T_{V,N}$. When h is chosen as the optimum of the following optimisation problem

$$\max_h \frac{h^H T_{S,N} h}{h^H T_{V,N} h}, \qquad (13.98)$$

what does it represent? Show that the optimum of this problem is defined up to a factor, and give the expression of its solution. With the additional normalising constraint $h^H d(f_0) = 1$, show that we get the solution to problem (13.96).

c) We now assume that V is Gaussian. Calculate the maximum likelihood estimator of ξ obtained when minimising the probability density function of $\mathbf{X_n}$ conditional to ξ. What conclusion can be drawn?

13.2 (Minimum variance spectral estimator) Let X denote any WSS process. We consider the spectrum estimator of X defined at point f when a model of the form $X_n = \xi_f e^{2i\pi n f} + V_n$ is assumed for the process X under consideration, where $V = (V_n)_{n \in \mathbb{Z}}$ is a white noise uncorrelated with the random variables ξ_f.

a) Estimating ξ_f as in the above exercise by $h_f^H \mathbf{X_n}$, where $\mathbf{X_n} = [X_n, \ldots, X_{n+N}]^T$, show that the spectrum estimator $S_{MV}(f)$, called the *minimum variance spectral estimator*, or *Capon spectrum estimator*, and given by $S_{MV}(f) = (N+1) \| h_f^H \mathbf{X_n} \|^2$, can be rewritten as

$$S_{MV}(f) = \frac{N+1}{d(f)^H \hat{T}_{X,N}^{-1} d(f)}, \qquad (13.99)$$

where $[\hat{T}_{X,N}]_{ab} = \hat{R}_{X,N}(a-b)$ and $d(f) = [1, e^{2i\pi f}, \ldots, e^{2i\pi(N-1)f}]^T$.

b) Calculate $S_{MV}(f)$ when X is a white noise and when X is of the form $X_n = \xi e^{2i\pi n f_0} + V_n$, where V is a white noise.

c) Express the minimum variance model of order p (that is for $N = p$), denoted by $S_{MV}^{(p)}(f)$. Conversely, express the AR spectrum models of orders at most p and the AR spectrum model of order p in terms of the minimum variance model of orders at most p.

d) Calculate the asymptotic variance of $S_{MV}^{(p)}(f)$.

(Hint: refer to the expression of the asymptotic variance of the AR spectrum estimators.)

13.3 (Maximum likelihood estimation of a noisy sinusoid) Let X denote a process of the form

$$X_n = Ae^{2i\pi n f_0} + V_n, \tag{13.100}$$

where A and f_0 are an unknown complex amplitude and an unknown frequency respectively, and $V = (V_n)_{n \in \mathbb{Z}}$ a white, Gaussian circular noise.

a) Show that the maximum likelihood estimator of (A, f_0) from $\mathbf{X}_N = [X_0, \dots, X_N]^T$, denoted by $(\hat{A}, \hat{f}_0)$, is given by

$$\begin{cases} \hat{f}_0 = \arg\max_f d(f)^H \hat{T}_{X,N} d(f) \\ \hat{A} = \| d(\hat{f}_0) \|^{-2} d(\hat{f}_0)^H \mathbf{X}_N, \end{cases} \tag{13.101}$$

where $[\hat{T}_{X,N}]_{ab} = \hat{R}_{X,N}(a - b)$ and $d(f) = [1, e^{2i\pi ft}, \dots, e^{2i\pi(N-1)ft}]^T$.

b) Interpret $(\hat{A}, \hat{f}_0)$ in terms of the Fourier transform and of the periodogram of X.

13.4 (Prony's method) Let $x = (x_n)_{n \in \mathbb{Z}}$ denote a signal that we want to model as a sum of p damped sinusoids, approximating it by

$$\hat{x}_n = \sum_{k=1,p} c_k z_k^n. \tag{13.102}$$

We want to calculate $(c_k)_{k=1,p}$ and $(z_k)_{k=1,p}$ from x.

a) Denoting $a(z) = \prod_{k=1,p}(z - z_k) = z^p + \sum_{l=1,p} a_l z^{p-l}$, show that

$$\begin{pmatrix} x_p & \cdots & x_1 \\ x_{p+1} & \cdots & x_2 \\ \cdots & \cdots & \cdots \\ x_{2p-1} & \cdots & x_p \end{pmatrix} \times \begin{pmatrix} a_1 \\ \cdot \\ \cdot \\ a_p \end{pmatrix} = - \begin{pmatrix} x_{p+1} \\ \cdot \\ \cdot \\ x_{2p} \end{pmatrix}. \tag{13.103}$$

Then, $(z_k)_{k=1,p}$ is obtained by solving $z^p + \sum_{l=1,p} a_l z^{p-l} = 0$.

b) Check that $(c_k)_{k=1,p}$ can be recovered from x and $(z_k)_{k=1,p}$ by solving the system

$$\begin{pmatrix} z_1 & \cdots & z_N \\ \cdots & \cdots & \cdots \\ z_1^p & \cdots & z_N^p \end{pmatrix} \times \begin{pmatrix} c_1 \\ \cdot \\ c_p \end{pmatrix} = \begin{pmatrix} x_1 \\ \cdot \\ x_n \end{pmatrix}, \tag{13.104}$$

that we denote $\mathbf{zc} = \mathbf{x}$ in the matrix form (here, we can let $N = p$).

c) If x does not exactly match model (13.102), $(c_k)_{k=1,p}$ and $(z_k)_{k=1,p}$ can be estimated from $(x_k)_{k=1,N}$ $(N \geq 2p)$ in the mean square sense: $(z_k)_{k=1,p}$ are calculated by solving $z^p + \sum_{l=1,p} a_l z^{p-l} 0$ where $(a_k)_{k=1,p}$ minimise

$$\sum_{n=p+1,N} \left| x_n + \sum_{k=1,p} a_k x_{n-k} \right|^2. \tag{13.105}$$

Explain this criterion, and show that the mean square estimate of $(c_k)_{k=1,p}$ is then given by

$$\hat{c} = (\mathbf{z}^H \mathbf{z})^{-1} \mathbf{z}^H \mathbf{x}. \tag{13.106}$$

d) We assume now that $X = (X_n)_{n \in \mathbb{Z}}$ is a process of the form $X_n = \sum_{k=1,p} \xi_k z_k^n + V_n$, where $V = (V_n)_{n \in \mathbb{Z}}$ is a white noise and $(\xi_k)_{k=1,p}$ are independent random variables, and $|z_k| < 1$ for $k = 1, \ldots, p$. Check that X is an ARMA process that satisfies the difference equation

$$X_n + \sum_{k=1,p} a_k X_{n-k} = \sum_{l=1,p} a_l V_{n-l}. \tag{13.107}$$

Propose a technique for estimating $(a_k)_{k=1,p}$ and then $(z_k)_{k=1,p}$ and the values of $(\xi_k)_{k=1,p}$ from a realisation of $(X_n)_{n=1,N}$ $(N \gg p)$.

13.5 We assume that X is an MA(q) Gaussian process.

a) Express the log-likelihood of this model for data $X_1, \ldots, X_N$ in terms of the coefficients of the MA model.

b) With a view to optimising the criterion, calculate the gradient of the log-likelihood criterion.

13.6 Show that the maximum likelihood estimator of the coefficient of an AR(1) model is the solution to a cubic equation.

13.7 We consider the spectrum estimation of an MA process for which we assume that an initial estimate of the PSD of the form $S(f) = \sum_{k=-q,q} b_k e^{-2i\pi k f}$ is available. Show that the algorithm presented in Exercise 12.7 can be used to obtain a modified estimate of the PSD of an MA(q) model that matches the positivity constraint.

13.8 (ARIMA processes) $X = (X_n)_{n \in \mathbb{Z}}$ is said to be an ARIMA(p, d, q) process (I for integrated) if Y defined by $Y_n = [(1 - z^{-1})^d] X_n$ is an ARMA(p, q) process.

a) Check that the process $X = (X_n)_{n \in \mathbb{N}}$ defined by $X_n = X_0 + \sum_{k=1,n} Y_k$, where $Y_n = \sum_{k>0} a^k V_{n-k}$, with $|a| < 1$ and with $V = (V_n)_{n \in \mathbb{Z}}$ a white noise is an ARIMA($1,\overline{1},0$) process.

13.9 (Polynomial trend) We consider $X_n = P(n) + Y_n$, where $P(x)$ is a degree d polynomial and Y an ARMA(p, q) process. Show that X is an ARIMA(p, d, q) process (see the definition of an ARIMA(p, d, q) process in the above exercise).

13.10 (Model order selection, AIC criterion) Let X denote an AR(p) model of the form $X_n - \sum_{k=1,p} a_k X_{n-k} = V_n$, with $\| V_n \|^2 = \sigma^2$. We denote by $\hat{a}$ the maximum likelihood estimator of $a = [a_1, \ldots, a_p]^T$. Let $Y = (Y_n)_{n \in \mathbb{Z}}$ denote an AR(p) process independent of X but with the same distribution.

a) Show that the final prediction error variance, defined by $\| Y_n - \sum_{k=1,p} a_k Y_{n-k} \|^2$, satisfies

$$\| Y_n - \sum_{k=1,p} a_k Y_{n-k} \|^2 = \sigma^2 + \mathbb{E}[(\hat{a} - a)^H T_{Y,p-1}(\hat{a} - a)], \qquad (13.108)$$

where $[T_{y,p-1}]_{ab} = R_Y(a - b)$.
(Hint: note that

$$\mathbb{E}[(\hat{a} - a)^H \mathbf{Y}_{n-1}^* \mathbf{Y}_{n-1}^T (\hat{a} - a)] = \mathbb{E}[\mathbb{E}[(\hat{a} - a)^H \mathbf{Y}_{n-1}^* \mathbf{Y}_{n-1}^T (\hat{a} - a)] | X], \qquad (13.109)$$

where $\mathbf{Y}_{n-1} = [Y_{n-1}, \ldots, Y_{n-p}]^T$.)
b) Accounting for the asymptotic distribution of $\sqrt{n}(\hat{a} - a)^H$, where n is the data length (see Section 13.7.2), show that under the hypothesis of an order k model for X and Y,

$$\min_{\alpha_1, \ldots, \alpha_k} \| Y_n - \sum_{l=1,k} \alpha_l Y_{n-l} \|^2 = \sigma_k^2 (1 + k/n) + o(1/n), \qquad (13.110)$$

where σ_k^2 is the variance of the linear prediction error of order k.
c) Check that the maximum likelihood estimator of σ^2, denoted by $\hat{\sigma}^2$, is asymptotically distributed as follows: $n\sigma^{-2}\hat{\sigma}^2 \sim \chi^2(n - k)$. Then, show that $\| Y_n - \sum_{l=1,k} a_l Y_{n-l} \|^2$ can be estimated by $\hat{\sigma}^2 \frac{n+k}{n-k}$ and explain why it is of interest to use the value

$$\hat{p} = \arg \min_{0 < k < n} \hat{\sigma}_k^2 \left(\frac{n+p}{n-p} \right) \qquad (13.111)$$

for the model order selection, where $\hat{\sigma}_k^2$ is the final error prediction for an AR(k) model assumption.

13.11 (Two indexed processes) Propose an extension of autoregressive, minimum variance and maximum entropy spectrum estimation concepts to the case of a two indexed WSS process $X = (X_{m,n})_{m,n \in \mathbb{Z}}$.

14. Higher Order Statistics

Purpose In this chapter, we present a few basic notions about higher order statistics that generalise notions defined above for second order statistics. We show, in particular, how the definition of the spectral representation of a process can be generalised for processes that are stationary at orders higher than two.

14.1 Introduction

Most conventional techniques for random signal processing only take into consideration knowledge of the second order statistics of the processes studied, that is, their autocovariance coefficients, or in a similar way, that of their spectrum.

We can see several explanations for this: on the one hand, in many situations the processes studied are Gaussian, or at least the Gaussian hypothesis is assumed to be reasonable. The distribution of the process is then entirely characterised by knowledge of its first and second order moments. On the other hand, many results of great practical utility, such as those of the linear prediction theory, are expressed from second order statistics. Second order statistics are also sufficient for characterising systems that can be seen as the output of a filter with minimum phase or known phase, and with white noise input.

However, for some problems, the observed signals are non-Gaussian. It may then be of interest to account for the information provided by estimated moments such as $\mathbb{E}[X_{k_1} \ldots X_{k_l}]$, with $l > 2$. Generally, the statistics involving estimators of such moments are called *higher order* statistics.

In addition, when processing non-Gaussian signals in the presence of additive Gaussian noise, it is possible to remove the contribution of the noise, up to estimation errors, when working with higher order statistics.

We also point out that some problems such as blind deconvolution, which involves deconvolving a random process filtered by an unknown filter, or such as estimating a non-minimum-phase transfer function, can often be addressed using higher order statistics, while this is not possible considering only second order statistics.

But, it must be pointed out that using higher order statistics is not without problems: the complexity of the methods is increased by the increasing number of dimensions of the mathematical objects at hand, and we are often led to solve non-linear problems. But the main handicap is certainly that of the higher variance of the estimators of higher order statistics compared to that of second order moments.

In fact, it is often preferable, when possible, to take into account all the statistical information about a process. This can be done by using an appropriate parametric model of the distribution of the process, rather than considering only the partial statistical information consisting of the knowledge of the moments up to a certain order. We may then use parametric models to describe the distribution of the processes brought into play, as will be seen in Chapter 15. Higher order statistics techniques, in fact, occupy an intermediate position between simple second order techniques and more general techniques that often require a high computational burden.

In this chapter, we shall begin by defining the cumulants, which are parameters often used in preference to moments. We shall then introduce the notions of cumulant functions and of cumulant spectra for processes that are stationary at orders higher than two. We shall illustrate why it is of interest to use these tools in the context of a few particular methods for characterising rational transfer functions.

14.2 Moments and Cumulants

14.2.1 Real-valued Random Variables

We recall that if X denotes a real scalar random variable, its first and second characteristic functions, denoted by $\Phi_X(u)$ and $\Psi_X(u)$, are defined by

$$\Phi_X(u) = \mathbb{E}[e^{iuX}],$$

$$\text{and } \Psi_X(u) = \log(\Phi_X(u)) \tag{14.1}$$

respectively. The moments and the cumulants of X are defined from the coefficients of the series development of the first and of the second characteristic function, or equivalently from the derivatives of these functions:

$$m_{X,(r)} = \mathbb{E}[X^r] = (-i)^r \left(\frac{d^r \Phi_X(u)}{du^r} \right)_{u=0},$$

$$\text{and } C_{X,(r)} = (-i)^r \left(\frac{d^r \Psi_X(u)}{du^r} \right)_{u=0}. \tag{14.2}$$

By using the relations $\Psi_X(u) = \log(\Phi_X(u))$, and $\Phi_X(u) = \exp(\Psi_X(u))$, we see that the cumulants can be expressed by means of the moments with lower or equal orders, and conversely ([61] p.33):

$$C_{X,(r)} = \sum_{\lambda_1+\ldots+\lambda_q=r} \frac{(-1)^{q-1}}{q} \frac{r!}{\prod_{i=1,q} \lambda_i!} \prod_{p=1,q} m_{X,(\lambda_p)},$$

$$(14.3)$$

$$m_{X,(r)} = \sum_{\lambda_1+\ldots+\lambda_q=r} \frac{1}{q!} \frac{r!}{\prod_{i=1,q} \lambda_i!} \prod_{p=1,q} C_{X,(\lambda_p)}.$$

Thus, for zero mean random variables,

$$C_{X,(k)} = m_{X,(k)} \qquad k = 1,3,$$

$$(14.4)$$

$$C_{X,(4)} = m_{X,(4)} - 3m_{X,(2)}^2.$$

We notice that in the Gaussian case

$$\Psi_X(u) = i.m_{X,(1)}u - \frac{1}{2}m_{X,(2)}\, u^2. \tag{14.5}$$

The cumulants of orders higher than two are therefore all equal to zero. This important remark is one of the two main reasons that often lead to working with cumulants rather than with moments, although moments are, in practice, simpler to estimate.

The other reason that justifies using cumulants is their additivity in the independent case. Indeed, if X and Y are independent random variables, it is clear that $\Phi_{X+Y}(u) = \Phi_X(u)\Phi_Y(u)$, and consequently, $\Psi_{X+Y}(u) = \Psi_X(u) + \Psi_Y(u)$.

We sometimes use standardised cumulants, defined by

$$K_{X,(r)} = \frac{C_{X-\mathbb{E}[X],(r)}}{C_{X-\mathbb{E}[X],(2)}^{r/2}} = C_{Y,(r)}, \tag{14.6}$$

with $Y = C_{X,(2)}^{-1/2}(X - \mathbb{E}[X])$. The coefficients $K_{X,(3)}$ and $K_{X,(4)}$ are called the *skewness* and *kurtosis* factors, by comparison of the shape of the distribution of X with that of a Gaussian distribution.

14.2.2 Real and Complex Random Vectors

For a real random vector $X = [X_1,\ldots,X_N]^T$, we define its characteristic functions, $\Phi_X(u)$ and $\Psi_X(u)$, by

$$\Phi_X(u) = \mathbb{E}[\exp(i\sum_{k=1,N} u_k X_k)]$$

$$= \mathbb{E}[e^{iu^T X}], \tag{14.7}$$

and $\Psi_X(u) = \log(\Phi_X(u)),$

where $u = [u_1, \ldots, u_N]^T$. Generalising the scalar case, we may define the moments and the cumulants of X by

$$m_{X,i_1,\ldots,i_r} = \mathbb{E}[X_{i_1} \ldots X_{i_r}]$$

$$= (-i)^r \left(\frac{\partial^r \Phi_X(u)}{\partial u_{i_1}, \ldots, \partial u_{i_r}} \right)_{u=0}, \tag{14.8}$$

and $C_{X,i_1,\ldots,i_r} = (-i)^r \left(\frac{\partial^r \Psi_X(u)}{\partial u_{i_1}, \ldots, \partial u_{i_r}} \right)_{u=0}.$

We shall also use the notation $C_{X,i_1\ldots i_r} = C(X_{i_1}, \ldots, X_{i_r})$. In particular, we will have

$$C(X_i, \ldots, X_i) = C_{X,ii\ldots i} \tag{14.9}$$

$$= C_{X_i,(r)}.$$

We indicate the expressions of the first cumulants as a function of the moments (see [61] p.33 for a general relation):

$$C_{X,ij} = m_{X,ij} - m_{X,i} m_{X,j}$$

$$C_{X,ijk} = m_{X,ijk} - [3] m_{X,i} m_{X,jk} + 2 m_{X,i} m_{X,j} m_{X,k} \tag{14.10}$$

$$C_{X,ijkl} = m_{X,ijkl} - [4] m_{X,i} m_{X,jkl} - [3] m_{X,ij} m_{X,kl}$$

$$+ 2[6] m_{X,i} m_{X,j} m_{X,kl} - 6 m_{X,i} m_{X,j} m_{X,k} m_{X,l},$$

where the notation $[k]$ means that we add the k terms obtained by reordering the indices of the expression that follows $[k]$. We find, in particular, that if X is Gaussian and has zero mean, $C_{X,ijkl} = 0$, that is

$$m_{X,ijkl} = [3] m_{X,ij} m_{X,kl}. \tag{14.11}$$

Formally, we can see a complex vector of size n as a real vector of size $2n$. We then naturally define the first characteristic function of a complex vector $X = X_1 + iX_2$, where X_1 and X_2 are real random vectors, at point $u = u_1 + iu_2$ ($u_1, u_2 \in \mathbb{R}^n$) by

$$\Phi_X(u) = \mathbb{E}[e^{i(u_1^T X_1 + u_2^T X_2)}]$$

$$= \mathbb{E}[e^{i\mathcal{R}e[u^H X]}]. \tag{14.12}$$

We define $\Psi_X(u)$ by $\Psi_X(u) = \log(\Phi_X(u))$. A straightforward consequence of the definition of $\Phi_X(u)$ is that

$$\forall a \in \mathbb{C}, \quad \Phi_{aX}(u) = \Phi_X(a^* u). \tag{14.13}$$

Taking into account the possible conjugation of certain random variables, the complex case is generally expressed by notations such as

$$C_{X,i_1\ldots i_p}^{i_{p+1}\ldots i_r} = C(X_{i_1}, \ldots, X_{i_p}, X_{i_{p+1}}^*, \ldots, X_{i_r}^*). \tag{14.14}$$

As in the scalar case, the cumulants may be standardised in the complex and vector cases. For this, we consider a factorisation LL^H of the covariance matrix R_X of the vector X: $R_X = \mathbb{E}[XX^H] - \mathbb{E}[X]\mathbb{E}[X]^H$. The standardised cumulants are then defined as those of the vector $Y = L^{-1}(X - \mathbb{E}[X])$, which has a covariance matrix equal to the identity. The relation $R_X = LL^H$ does not define the matrix L uniquely (it is defined up to one unitary matrix factor), and we generally take for L the matrix obtained from the eigenvalue decomposition of R_X: $R_X = U\Lambda U^H$ and $L = U\Lambda^{1/2}$.

14.2.3 Properties of Moments and Cumulants

Moments and cumulants define multi-linear operators. Thus, if $Y = AX$, where A is a matrix, we will have, for example,

$$C_{Y,ijkl} = \sum_{a,b,c,d} A_{ia} A_{jb} A_{kc} A_{ld} C_{X,abcd}. \tag{14.15}$$

In the complex case, we obtain similar results, but with taking into account the possible conjugation of the coefficients. Thus,

$$C_{X,ik}^{jl} = \sum_{a,b,c,d} A_{ia} A_{jb}^H A_{kc} A_{ld}^H C_{X,ac}^{bd}. \tag{14.16}$$

These multi-linear operators are called *tensors*.

We have seen that if X and Y are independent random vectors, $C_{X+Y,(r)} = C_{X,(r)} + C_{Y,(r)}$. In addition, if $\{X_1, ..., X_p\}$ and $\{X_{p+1}, ..., X_r\}$ are two independent families of random variables, it is clear that $C(X_1, \ldots, X_r) = 0$, since $\Phi_{\{X,Y\}}([u^T, v^T]^T) = \Phi_X(u)\Phi_Y(v)$ and thus $\Psi_{\{X,Y\}}([u^T, v^T]^T) = \Psi_X(u) + \Psi_Y(v)$.

14.2.4 Empirical Estimation of Moments and Cumulants

The empirical moments of a zero mean random variable X calculated from the observations $(x_n)_{n=1,N}$ are of the form $N^{-1}\sum_{n=1,N} X_n^r$. For independent realisations, these estimators are unbiased and strongly consistent, as we saw in the chapter relating to non-parametric spectral estimation.

Unfortunately, the estimators of the cumulants obtained by replacing the moments by their empirical estimators in the expression of the cumulants as a linear combination of products of moments (relation (14.3)) are biased estimators.

We can obtain unbiased estimators of the cumulants by a suitable weighting of the terms of this sum. These estimators are called *k-statistics*. A study of these statistics can be found in [25]. We note that the variance of these estimators decreases in N^{-1}.

Unfortunately, for fixed N, the variance of the estimated moments and cumulants tends to increase rapidly with the order r under consideration. To illustrate this point, let us consider a simple example. For a data sample of size N, made up of independent random variables with the same distributions, the variance of the empirical estimator of $\mathbb{E}[X^r]$ is equal to $\sigma^2_{X^{2r}} = N^{-1}(\mathbb{E}[X^{2r}] - \mathbb{E}^2[X^r])$. In the case of Gaussian random variables with variance σ^2, we calculate for $r = 2k$ the standard deviation of the empirical estimator of $\mathbb{E}[X^r]$ divided by $\mathbb{E}[X^r]$. Noting that $\mathbb{E}[X^{2k}] = (2k)!(2^k k!)^{-1}\sigma^{2k}$ and using Stirling's formula: $n! \sim n^n e^{-n}\sqrt{2\pi n}$, we obtain for the normalised standard deviation $\sigma_{X^{2r}}/\mathbb{E}^2[X^r]$

$$\frac{1}{\mathbb{E}[X^r]}\sqrt{N^{-1}(\mathbb{E}[X^{2r}] - \mathbb{E}^2[X^r])} \sim \frac{2^k}{\sqrt{\sqrt{2}N}}. \tag{14.17}$$

This simple example shows that we will have to increase considerably the length N of the data sequences used to estimate the moments and the cumulants reliably when r increases. This increasing variance property is one of the main handicaps of using higher order techniques.

14.3 Cumulants of a Process and Cumulant Spectra

The second order properties of stationary processes are often expressed in terms of an autocovariance function or a spectrum function. Similarly, for order r stationary processes, we can define the notions of cumulant function and cumulant spectra, often also called *polyspectra*. This enables filtering relations in the time domain and in the frequency domain to be expressed at higher orders.

14.3.1 Cumulants of a Process

We consider a real-valued process X. In this case, we define the order $p-1$ cumulant sequence by

$$C_{X,p-1}(n, k_1, \ldots, k_{p-1}) = C(X_n, X_{n+k_1}, \ldots, X_{n+k_{p-1}}). \tag{14.18}$$

For a complex-valued process, we define the cumulant function of order $(p-1, q)$, with $p \geq q$, by

$$C_{X,p-1,q}(n, k_1, \ldots, k_{p-1})$$

$$= C(X_n, X_{n+k_1}, \ldots, X_{n+k_{q-1}}, X^*_{n+k_q}, \ldots, X^*_{n+k^*_{p-1}}). \tag{14.19}$$

In the case of a p-th order stationary process, these cumulants do not depend on n and they will simply be denoted by $C_{X,p-1}(k_1, \ldots, k_{p-1})$ or by $C_{X,p-1,q}(k_1, \ldots, k_{p-1})$. In what follows, we shall assume that the processes considered are stationary up to the orders under consideration.

We note that there are many symmetries that limit the number of cumulants to be computed in practice. Thus, in the real case, we have, for example,

$$C_{X,2}(k_1, k_2) = C_{X,2}(k_2, k_1)$$

$$= C_{X,2}(-k_2, k_1 - k_2) \tag{14.20}$$

$$= C_{X,2}(-k_1, k_2 - k_1).$$

14.3.2 Cumulant Spectra

Assuming that $\sum_{(k_1,\ldots,k_{p-1}) \in \mathbb{Z}^{p-1}} |C_{X,p-1}(k_1, \ldots, k_{p-1})| < \infty$, we define the order $p-1$ cumulant spectrum of a real process by

$$S_{X,p-1}(f_1, \ldots, f_{p-1})$$

$$= \sum_{(k_1,\ldots,k_{p-1}) \in \mathbb{Z}^{p-1}} C_{X,p-1}(k_1, \ldots, k_{p-1}) \, e^{-2i\pi \sum_{j=1,p-1} k_j f_j}. \tag{14.21}$$

Conversely, the cumulants can then be recovered as the Fourier coefficients of $S_{X,p-1}(f_1, \ldots, f_{p-1})$. The function $S_{X,p-1}(f_1, \ldots, f_{p-1})$ is called the *cumulant spectral density* of X. But, in the same way that the PSD of a process may not exist at the second order and that the more general notion of spectral measure has to be introduced, a process X stationary up to the p-th order, does not always have an order $p-1$ cumulant spectral density.

Intuitively, the cumulant spectral measure of X can be constructed from the spectral representation of X:

$$X_n = \int_{\mathcal{I}} e^{2i\pi n f} d\hat{X}(f). \tag{14.22}$$

From the definition of $C_{p-1}(k_1, \ldots, k_{p-1})$ and using (14.22), we obtain

$$C_{X,p-1}(k_1, \ldots, k_{p-1})$$

$$= C(X_n, X_{n+k_1}, \ldots, X_{n+k_{p-1}})$$

$$= \int_{\mathcal{I} \times \cdots \times \mathcal{I}} e^{2i\pi \sum_{j=1,p-1} k_j f_j} \, e^{2i\pi n (f_0 + \sum_{i=1,p-1} f_i)} \tag{14.23}$$

$$\times C\left(d\hat{X}(f_0), d\hat{X}(f_1), \ldots, d\hat{X}(f_{p-1})\right),$$

where we can show that $C\left(d\hat{X}(f_0), d\hat{X}(f_1), ..., d\hat{X}(f_{p-1})\right)$ defines a measure on $(\mathbb{R}^p, \mathcal{B}(\mathbb{R}^p))$ for a large class of processes (see [61] p.167). This measure is defined by the relation

$$C\left(d\hat{X}(f_0), ..., d\hat{X}(f_{p-1})\right)(\Delta_0 \times ... \times \Delta_{p-1})$$

$$= C\left(\hat{X}(\Delta_0), ..., \hat{X}(\Delta_{p-1})\right), \tag{14.24}$$

for $\Delta_0, ..., \Delta_{p-1} \in \mathcal{B}(\mathbb{R})$.

The independence of the cumulant with respect to n implies that in relation (14.23) we have $f_0 = -\sum_{i=1,p-1} f_i$. We define the cumulant spectral measure of X, denoted by $\mu_{X,p-1}$, as the measure whose Fourier coefficients are the coefficients $C_{X,p-1}(k_1, ..., k_{p-1})$. Therefore,

$$d\mu_{X,p-1}(f_1, ..., f_{p-1}) = C\left(d\hat{X}(-\sum_{i=1,p-1} f_i), d\hat{X}(f_1), ..., d\hat{X}(f_{p-1})\right). \tag{14.25}$$

Moreover, in the case where $\mu_{X,p-1}$ is absolutely continuous,

$$d\mu_{X,p-1}(f_1, ..., f_{p-1}) = S_{X,p-1}(f_1, ..., f_{p-1})df_1 ... df_{p-1}. \tag{14.26}$$

Similarly, in the complex case, it is better to define the order $(p-1, q)$ cumulant spectral measures by

$$d\mu_{X,p-1,q}(f_1, ..., f_{p-1}) = C(d\hat{X}(-\sum_{i=1,q-1} f_i + \sum_{j=q,p-1} f_j), d\hat{X}(f_1),$$

$$..., d\hat{X}(f_{q-1}), d\hat{X}(f_q)^*, ..., d\hat{X}(f_{p-1})^*). \tag{14.27}$$

For $(p-1, q) = (1, 1)$, $d\mu_{X,1,1}(f_1) = \mathbb{E}[d\hat{X}(f_1)d\hat{X}(f_1)^*]$, which corresponds to the spectral measure as it was defined at the second order.

14.3.3 Higher Order White Noise

A process stationary up to the order p is said to be white up to the order p if its cumulants are of the form

$$C_{X,q-1}(k_1, ..., k_{q-1}) = C_{X_n,(q)}\delta_{0,k_1,...,k_{q-1}}, \quad q \leq p, \tag{14.28}$$

where δ here represents Kronecker's symbol ($\delta_{0,k_1,...,k_{q-1}} = 1$ if $k_1 = ... = k_{q-1} = 0$, and 0 otherwise). We note that second order white noise is made up of uncorrelated samples, and that a process white at any order is made up of independent random variables. We again find the equivalence between uncorrelation and independence in the Gaussian case for which all the cumulants of orders higher than two are equal to zero.

14.3.4 Estimation of Cumulants and of Cumulant Spectra

A cumulant spectrum can be estimated by the multidimensional Fourier transform of the estimated cumulants. As for the second order periodogram estimator, smoothing techniques can be considered to decrease the variance of estimated cumulant spectra.

Another approach involves using relation (14.25):

$$d\mu_{X,p-1}(f_1, \ldots, f_{p-1}) = C(d\hat{X}(- \sum_{i=1,p-1} f_i), d\hat{X}(f_1), \ldots, d\hat{X}(f_{p-1})).$$

$$(14.29)$$

In practice, the components $\hat{X}(f_k)$ are obtained by a narrow band filtering of X around f_k, which can be provided by the discrete Fourier transform.

14.3.5 Linear Filtering

We shall study the way in which the cumulant functions of order p are transformed by a linear filter. First, we consider the scalar case. The input X and the output Y of a filter with impulse response $(h_n)_{n \in \mathbb{Z}}$ and with frequency response $H(f)$ are linked by the convolution relation $Y_n = \sum_m h_m X_{n-m}$. From the definition of the cumulants of a process

$$C_{Y,p-1}(k_1, \ldots, k_{p-1}) = C(Y_n, Y_{n+k_1}, \ldots, Y_{n+k_{p-1}}), \qquad (14.30)$$

it immediately results that the order $p-1$ cumulants of X and of Y are linked by the relation

$$C_{Y,p-1}(k_1, \ldots, k_{p-1}) =$$

$$\sum_{l_1} \cdots \sum_{l_{p-1}} C_{X,p-1}(k_1 - l_1, \ldots, k_{p-1} - l_{p-1})(\sum_l h_l h_{l+l_1} \ldots h_{l+l_{p-1}}).$$

$$(14.31)$$

Since $d\hat{Y}(f) = H(f)d\hat{X}(f)$, it results from relation (14.25) and the linearity of the cumulants that the spectral measure of Y is given by

$$d\mu_{Y,p-1}(f_1, \ldots, f_{p-1}) =$$

$$H(- \sum_{i=1,p-1} f_i) H(f_1) \ldots H(f_{p-1}) \, d\mu_{X,p-1}(f_1, \ldots, f_{p-1}).$$

$$(14.32)$$

The generalisation of this result to the complex case and to the vector case is straightforward.

Finally, let us remark that for systems with rational spectra, it can be useful to express the relations between the input and the output cumulants by using the parameters of the state space representation associated with the system [59].

14.4 Estimation of a Transfer Function

14.4.1 Model

In certain problems we observe a process X, which can be modelled as the output of a causal filter with an input noise process, denoted by V, that is, white up to the order r. Here, we consider the real scalar case, and we assume that $X_n = [h(z)]V_n$, where $h(z)$ is modelled as a rational transfer function,

$$h(z) = \frac{b(z)}{a(z)} = \frac{\sum_{l=0,q} b_l z^{-l}}{1 + \sum_{k=1,p} a_k z^{-k}}. \tag{14.33}$$

X is therefore an ARMA process (we assume that the zeroes of $a(z)$ are located in the open unit disk). The cumulants of V are of the form

$$C_{V_n,q-1}(k_1, ..., k_{q-1}) = C_{V_n,(q)}\,\delta_{k_1,...,k_{q-1}}, \quad q \leq r. \tag{14.34}$$

When $h(z)$ is minimum-phase or has a known phase, identifying the coefficients of $h(z)$ can be done simply from the knowledge of the second order moments of X. Indeed, in this case, the phase of $h(z)$ can be written as the discrete Hilbert's transform of $\log|h(z)|$ (see Appendix W). Consequently, as $S_{X,1}(f) = |h(e^{2i\pi f})|^2 C_{V_n,(2)}$, the knowledge of $S_{X,1}(f)$ is then sufficient to identify $h(z)$ up to one factor with modulus 1.

14.4.2 Interest of Higher Order Statistics

We now want to take into account the fact that the system that we wish to model can be represented by a realisable filter, that is, causal and stable, but not necessarily minimum-phase. This kind of situation appears, for example, in modelling seismic data used to analyse the signals backscattered at the underground layers interface, or in modelling multipath propagation channels in telecommunications. We shall see how using higher order statistics makes it possible to remove the ambiguity that then remains at the second order.

We first show that the ambiguity, in fact, concerns the MA part of the transfer function. We factorise $b(z)$ in the form

$$b(z) = b_0 \prod_{k=1,p} (1 - \beta_k z^{-1}). \tag{14.35}$$

The output of a transfer function filter $b(z)$ with a white noise input process of variance 1 is an MA process, denoted by Y, with PSD

$$S_Y(f) = |b_0 \prod_{k=1,p} (1 - \beta_k e^{-2i\pi f})|^2. \tag{14.36}$$

$b(z)$ is therefore a causal factorisation of degree p of $S_Y(f)$. But, any causal factorisation of degree p is a polynomial of the form $c\prod_{k=1,p}(1 - \beta'_k z^{-1})$, with $\beta'_k = \beta_k$, or $\beta'_k = (\beta^*_k)^{-1}$, for

$$|1 - \beta_k e^{-2i\pi f}| = |\beta_k| \times |1 - (\beta^*_k)^{-1} e^{-2i\pi f}|. \tag{14.37}$$

There are therefore exactly 2^p MA filters whose frequency response modulus is equal to $|b(e^{-2i\pi f})|$. Of course, all these filters are causal and stable. In addition, the stability constraint means that the zeroes of the AR part lie inside the unit circle, which determines the latter uniquely from its modulus.

The problem of the estimation of $h(z)$ by means of higher order techniques that we consider here, therefore seems to concern only the MA part of $h(z)$, the AR part being entirely characterised by the sole knowledge of its modulus due to the stability constraint. In fact, higher order techniques are also interesting for the AR estimation of a system when there is an additive Gaussian noise at its output; indeed, in this case the cumulants of orders higher than two of the interfering noise are equal to zero, up to the estimation errors, and their contribution disappears when estimating the coefficients of the filter.

We shall indicate two types of approaches proposed in the literature. The first one aims at testing which of the 2^p possible filters must be retained by minimising a distance criterion between parametric expressions of higher order parameters and their empirical counterparts. The second one directly estimates the parameters of the model. Only a few significant methods are presented here. A summary of existing methods can be found in [57].

14.4.3 Distance Criteria

Spectral Domain Approach Since the modulus of $H(f) = h(e^{2i\pi f})$ can be estimated at the second order, the first idea involves finding only its phase at higher orders. From (14.32), we have, in particular,

$$S_{X,2}(f_1, f_2) = H(-f_1 - f_2)H(f_1)H(f_2)\,C_{V_n,(3)}, \tag{14.38}$$

and the phase φ_H of H is related to the phase $\varphi_{X,2}$ of $S_{X,2}$ by the relation

$$\varphi_{X,2}(f_1, f_2) = \varphi_H(-f_1 - f_2) + \varphi_H(f_1) + \varphi_H(f_2). \tag{14.39}$$

By identifying this relation for a sufficient number of values of f_1 and of f_2, we can select which of the 2^p possible filters for $b(z)$ best represents the system studied. We note that this approach requires $S_{X,2}(f_1, f_2)$ to be estimated. If $C_{V_n,(3)} = 0$, we have to work at an order $k > 2$ for which $C_{V_n,(k)}$ is non-zero.

Time Domain Approach In the time domain, it is possible to choose which of the 2^p filters is suitable from the comparison of the distance between the empirical estimates of cumulants and the values of the cumulants associated with the 2^p possible filters obtained at the second order [58].

14.4.4 Direct Identification of the Filter Coefficients

MA Filters To identify an MA model given by the difference equation

$$X_n = \sum_{k=0,q} b_k V_{n-k},$$ (14.40)

we can consider an approximation of the inverse filter relation of the form

$$V_n = \sum_{l=-l_1,l_2} c_l X_{n-l},$$ (14.41)

where the presence of negative coefficients accounts for the fact that $h(z)$ may not be causal.

Relations (14.40) and (14.41) make it possible to compute the cumulants $C(V_n, X_{n+k})$ and $C(V_n, X_{n+k}^2)$ in two different ways, which leads easily to the relations

$$\frac{1}{C(V_n, V_n)} \sum_{l=-l_1,l_2} c_l C(X_{n-l}, X_{n+k}) = b_k$$ (14.42)

$$\frac{1}{C(V_n, V_n, V_n)} \sum_{l=-l_1,l_2} c_l C(X_{n-l}, X_{n+k}^2) = b_k^2,$$ (14.43)

with $b_k = 0$ for $k > q$ or $k < 0$. There are several ways of exploiting these relations. We can, in particular, limit ourselves to exploiting only relation (14.42), and relation (14.43) for which $k > q$ and $k < 0$. More precisely, we consider a linear system of equations of the form $Ac = b$, where the vectors b and c contain the coefficients b_k and c_l respectively, relation (14.42) being taken for $k = -q - l_2, \ldots, q + l_1$, and relation (14.43) for $k = -q - l_2, \ldots, -1, q+1, \ldots, q + l_1$. We may solve this system in the least squares sense after setting c_0 to 1.

AR Filters For an AR process defined by the recurrence equation

$$\sum_{k=0,p} a_k X_{n-k} = V_n,$$ (14.44)

with $a_0 = 1$, it is clear that we may generalise the higher order Yule-Walker equations by considering relations of the type

$$\sum_{k=0,p} a_k C(X_{n-k}, X_{n-l}, \ldots, X_{n-l}) = \sum_{k=0,p} a_k C_{X,m-1}(l - k, 0, \ldots, 0)$$

$$= 0.$$ (14.45)

These relations are valid for $l > 0$ if X is an AR process and more generally for $l \geq q$ if X is an ARMA(p,q) process. They are also satisfied for any order m at which V is a white noise. By concatenating a certain number of relations of this form, we can obtain a linear system of equations that can be solved in the least squares sense to estimate the coefficients a_k. The choice of the equations that should be considered in order to obtain such a system is studied in [63].

ARMA Models It is possible to estimate the coefficients of an ARMA model by first estimating those of the AR part and then those of the MA part. Furthermore, when the AR coefficients of an ARMA model defined by the transfer function (14.33) have been identified, the coefficients of the impulse response can be computed [62] by the relations

$$h_n = h_0 \frac{\sum_{k=0,p} a_k C_{X,p-1}(q - k, n, 0, ..., 0)}{\sum_{k=0,p} a_k C_{X,p-1}(q - k, 0, 0, ..., 0)}. \tag{14.46}$$

Indeed,

$$\sum_{k=0,p} a_k C_{X,p-1}(q - k, n, 0, ..., 0)$$

$$= C(\sum_{k=0,p} a_k X_{m-k}, X_{m+n-q}, X_{m-q}, \ldots, X_{m-q})$$

$$= C(\sum_{l=0,q} b_l V_{m-k}, \sum_{j=-n,\infty} h_{n+j} V_{m-q-j}, X_{m-q}, \ldots, X_{m-q})$$

$$= C(\sum_{l=0,q} b_l V_{m-k}, h_n V_{m-q}, X_{m-q}, \ldots, X_{m-q})$$

$$= h_n h_0^{-1} C(\sum_{l=0,q} b_l V_{m-k}, h_0 V_{m-q}, X_{m-q}, \ldots, X_{m-q})$$

$$= h_n h_0^{-1} C(\sum_{k=0,p} a_k X_{m-k}, X_{m-q}, X_{m-q}, \ldots, X_{m-q})$$

$$= h_n h_0^{-1} \sum_{k=0,p} a_k C_{X,p-1}(q - k, 0, \ldots, 0). \tag{14.47}$$

Other methods make it possible to envisage the joint estimation of the AR and MA parts.

Remarks

1) In this section, we have restricted ourselves to the real scalar case. The techniques considered here can be extended to the complex case and to the vector case.

2) Here, we have only considered the problem of linear filtering. However, it is to be noted that higher order statistics appear fairly naturally in the context of non-linear filtering. In particular, dropping the Gaussian hypothesis often leads to non-linear systems being envisaged. An important category of non-linear filters is the class of the Volterra filters, which appear as a generalisation of linear filters. For such an order p filter, the output Y is linked to the input X by a relation of the type

$$Y(n) = h_0 + \sum_{\substack{\{l_{m,1},...,l_{m,m}\} \\ m=1,p}} h_m\,(l_{m,1},...,l_{m,m})\, X\,(n - l_{m,1})\,...X\,(n - l_{m,m})$$

$$(14.48)$$

and we can easily understand that estimation of the coefficients $h_m\,(l_{m,1},...,$ $l_{m,m})$ involves higher order statistics of X.

Exercises

14.1 (Bispectrum) Let $X = (X_t)_{t \in \mathbb{R}}$ denote a third order real-valued stationary process. We assume that the third order spectrum is absolutely continuous with respect to Lebesgue's measure, and we note $S_{X,2}(f_1, f_2)$ as its PSD.

a) Check that $\mathbb{E}[d\hat{X}(f_1)d\hat{X}(f_2)d\hat{X}(f_3)] = S_{X,2}(f_1, f_2)\delta_{f_1 + f_2 - f_3}$.

b) Show that for $f_{2k} > f_{2k-1}$, $k = 1, 2, 3$,

$$\mathbb{E}[\hat{X}([f_1, f_2])\hat{X}([f_3, f_4])\hat{X}([f_5, f_6])] = \int_A S_{X,2}(u, v)dudv, \qquad (14.49)$$

where A is the convex set of the plane limited by the lines of equations $u = f_1$, $u = f_2$, $v = f_3$, $v = f_4$, $v = f_5 - u$, $v = f_6 - u$.

The third order spectrum is often referred to as the bispectrum since it involves two frequency variables f_1 and f_2.

14.2 Let $X = (X_n)_{n \in \mathbb{Z}}$ denote a zero mean, Gaussian, WSS process. Express the fourth order cumulant function of X in terms of its autocovariance function.

14.3 (Skewness and kurtosis) Calculate the skewness and the kurtosis of the random variable X when the probability density function of X denoted by $f_X(x)$ is of the form $f_X(x) = \lambda e^{-\lambda x}\mathbb{I}_{\mathbb{R}_+}(x)$, $f_X(x) = (\lambda/2)e^{-\lambda|x|}$, and when X is a Bernoulli random variable.

14.4 (Edgeworth's development) Let X denote a random variable with mean and covariance equal to 0 and 1 respectively. Let us denote by $\Psi_0(u) = -u^2/2$ the second characteristic function of the $\mathcal{N}(0,1)$ distribution, and $\Psi_X(u)$ that of X. We are looking for a development of $\Psi_X(u)$ around $\Psi_0(u)$. To this end, we note

$$\exp[\Psi_X(u) - \Psi_0(u)] = \sum_{n=0,\infty} \frac{b_n}{n!}(iu)^n, \qquad (14.50)$$

assuming that this series expansion exists for the random variable X.

a) Show that the probability density function of X, denoted by $f_X(x)$ satisfies

$$f_X(x) = \frac{1}{\sqrt{2\pi}} e^{-x^2/2} [\sum_{n=0,\infty} \frac{b_n}{n!} h_n(x)], \qquad (14.51)$$

where $(h_n)_{n=0,\infty}$ are the Hermite polynomials defined by the recursion $h_0(u) = 1$ and

$$h_n(x) = x h_{n-1}(x) - h'_{n-1}(x) \qquad (n > 0). \qquad (14.52)$$

(Hint: check that $h_n(x) = e^{u^2/2} \frac{i^n}{\sqrt{2\pi}} \int_{\mathbb{R}} e^{-t^2/2} t^n e^{-itx} dt$.)

b) Example: Calculate Edgeworth's development of a Bernoulli random variable X with distribution $P(X = +1) = P(X = -1) = 1/2$

14.5 (Cumulant estimation) Let X denote a zero mean random variable and let us consider the estimator of $C_{X,4}$ of the form

$$\hat{C}_{X,4} = (N^{-1} \sum_{n=1,N} X_n^4) - 3(N^{-1} \sum_{n=1,N} X_n^2)^2, \qquad (14.53)$$

where $(X_n)_{n=1,N}$ are independent random variables with the same distribution as X.

a) Check that $\mathbb{E}[\hat{C}_{X,4}] = C_{X,4} - 3N^{-1}[C_{X,4} + 2C_{X,2}^2]$.

b) In order to find an unbiased estimator of $C_{X,4}$, we are looking for an estimator of the form $\alpha \sum_{n=1,N} X_n^4 + \beta \sum_{m,n=1,N} X_m^2 X_n^2$. Calculate α and β that yield an unbiased estimator of $C_{X,4}$.

c) Show that the variance of this estimator is equal to

$$N^{-1}[C_{X,8} + 16C_{X,6}C_{X,2} + 48C_{X,5}C_{X,3} + 34C_{X,4}^2$$

$$+ 72C_{X,4}C_{X,2}^2 + 144C_{X,3}^2 C_{X,2} + 24C_{X,2}^4]. \qquad (14.54)$$

14.6 Let $X = (X_n)_{n \in \mathbb{Z}}$ denote a real-valued third order stationary process and $S_{X,3}(f_1, f_2)$ its third order spectrum. Check the following relationships:

$$S_{X,3}(f_1, f_2) = S_{X,3}(f_2, f_1) = S_{X,3}(f_1, -f_1 - f_2) = S_{X,3}(f_2, -f_1 - f_2). \qquad (14.55)$$

14.7 (Sampling) Let $X = (X_t)_{t \in \mathbb{R}}$ denote a third order stationary process with spectral measure $d\mu_X(f)$ carried by $[-B, B]$.

a) Check that the bispectral measure $d\mu_{X,2}(f_1, f_2)$ is carried by the hexagonal set

$$\{(f_1, f_2); -B \le f_1, f_2 \le B, -B \le f_1 + f_2 \le B\}. \qquad (14.56)$$

(Hint: note that $d\hat{X}(f) = \mathbb{I}_{[-B,B]}(f) d\hat{X}(f)$ and use the third order filtering equations.)

b) X is sampled with period T, yielding $X_e = (X_{nT})_{n \in \mathbb{Z}}$. Express $d\mu_{X_e,2}(f_1, f_2)$ in terms of the spectral measure of X and check that a sampling condition to avoid third order spectrum aliasing is given by $T < (3B)^{-1}$.

14.8 (Higher order periodogram) Let $X = (X_t)_{t \in \mathbb{Z}}$ denote a zero mean real-valued p^{th} order stationary process. We define the pth order periodogram as

$$\hat{S}_{X,p-1,n}(f_1, \ldots, f_{p-1}) = F_{X,n}(- \sum_{k=1,p-1} f_k) \prod_{k=1,p-1} F_{X,n}(f_k), \quad (14.57)$$

where $F_{X,n}(f) = \sum_{k=1,n} X_k e^{-2i\pi k f}$. Similarly the pth order cumulant periodogram would be defined as

$$\hat{S}^C_{X,p-1,n}(f_1, \ldots, f_{p-1})$$

$$= C(F_{X,n}(-\sum_{k=1,p-1} f_k), F_{X,n}(f_1), \ldots, F_{X,n}(f_p - 1). \quad (14.58)$$

Calculate the bias and the variance of $\hat{S}_{X,2,n}(f_1, f_2)$.

14.9 (ARMA process identification) In this section, we consider the system identification problem for known input and known noisy output. Let $X = (X_n)_{n \in \mathbb{Z}}$ denote an ARMA process defined by $[a(z)]X_n = [b(z)]V_n$, where $a(z) = 1 + \sum_{k=1,p} a_k z^{-k}$, $b(z) = 1 + \sum_{l=1,q} b_l z^{-l}$, and $V = (V_n)_{n \in \mathbb{Z}}$ is a white noise up to the third order. We observe the process $Y = (Y_n)_{n \in \mathbb{Z}}$ with $Y_n = X_n + W_n$, where $W = (W_n)_{n \in \mathbb{Z}}$ is a Gaussian noise. Show that $a(z)$ and $b(z)$ can be identified from the knowledge of V and Y through relationships involving cumulants of the form $C(Y_n, V_m, V_m)$ and $C(V_n, V_m, V_m)$.

14.10 (Frequency estimation) Let $X = (X_n)_{n \in \mathbb{Z}}$ denote a WSS process with $X_n = \sum_{k=1,p} \xi_k e^{2i\pi n f_k} + V_n$, where the random variables ξ_k are zero mean, non-Gaussian and independent, and $V = (V_n)_{n \in \mathbb{Z}}$ a Gaussian white noise. Let us consider the matrix $Q_{X,N}$ of size $N^2 \times N^2$ with general term $[Q_{X,N}]_{Na+b, Nc+d} = C(X_a, X_b, X_c^*, X_d^*)$.

a) Show that $Q_{X,N} = \sum_{k=1,p} C_{\xi_k, 4} \{d(f_k) \otimes d(f_k)\} \{d(f_k) \otimes d(f_k)\}^H$, where $\otimes$ represents here the Kronecker product[1].

b) Show that using the Caratheodory theorem (Theorem 11.2) it is possible to identify $(f_k)_{k=1,p}$ from the knowledge of $Q_{X,N}$ provided that $p < 2N - 1$.
(Hint: the space spanned by vectors of the form $u \otimes u$ where $u \in \mathbb{R} \times \mathbb{C}^{N-1}$ is a real vector space of dimension $2N - 1$.)

c) What might be the advantages of identifying $(f_k)_{k=1,p}$ from $Q_{X,N}$ rather than the $N \times N$ covariance matrix $T_{X,N}$ with general term $[T_{X,N}]_{ab} = \mathbb{E}[X_a X_b^*]$?

14.11 (Frequency estimation) Let $X_n = \xi_1 e^{2i\pi n f_1} + \xi_2 e^{2i\pi n f_2}$, where ξ_1 and ξ_2 are zero mean independent non-Gaussian random variables. Discuss the advantage of using the fourth order periodogram rather than the second order periodogram to estimate f_1 and f_2 when $|f_1 - f_2|$ is small. Calculate the variance of the fourth and second order periodograms.

---

[1] The Kronecker product of two matrices A and B of respective size $m_a \times n_a$ and $m_b \times n_b$ is the matrix $A \otimes B$ of size $m_a m_b \times n_a n_b$ with general term $[A \otimes B]_{im_a+k, jn_a+l} = [A]_{ij}[B]_{kl}$.

14.12 .(**Price's formula**) Let $g_k(t)$, $k = 1, 2$, be two differentiable functions and $[X_1\ X_2]^T$ a zero mean Gaussian vector. We note $c = cov[X_1, X_2]$.

a) Prove that under certain conditions upon the functions g_1 and g_2 we obtain Price's formula:

$$\frac{\partial}{\partial c}\mathbb{E}[g_1(X_1)g_2(X_2)] = \mathbb{E}[g_1'(X_1)g_2'(X_2)]. \tag{14.59}$$

(Hint: express $\mathbb{E}[g_1(X_1)g_2(X_2)]$ in terms of the Fourier transforms $\hat{g}_k(t) = \int_{\mathbb{R}} e^{-2i\pi ft} g_k(t)dt$, $k = 1, 2$.)

b) Application: Let $X = (X_t)_{t\in\mathbb{R}}$ be a WSS Gaussian process with autocovariance function $R_X(t)$. Calculate the autocovariance functions of the processes $(X_t^2)_{t\in\mathbb{R}}$ and $(X_t^3)_{t\in\mathbb{R}}$ in terms of $R_X(t)$.

14.13 (Source separation) In this exercise, we address the principle of blind source separation in the presence of an instantaneous mixture. Let $Y = (Y_n)_{n\in\mathbb{Z}}$ denote a multivariate process of size p, $Y_n = AX_n + V_n$, where A is a full rank matrix of size $N \times p$ ($N > p$), the p coordinates of $X = (X_n)_{n\in\mathbb{Z}}$ are independent random processes and $V = (V_n)_{n\in\mathbb{Z}}$ is a Gaussian noise vector with covariance matrix $\sigma_V^2 I$.

a) Let M_1 and M_2 denote two matrices and assume that M_1 and M_2 have representations of the form $M_1 = AD_1A^H$ and $M_2 = AD_2A^H$ respectively, where D_1 and D_2 are diagonal with non-zero diagonal entries. Show that A can be identified from M_1 and M_2 up to one permutation of its columns and to a right diagonal matrix factor if and only if all the eigenvalues of the matrix pencil (M_1, M_2), that is, the eigenvalues of $M_2^{-1}M_1$ are distinct.

b) How could this result be exploited to estimate the matrix A from the observation of Y?

(Hint: consider, for instance, second and fourth order matrix cumulants of Y to build two matrices of the form ADA^H.)

c) Let $\hat{A}$ be an estimate of A. Calculate then the best linear unbiased estimate of X_n from the knowledge of Y_n.

15. Bayesian Methods and Simulation Techniques

Purpose Here, we present a general methodology for solving problems of filtering, smoothing, and prediction, along with that of identifying transfer functions. For this, we use a Bayesian approach, which allows possible *a priori* information about the desired parameters to be incorporated. In order to be able to perform the numerical computation of the estimators, we use Monte Carlo techniques to solve integration and maximisation problems that appear in Bayesian estimation.

15.1 Introduction

Many parametric estimation techniques and, in particular, maximum likelihood estimation considered in Chapter 13, have their drawbacks. In particular, these methods do not incorporate possible available information about the parameters of interest, denoted by θ. Prior knowledge about θ can be taken into account by considering a probability distribution for θ, called the prior distribution. In order to simplify the notations, θ will represent either a vector of random parameters or a realisation of this random vector.

In addition, maximum likelihood methods tend to over-parameterise the order of the models. Indeed, it is clear that by increasing the number of parameters of a model, we generally obtain a new model, for which the value of the maximum of the likelihood is larger. As the maximum likelihood method does not allow a reasonable size for the model to be defined, a penalty term is sometimes added to the likelihood criterion (e.g. AIC criterion, [10] Chap. 9).

The Bayesian approach involves seeing the parameters of interest as random variables for which probability distributions, called *prior distributions*, are available. Using such distributions makes it possible to incorporate the *a priori* information relative to the parameters and to efficiently manage problems of model size. This approach is quite general since it is always possible to use prior distributions or parameterised prior distributions that involve little information about θ, such as uniform distributions. The parameters that may possibly appear in the description of prior distributions are called *hyperparameters*.

In this chapter, we shall begin by recalling the principle of Bayesian estimation, then that of Monte Carlo methods, which makes it possible to solve problems of integration and optimisation encountered in Bayesian estimation by means of random variables simulation. The techniques for generating independent random variables are often insufficient, so we then use Monte Carlo methods based on the simulation of Markov chains, so-called *Monte Carlo Markov Chain* (MCMC) methods. Therefore, we shall recall some of the basics concerning Markov chains, before presenting conventional Metropolis-Hastings and Gibbs simulation algorithms. We shall illustrate why it is of interest to use these tools for problems of filtering and estimating rational transfer functions.

15.2 Elements of Bayesian Estimation

In the following, to simplify the notations and without loss of generality we shall assume that the probability distributions considered are absolutely continuous (with respect to Lebesgue's measure). The density of the prior distribution of θ will be denoted by $\pi(\theta)$, that of the distribution of observations parameterised by θ $f(x|\theta)$, and that of the distribution of the variables observed for the prior distribution $\pi(\theta)$ of the parameters, $f_\pi(x)$. We shall denote by $f(\theta|x)$ the density of the distribution of θ conditional to the value x taken by the vector X of observed random variables. $f(\theta|x)$ is called the *posterior distribution*. E and Θ will represent the spaces in which X and θ take their values. We note that

$$f_\pi(x) = \int_\Theta f(x|\theta)\pi(\theta)d\theta, \tag{15.1}$$

and that from Bayes' rule,

$$f(\theta|x) = \frac{f(x|\theta)\pi(\theta)}{f_\pi(x)}$$
$$= \frac{f(x|\theta)\pi(\theta)}{\int_\Theta f(x|\theta)\pi(\theta)d\theta}. \tag{15.2}$$

In estimation procedures where θ is a deterministic vector of unknown parameters, we generally estimate θ by minimising a certain mean cost function of the form $\mathbb{E}[C(\theta, T(X))]$. In Bayesian estimation, θ is a random vector, and the cost function takes the form

$$\mathbb{E}[C(\theta, T(X))] = \mathbb{E}[\mathbb{E}[C(\theta, T(X))|\theta]]$$
$$= \int_\Theta \mathbb{E}[C(\theta, T(X))|\theta]\pi(\theta)d\theta. \tag{15.3}$$

For a quadratic cost function, we obtain the estimator $\mathbb{E}[\theta|X]$. Its value $\mathbb{E}[\theta|X = x]$ obtained for $X = x$, will simply be denoted by $\mathbb{E}[\theta|x]$. We note the following important particular case:

Theorem 15.1 *When $f(x|\theta)$ is of the form $h(x)\exp(\theta^T x - \phi(\theta))$,*

$$\mathbb{E}[\theta|x] = \nabla_x \log f_\pi(x) - \nabla_x \log h(x). \tag{15.4}$$

Proof

$$\mathbb{E}[\theta_i|x] = \frac{1}{f_\pi(x)} \int_\Theta \theta_i h(x) e^{\theta^T x - \phi(\theta)} \pi(\theta) d\theta$$

$$= \frac{1}{f_\pi(x)} \frac{\partial}{\partial x_i} \int_\Theta h(x) e^{\theta^T x - \phi(\theta)} \pi(\theta) d\theta - \frac{1}{h(x)} \frac{\partial h(x)}{\partial x_i} \tag{15.5}$$

$$= \frac{\partial}{\partial x_i} [\log f_\pi(x) - \log h(x)]. \square$$

The Maximum A Posteriori (MAP) estimator, that is, the estimator that maximises $f(\theta|x)$, is obtained by considering the cost functions equal everywhere to 1 except over a small neighbourhood of the values of θ where $f(\theta|x)$ is maximum and where the cost function is set to zero. We obtain the MAP estimator by letting the size of this neighbourhood decrease to 0.

Concerning the choice of the prior distribution $\pi(\theta)$, a possible technique involves using parametric families of so-called conjugate distributions, that is, families of parameterised distributions, such that if the prior distribution $\pi(\theta)$ belongs to this parametric family, then the same is true for the posterior distribution. Obviously, the choice for the parametric family of $\pi(\theta)$ is then determined by $f(x|\theta)$. Another technique involves taking into account certain invariance properties that can be natural to consider for the distribution of θ, such as invariance by translation of the prior density of θ.

The above choices of prior distributions are not always either possible or justified. Even when they are, it often appears that the computation of integral expressions of the form $\int_\Theta \mathbb{E}[C(\theta, T(X))|\theta]\pi(\theta)d\theta$, or the maximisation of the posterior probability density function $f(\theta|x)$ cannot be performed analytically. Monte Carlo simulation methods then make it possible to solve this kind of problem numerically.

15.3 Monte Carlo Methods

Very often, the problems of integration and optimisation encountered in Bayesian estimation do not have an analytical solution. Therefore, we resort to numerical methods. The numerical methods that today seem to be the most satisfactory in many cases are simulation methods based on the generation of random variables. Their use is particularly adapted for problems of integration, with respect to a large number of variables, or to maximise functions that have a large number of local optima.

In this section, we shall briefly point out the principle of Monte Carlo methods for integration and optimisation. These methods allow problems of integration and optimisation to be solved in an approximate way, by using the simulation of random variables to generate estimators of the desired quantities. Therefore, we shall also present some elementary techniques for generating independent random variables, useful for implementing Monte Carlo methods. The techniques based on the generation of correlated random variables will be addressed later in a specific section.

15.3.1 Monte Carlo Integration

In Bayesian estimation, integral expressions such as

$$f_\pi(x) = \int_\Theta f(x|\theta)\pi(\theta)d\theta, \tag{15.6}$$

or the mean cost function

$$\int_{E,\Theta} C(\theta, T(x))f(x|\theta)\pi(\theta)dx d\theta, \tag{15.7}$$

cannot generally be computed analytically. In order to solve this kind of problem, we consider the general expression

$$I = \int_E h(x)f(x)dx, \tag{15.8}$$

where f is a Probability Density Function (PDF). We can compute a numerical approximation of I by simulating n realisations of independent random variables $X_1, \ldots, X_n$, with the same PDF $f(x)$. We then consider the estimator of I defined by

$$\hat{I} = \frac{1}{n} \sum_{k=1,n} h(X_k). \tag{15.9}$$

This is an unbiased estimator, and it is clear that its variance is given by

$$\mathrm{var}[\hat{I}] = \frac{1}{n} \left(\int_E h^2(x)f(x)dx - (\int_E h(x)f(x)dx)^2 \right). \tag{15.10}$$

We can try to reduce the variance of this estimator by using the so-called *importance sampling method*, also called the *weighted sampling method*. For this, we simulate n independent random variables $Y_1, \ldots, Y_n$, distributed according to a certain PDF denoted by g. g is called the *importance distribution*. We then use the estimator

$$\tilde{I} = \frac{1}{n} \sum_{k=1,n} h(Y_k) \frac{f(Y_k)}{g(Y_k)}, \tag{15.11}$$

which is again an unbiased estimator of I. This method also enables a density g, which is simpler to simulate than f, to be used.

Theorem 15.2 *The PDF g, which minimises the variance of $\tilde{I}$, is*

$$\hat{g}(x) = \frac{|h(x)|f(x)}{\int_E |h(y)|f(y)dy}. \tag{15.12}$$

Proof We first note that from Cauchy-Schwarz's inequality, for any PDF g,

$$\left(\int_E |h(y)|f(y)dy \right)^2 \le \int_E h^2(y) \frac{f^2(y)}{g(y)} dy \int_E g(y)dy \tag{15.13}$$

$$\le \int_E h^2(y) \frac{f^2(y)}{g(y)} dy,$$

as long as the previous integrals are defined. Consequently, the variance

$$\frac{1}{n} \left(\int_E \frac{h^2(y)f^2(y)}{g(y)} dy - \left(\int_E h(y)f(y)dy \right)^2 \right) \tag{15.14}$$

of $\tilde{I}$ is minimal when g is proportional to $|h(y)|f(y)$, that is, taking into account the normalisation condition $\int_E g(y)dy = 1$, $g = \hat{g}$. $\square$

We shall come back to the importance sampling method later on, when considering the use of Monte Carlo methods for the problem of state space model filtering.

Another approach, known as Riemann's simulation by sums, involves approaching the integral I by an expression of the type

$$\sum_{k=0,n-1} h(z_k)(z_{k+1} - z_k), \tag{15.15}$$

where the coefficients z_k are obtained by performing n independent simulations using the PDF f, and rearranging the values obtained in increasing order. This method permits an estimator of I with reduced variance to be obtained for simple integrals, but becomes less interesting for multiple integrals.

15.3.2 Optimisation

For an optimisation problem of the form

$$\max_{\theta \in \Delta} h(\theta), \tag{15.16}$$

it may be more efficient to use a simulation method than conventional descent techniques (which are introduced at the beginning of Chapter 16). This is particularly true when the function h or the domain Δ are not regular, or when h has local optima.

Optimisation of functions of the form $h(\theta) = \mathbb{E}[H(X,\theta)]$ will be more specifically studied in the next chapter.

Before considering the simulated annealing method, which represents a general simulation-based optimisation tool, we indicate a technique for maximising likelihood functions, which involves the notion of a complete-data model.

EM (Expectation Maximisation) Algorithms We shall now present an important method known as the EM algorithm. When we want to maximise the probability density $f(x|\theta)$, the optimisation is sometimes simplified when $f(x|\theta)$ can be seen as a marginal density of a PDF $f(x,y|\theta)$, which is called the *complete* PDF. We shall denote by F the space in which y takes its values. In this kind of situation, we can search for a maximum of $f(x|\theta)$ by means of the following algorithm, called the EM algorithm [68]:

1. compute $Q(\theta, \theta_n) = \mathbb{E}[\log(f(x,y|\theta))|x, \theta_n]$:

$$Q(\theta, \theta_n) = \int_F \log f(x, y|\theta) f(y|x, \theta_n) dy, \tag{15.17}$$

2. maximise $Q(\theta, \theta_n)$: $\theta_{n+1} = \arg\max_\theta Q(\theta, \theta_n)$.

Theorem 15.3 *For a strictly positive bounded PDF $f(x|\theta)$, the EM algorithm converges towards a local maximum of the likelihood.*

Proof We note that

$$Q(\theta, \theta') = \int_F \log f(x|\theta) f(y|x, \theta') dy + \int_F \log f(y|x, \theta) f(y|x, \theta') dy$$

$$= \log f(x|\theta) + H(\theta, \theta'). \tag{15.18}$$

Since $f(x|\theta)$ is bounded, it is sufficient to show that

$$f(x|\theta_{n+1}) \geq f(x|\theta_n). \tag{15.19}$$

Since $\log f(x|\theta) = Q(\theta, \theta') - H(\theta, \theta')$, this amounts to showing that

$$Q(\theta_{n+1}, \theta_n) - H(\theta_{n+1}, \theta_n) \geq Q(\theta_n, \theta_n) - H(\theta_n, \theta_n). \tag{15.20}$$

As $Q(\theta_{n+1}, \theta_n) \geq Q(\theta_n, \theta_n)$, it therefore suffices to check that $H(\theta_{n+1}, \theta_n) - H(\theta_n, \theta_n) \leq 0$. But from Jensen's inequality (see Appendix A),

$$H(\theta_{n+1}, \theta_n) - H(\theta_n, \theta_n) = \int_F \log \frac{f(y|x, \theta_{n+1})}{f(y|x, \theta_n)} f(y|x, \theta_n) dy$$

$$\leq \log \left(\int_F \frac{f(y|x, \theta_{n+1})}{f(y|x, \theta_n)} f(y|x, \theta_n) dy \right) \qquad (15.21)$$

$$\leq 0.$$

Moreover, we note that this is in fact a strict inequality, except if we have for almost every y, with respect to the measure with density $f(y|x, \theta_n)$, $f(y|x, \theta_{n+1}) = f(y|x, \theta_n)$, which completes the proof. $\square$

There are modified versions of this algorithm that make use of the simulation of random variables. In particular, when the direct calculation of the integral $Q(\theta_n, \theta)$ is not possible or difficult, step 1) can be replaced by a simulation step of the complete data y conditionally to θ_n and to the observation x. Let y_n be the vector thus simulated. The function optimised in step 2) is then the likelihood of the complete data $f(x, y_n|\theta)$. The principle of this approach is to construct a Markov chain whose mean is asymptotically equal to the maximum likelihood estimator of the parameters. This method is known as the SEM (Stochastic EM) algorithm [66].

For exponential models, the opposite of the log-likelihood of the complete data model is of the form

$$L(z|\theta) = -\psi(\theta) + < S(x, y), \phi(\theta) >, \qquad (15.22)$$

As

$$\mathbb{E}[L(X, Y|\theta)] = -\psi(\theta) + < \mathbb{E}[S(X, Y)], \phi(\theta) >, \qquad (15.23)$$

the idea here is to construct a stochastic approximation of $\mathbb{E}[S(X, Y)]$: at the $n + 1^{\text{th}}$ iteration of the algorithm, Y is simulated conditionally to θ_n and to $X = x$, and the value y_n obtained is used to update the estimator of $\mathbb{E}[S(X, Y)]$ by the formula

$$S_{n+1} = (1 - \gamma_n) S_n + \gamma_n S(x, y_n). \qquad (15.24)$$

The coefficients $(\gamma_n)_{n \in \mathbb{N}}$ here represent a sequence of real step coefficients, decreasing towards 0. θ is then updated by

$$\theta_{n+1} = \arg \min_\theta [-\psi(\theta) + < S_{n+1}, \phi(\theta) >]. \qquad (15.25)$$

This method is known as the SAEM (Stochastic Approximation EM) algorithm [67]. The iterative algorithms of the form (15.24) will be studied with more details in the next chapter. A study of the convergence of the SAEM algorithm can also be found in [73].

Simulated Annealing We now consider criteria to be optimised that can be more general than likelihood functions. The simulated annealing method makes it possible to avoid convergence towards the local minima of the criterion of the function to be minimised, denoted by $h(\theta)$, which we shall assume has positive values. For this, the algorithm generates a sequence of values θ_n such that the sequence $h(\theta_n)$ can grow locally, in order to escape from local minima and converge towards a global minimum of $h(\theta)$. In order to do this, we choose a prior distribution $g(\theta)$, and θ_{n+1} is chosen from θ_n in the following way:

1. simulate x from $g(x - \theta_n)$,
2. take $\theta_{n+1} = x$ with the probability

$$\min\{\exp(-T_n^{-1}[h(x) - h(\theta_n)]), 1\}, \tag{15.26}$$

   and $\theta_{n+1} = \theta_n$, otherwise,
3. decrease T_n in T_{n+1}.

Results exist, which specify the conditions of decreasing towards 0 of sequence $(T_n)_{n \in \mathbb{N}}$ that ensure the convergence towards a global optimum of the problem [77].

In order to understand the way this algorithm works, we shall indicate below the link existing between the simulated annealing method, and the simulation of random variables by Markov chain algorithms.

15.3.3 Simulation of Random Variables

Uniform Random Variables Techniques for the simulation of random variables of general distributions make use of simulation of uniform random variables. In order to generate a random variable with a uniform distribution on $[0, 1]$, denoted by $\mathcal{U}_{[0,1]}$, we can basically use two methods. The first one involves constructing a sequence of the form

$$x_{n+1} = ax_n + b \bmod[M + 1]. \tag{15.27}$$

We thus obtain a periodic sequence with integer values in $[0, M]$. The realisations of the variable of distribution $\mathcal{U}_{[0,1]}$ are then simulated by the values $M^{-1}x_n$.

The other method involves constructing a sequence of binary vectors $\mathbf{x}_n$ by the relations

$$\mathbf{x}_{n+1} = T\mathbf{x}_n, \tag{15.28}$$

where T is a binary matrix with input values 0 or 1. The addition and multiplication operations are carried out in base 2. We often use matrices T that have a Toeplitz structure, that is, whose terms are constant along a same parallel to the diagonal. If l is the size of the vectors $\mathbf{x}_n$, the uniform variables are simulated by $2^{-l}x_n$, where x_n is the scalar decimal representation associated with the binary vector $\mathbf{x}_n$.

Of course, to obtain a good generator, the periodicity of the sequences created must be as long as possible, which is obtained by an appropriate choice of the values a or b in the first case, or of T in the second. We can increase the period of the sequences generated by mixing the two above-mentioned approaches [77].

Simulation of General Distributions To simulate more general distributions, the first idea involves noting that if X is a random variable, with bijective distribution function F from $\mathbb{R}$ (or an interval of $\mathbb{R}$) onto $]0, 1[$, then if $U \sim \mathcal{U}_{[0,1]}$ the random variable $F^{-1}(U)$ has the same distribution as X, that is, its distribution function is equal to F.

However, the calculation of F^{-1} is not always possible. Therefore, we often use the so-called acceptance-rejection method, of more general use, described in the following theorem:

Theorem 15.4 *Let X be a random variable with PDF $f(x)$, and let us assume that there exists an auxiliary PDF g and a constant M such that*

$$\forall x \in E, \ f(x) \le Mg(x). \tag{15.29}$$

Then, the acceptance-rejection procedure, defined by

1. generate $y \sim g$, and $u \sim \mathcal{U}_{[0,1]}$,

2. accept $z = y$ if $u \le f(y)(Mg(y))^{-1}$, and return to 1), otherwise,

provides a realisation z of a random variable Z, whose distribution is the same as that of X.

Proof We first note that

$$P(\{Y_k \text{ accepted}\}) = \mathbb{E}[\mathbb{E}[\mathbb{1}_{\{U \le f(Y_k)(Mg(Y_k))^{-1}\}}|Y_k]]$$
$$= \frac{1}{M}, \tag{15.30}$$

and

$$P(\{Y_k \text{ accepted}\} \cap \{Y_k \le z_0\}) = \mathbb{E}[\mathbb{E}[\mathbb{1}_{\{U \le f(Y_k)(Mg(Y_k))^{-1}\} \cap \{Y_k \le z_0\}}|Y_k]]$$
$$= \mathbb{E}[\mathbb{1}_{\{Y_k \le z_0\}} \frac{f(Y_k)}{Mg(Y_k)}]$$
$$= \frac{1}{M} P(X \le z_0). \tag{15.31}$$

Therefore,

$$P(Z \leq z_0)$$

$$= \sum_{i=0,\infty} P(\{Y_1, \ldots, Y_{i-1} \text{ rejected}\}) P(\{Y_i \text{ accepted}\} \cap \{Y_i \leq z_0\})$$

$$= \sum_{i=0,\infty} \left(1 - \frac{1}{M}\right)^i \frac{1}{M} P(X \leq z_0)$$

$$= P(X \leq z_0) \cdot \square$$

$$(15.32)$$

This method suffers from several limitations. In particular, the constant M^{-1} must sometimes be too small for the algorithm to be of practical interest. Indeed, if the rate of acceptance M^{-1} of the variables generated is very small, the algorithm will be too slow to be able to provide a sample of sufficient size.

Moreover, it is not always possible to compute the value of M, and therefore to implement the algorithm. Thus, in the context of Bayesian estimation, if we wish to simulate according to the density $f(\theta|x)$, and we know how to simulate according to the density $\pi(\theta)$, we can take $g = \pi$. Then, $M \geq f(\theta|x)\pi^{-1}(\theta)$, that is $M \geq f(x|\theta)f_\pi^{-1}(x)$, since $f(\theta|x) = f(x|\theta)\pi(\theta)f_\pi^{-1}(x)$. But, unfortunately, $f_\pi(x)$ is often unknown since we have seen that it involves the calculation of an integral normalisation factor, which prevents the computation of M.

15.4 MCMC (Monte Carlo Markov Chain) Methods

We have underlined our interest in the simulation of random variables for integration and optimisation in Bayesian estimation problems. MCMC simulation techniques have been developed to overcome the limitations of the the simulation of independent random variables. These techniques provide sequences $(x_n)_{n \in \mathbb{N}}$, which represent realisations of a Markov chain. In these conditions we may, for example, ask ourselves under what conditions estimators like $n^{-1} \sum_{k=1,n} h(X_n)$ converge towards $\int_E h(x)f(x)dx$, where f is the PDF that we wish to simulate.

We shall now recall some basics about Markov chains that will allow us to specify these conditions. Then, we shall present two important simulation techniques, the Metropolis-Hastings and the Gibbs algorithms. When the sequences generated provide convergent estimators for Monte Carlo methods, this kind of algorithm is known as an *MCMC* method.

15.4.1 Markov Chains

We shall assume that the random variables under consideration have values in E (often, $E = \mathbb{R}$ or $E = \mathbb{R}^m$). We recall that a Markov chain is a process

$X = (X_n)_{n \in \mathbb{N}}$, defined on a probability space $(\Omega, \mathcal{A}, P)$, which satisfies $\forall A \in \mathcal{B}(E)$,

$$P(X_{n+1} \in A | X_n = x_n, \ldots, X_0 = x_0) = P(X_{n+1} \in A | X_n = x_n). \quad (15.33)$$

Before specifying the convergence properties of Monte Carlo algorithms based on the simulation of Markov chains, we shall recall some important notions about Markov chains. In particular, these results make it possible to give conditions under which, for a Markov chain $X = (X_n)_{n \in \mathbb{N}}$, the distribution of X_n converges in some sense towards a fixed distribution π. We shall also look at the convergence of sequences of the form $n^{-1} \sum_{k=1,n} h(X_k)$ towards $\int_E h(x) d\pi(x)$.

It would take far too long to present the proofs of the results mentioned in this section. Reference [75] can be consulted for a detailed presentation.

Homogeneous processes When $P(X_{n+1} \in A | X_n = x_n)$ does not depend on n, we speak of a homogeneous Markov chain. Unless specifically stated, we shall always assume that the Markov chains considered are homogeneous.

Transition kernel For a homogeneous Markov chain X, the function K defined by

$$\forall x \in E, \forall A \in \mathcal{B}(E), \; K(x, A) = P(X_{n+1} \in A | X_n = x) \quad (15.34)$$

is called the *transition kernel* of X. A homogeneous Markov chain is characterised by knowledge of K and of the probability measure of X_0.

Stationary distribution A transition kernel K has a stationary measure π if

$$\forall A \in \mathcal{B}(E), \; \pi(A) = \int_E K(x, A) d\pi(x). \quad (15.35)$$

If π is a probability measure, we speak of a stationary distribution. In this case, if K is the transition kernel of X and if π represents the distribution of X_n, this will also be the distribution of X_{n+1}.

Reversibility We say that K is π-reversible if

$$\forall A, B \in \mathcal{B}(E), \; \int_B K(x, A) d\pi(x) = \int_A K(x, B) d\pi(x). \quad (15.36)$$

We note that in this case, π is a stationary measure of K; indeed, letting $B = E$ yields relation (15.35). Consequently, if $X_0 \sim \pi$, we then have for all $n \geq 0$

$$P(\{X_{n+1} \in A\} \cap \{X_n \in B\}) = P(\{X_n \in A\} \cap \{X_{n+1} \in B\}), \quad (15.37)$$

and

$$P(X_{n+1} \in A | X_n \in B) = P(X_n \in A | X_{n+1} \in B). \quad (15.38)$$

It then appears that the distribution of X is not modified when considering the chain in the reverse direction.

Irreducibility Let φ be a probability measure. We say that X is φ-irreducible if

$$\forall x \in E, \forall A \in \mathcal{B}(E), \ \varphi(A) > 0 \Rightarrow \exists n > 0, \ P(X_n \in A | X_0 = x) > 0. \tag{15.39}$$

φ-irreducibility means that all the non-zero measure sets for φ can be reached by X, independently of the value taken by X_0.

The following result shows that we can construct measures, called *maximal irreducibility measures*, which characterise the irreducibility property of a Markov chain X more systematically than any particular measure φ for which X is φ-irreducible.

Theorem 15.5 *There exists a measure ψ such that*

- *X is ψ-irreducible,*
- *every measure φ such that X is φ-irreducible is absolutely measurable with respect to ψ,*
-

$$\forall A \in \mathcal{B}(E), \ \psi(A) = 0 \Rightarrow \psi(\{y; \exists n, P(X_n \in A | X_0 = y) > 0\}) = 0. \tag{15.40}$$

Proof See [75] p.88.

In what follows, ψ will always denote any maximum irreducibility measure of X.

Aperiodicity A ψ-irreducible Markov chain X is aperiodic if there exists no partition of E of the form $\{D_1, \ldots, D_p, N\}$, with $\psi(N) = 0$, such that

$$\forall n > 0, \ P(X_{n+1} \in D_{k+1} | X_n \in D_k) = 1, \tag{15.41}$$

with the notation $D_k = D_l$, for $k = l \ \text{mod}[p]$.

The aperiodicity hypothesis is important to ensure the convergence of the sequence of the distributions of the variables X_k. Indeed, consider, for instance, the case where conditionally to $X_0 \in D_i$ the support of the distribution of X_k is in D_{i_k}, where $i_k = (i + k) \ \text{mod}[p]$. Then, the distribution of X_k does not converge.

We can show that if for a certain value $x \in E$, $K(x, \{x\}) > 0$, or if the measure $K(x, .)$ has a strictly positive density on a neighbourhood V_x of x such that $\psi(V_x) > 0$, then X is aperiodic ([75] pp.116-118).

We note that if a kernel K is not aperiodic, it is, however, possible to construct an aperiodic kernel K_ε, defined by

$$\forall x \in E, \forall A \in \mathcal{B}(E), \ K_\varepsilon(x, A) = (1 - \varepsilon) \sum_{k=0,\infty} \varepsilon^k P(X_k \in A | X_0 = x), \tag{15.42}$$

with $0 < \varepsilon < 1$ ([75] p.118). Note that π is a stationary distribution of K if and only if π is a stationary distribution of K_ε ([75] p.241).

The few properties that we have just presented and which are generally fairly easy to satisfy when simulating Markov chains allow us to present the following result:

Theorem 15.6 *Let X be a ψ-irreducible Markov chain with transition kernel K and invariant distribution π. Then, for any π-measurable function h,*

$$\forall x_0 \in E, \; P[\lim_{n \to \infty} (n^{-1} \sum_{k=1,n} h(X_k) = \int_E h(x)d\pi(x)) | X_0 = x_0] = 1,$$

(15.43)

π-almost surely for any value of x_0.

Proof See [75] p.411.

In order to specify the convergence of the distribution of X_k, we introduce the notion of recurrence, which is stronger than the notion of irreducibility since it requires that in a certain sense the trajectories of X reach infinitely many times any set of A of $\mathcal{B}(E)$ such that $\psi(A) > 0$.

Recurrent chains A ψ-irreducible Markov chain X is said to be recurrent if

$$\forall A \in \mathcal{B}(E), \; \forall x \in E, \; \psi(A) > 0 \Rightarrow \mathbb{E}[\sum_{k=1,\infty} \mathbb{I}_A(X_k) | X_0 = x] = +\infty,$$

(15.44)

and Harris recurrent if

$$\forall A \in \mathcal{B}(E), \; \forall x \in A, \; \psi(A) > 0 \Rightarrow P[\sum_{k=1,\infty} \mathbb{I}_A(X_k) = +\infty | X_0 = x] = 1.$$

(15.45)

In fact, it can be shown that for the Harris recurrence, the implication (15.45) is satisfied $\forall x \in E$ as long as it is satisfied $\forall x \in A$ (see [75] p.204). The recurrence property indicates that the trajectories of X reach A infinitely many times on average, and the Harris recurrence that they reach A infinitely many times almost surely.

An important property of recurrent chains is that they have an invariant measure, unique up to one factor ([75] p.242), which defines a probability measure equivalent to any maximum irreducible measure ψ ([75] p.245). Conversely, if X is ψ-irreducible and has an invariant probability measure, then X is recurrent ([75] p.231). Consequently, this invariant probability measure is unique.

We now remark that if the distribution of X_0 is the measure μ_0, the distribution of X_n is the measure μ_n defined by

$$\mu_n \ : \ A \in \mathcal{B}(E) \to \int_E P(X_n \in A | X_0 = x) d\mu_0(x). \tag{15.46}$$

In particular, for $\mu_0 = \delta_{x_0}$, $\mu_n(A) = P(X_n \in A | X_0 = x_0)$. We then have the following result:

Theorem 15.7 *Let* X *be a Harris recurrent Markov chain with invariant distribution* π. *For any choice of the distribution* μ_0 *of* X_0, *the distribution* μ_n *of* X_n *converges towards* π *in the sense of the total norm variation:*

$$\lim_{n \to \infty} \left(\sup_{A \in \mathcal{B}(E)} |\mu_n(A) - \pi(A)| \right) = 0. \tag{15.47}$$

Proof See [75] p.383.

Note that there also exist weaker convergence results such that:

Theorem 15.8 *If* X *is irreducible, with invariant distribution* π,

$$\lim_{n \to \infty} \left(\sup_{A \in \mathcal{B}(E)} |\frac{1}{n} \sum_{k=1,n} \mu_k(A) - \pi(A)| \right) = 0. \tag{15.48}$$

If, moreover, the chain is aperiodic, the relation

$$\lim_{n \to \infty} \left(\sup_{A \in \mathcal{B}(E)} |P(X_n \in A | X_0 = x_0) - \pi(A)| \right) = 0 \tag{15.49}$$

is satisfied π-*almost surely for any choice* x_0 *of* X_0.

Proof See [80] pp.64-65.

Ergodicity Unfortunately, the above-mentioned theorems do not inform us about the speed of the convergence of the approximations $n^{-1} \sum_{k=1,n} h(X_k)$ towards $\int_E h(x) \pi(x) dx$, or of the distribution μ_n of X_n towards π, and we often want to check whether X has a geometrical ergodic property, that is, if there exist a function C and a constant α $(0 \le \alpha < 1)$, such that

$$\forall x_0 \in E, \forall n \in \mathbb{N}, \ \sup_{A \in \mathcal{B}(E)} |P[X_n \in A | X_0 = x_0] - \pi(A)| \le C(x_0) \alpha^n. \tag{15.50}$$

When a Markov chain X has an invariant probability measure, is Harris recurrent, and satisfies the geometrical ergodicity condition (15.50), it is said to be *geometrically ergodic*. If, in addition, $C(x)$ does not depend on x, it is said to be *uniformly ergodic*.

These properties are often difficult to prove, in which case in practice we limit ourselves to testing the convergence by verifying that the estimates of the desired quantities no longer evolve significantly after a certain number of iterations. Conditions under which geometrical convergence is guaranteed can be found in Chapter 15 of [75].

When geometrical ergodicity is satisfied, it enables the following central limit theorem to be presented:

Theorem 15.9 *Let X be a geometrically ergodic Markov chain with invariant distribution π and h any function such that $\int |h(x)|^{2+\varepsilon} d\pi(x) < \infty$, for some $\varepsilon > 0$. We define*

$$\sigma_h^2 = \text{var}_\pi[h(X_0)] + 2 \sum_{k=0,\infty} \text{cov}_\pi[h(X_0), h(X_k)], \tag{15.51}$$

where var_π and cov_π denote the variance and autocovariance under the hypothesis $X_0 \sim \pi$. If $\sigma_h^2 < \infty$,

$$\lim_{n \to \infty} \sqrt{n}[n^{-1} \sum_{k=1,n} h(X_k) - \int_E h(x) d\pi(x)] \sim \mathcal{N}(0, \sigma_h^2). \tag{15.52}$$

Proof See [80] p.68.

If the ergodicity is uniform, we have the same result under the weaker condition,

$$\int |h(x)|^2 d\pi(x) < \infty. \tag{15.53}$$

We notice that σ_h^2 is the power spectral density at the frequency 0 of the process Y defined by $Y_k = h(X_k)$ under the hypothesis $X_0 \sim \pi$. This makes it possible, for example, to estimate σ_h^2 by considering a smooth estimator of the periodogram of Y taken at the frequency 0.

It should be noted that the hypotheses of the above-mentioned theorem are very strong. For reversible Markov chains, the hypotheses of Theorem 15.9 can be weakened; indeed, if X has a stationary probability distribution π, and is reversible, irreducible, and aperiodic, Theorem 15.9 applies [77].

15.4.2 Metropolis-Hastings Algorithm

The Metropolis-Hastings algorithm involves a conditional distribution, called the proposal distribution, denoted by q. The algorithm is written in the form

1. generate $y_k \sim q(y|x_k)$,
2.

$$x_{k+1} = \begin{cases} y_k \text{ with the probability } \rho(x_k, y_k) \\ x_k \text{ with the probability } 1 - \rho(x_k, y_k), \end{cases} \tag{15.54}$$

$$\text{with } \rho(x_k, y_k) = \min\{\frac{f(y_k)q(x_k|y_k)}{f(x_k)q(y_k|x_k)}, 1\}.$$

This algorithm makes it possible to generate a realisation $(x_k)_{k\in\mathbb{N}}$ of a Markov chain $X = (X_k)_{k\in\mathbb{N}}$ whose stationary distribution is f:

Theorem 15.10 *Whatever the choice of q, the target distribution f is a stationary distribution of the sequence* $X = (X_k)_{k \in \mathbb{N}}$ *generated. We assume, moreover, that f is bounded and strictly positive on any compact connected set of E. Then, if there exists* $\varepsilon, \alpha > 0$ *such that*

$$|x - y| < \alpha \Rightarrow q(y|x) > \varepsilon,\tag{15.55}$$

the Markov chain X is f-irreducible and aperiodic.

Proof See Appendix X.

It appears clearly that the Metropolis-Hastings algorithm only requires knowledge of f up to one constant factor. This remarkable property is very useful in Bayesian estimation for simulating posterior distributions that are known up to one integral normalising factor.

We note that it is easy to construct distributions q that satisfy condition (15.55) when $q(y|x)$ is of the form $g(y - x)$.

In addition, when q does not depend on x, that is, $q(y|x) = g(y)$, and when there exists M such that $f(x) \leq Mg(x)$ on the support of f, then it can be shown that X is uniformly ergodic [77]:

$$\sup_{A \in \mathcal{B}(E)} |P(X_n \in A|X_0 = x_0) - \int_A f(x)dx| \leq (1 - \frac{1}{M})^n.\tag{15.56}$$

Acceptance Rate A good behaviour of the Metropolis-Hastings algorithm is the result of a compromise in the choice of q. Thus, if, for example,

$$q(y|x) = g(y - x)$$
$$= g(x - y),\tag{15.57}$$

it appears that a high acceptance rate, necessary for fast convergence of the algorithm, can be achieved if g is such that y_k and x_k are close. In this case, when f is continuous, we do have

$$\rho(x_k, y_k) = \min(f(x_k)f(y_k)^{-1}, 1)$$
$$\approx 1,\tag{15.58}$$

which means that the whole support of π is covered very slowly by the sequence of realisations x_k. In fact, a heuristic rule leads to choosing an acceptance rate of the order of $1/2$ or $1/4$ for models with dimensions higher than 2 (see, for example, [77]).

Links with Simulated Annealing It is easy to see that if we leave the temperature parameter T constant in the simulated annealing algorithm described above, we obtain a particular Metropolis-Hastings algorithm for which $q(y|x) = q(x|y)$, and with corresponding target PDF

$$f(x) = C_T \exp(-T^{-1}h(x)),\tag{15.59}$$

where C_T is a constant that ensures that f is a PDF. For small values of T, the significant values of f tend to be concentrated in the neighbourhood of the maxima of $h(x)$. Therefore, when T is very small the simulated values tend to be distributed close to the maxima of $h(x)$. However, the convergence results of the Metropolis-Hastings algorithm do not directly apply to the simulated annealing algorithm: since T varies when running the algorithm, the simulated Markov chain is not homogeneous.

15.4.3 Gibbs Sampler

In this section, we shall denote $\{y_a, y_{a+1}, \ldots, y_b\}$ by $y_{a:b}$. The principle of the Gibbs algorithm is as follows: let f denote the PDF to be simulated, and g a PDF such that

$$\int g(x, z)dz = f(x). \tag{15.60}$$

We note $Y = (X, Z) = [Y_1, \ldots, Y_p]^T$, and we assume that we know how to easily simulate the conditional distributions

$$g_1(y_1|y_{2:p}), g_1(y_2|y_1, y_{3:p}), \ldots, g_p(y_p|y_{1:p-1}). \tag{15.61}$$

The Gibbs sampling algorithm involves constructing a sequence of realisations $(y_{k,1:p})_{k \in \mathbb{N}}$ in the following way:

$$y_{k+1,1} \sim g_1(y_1|y_{k,2:p}),$$

$$y_{k+1,2} \sim g_2(y_2|y_{k+1,1}, y_{k,3:p}),$$

$$\ldots \qquad \ldots, \tag{15.62}$$

$$y_{k+1,p} \sim g_p(y_p|y_{k+1,1:p-1}).$$

Here, we can see that the choice of g is restricted by the fact that f must be a marginal density of g. Note furthermore that the components $y_{k,i}$ of y can be vectors.

Theorem 15.11 *Whatever the choice of g, the Gibbs algorithm simulates a Markov chain $Y = ([Y_{k,1}, \ldots, Y_{k,p}]^T)_{k \in \mathbb{N}}$ of which g is the stationary distribution. f therefore represents the stationary distribution of the subchain $X = (X_k)_{k \in \mathbb{N}}$. Moreover, if there exist $\varepsilon, \alpha > 0$ such that, $\forall i = 1, p$,*

$$|y - y'| < \alpha \Rightarrow g_i(y_i|y_{1:i-1}, y'_{i+1:p}) > \varepsilon, \tag{15.63}$$

then the Markov chain Y is g-irreducible and aperiodic (X is therefore f-irreducible and aperiodic). Furthermore, if $g_k(y_k|y_{i \neq k}) > 0$ for any value of y and of k, the chain is reversible.

Proof See Appendix Y.

It appears, therefore, that the Gibbs sampler allows us to simulate f as long as we know how to simulate the conditional densities $(g_k)_{k=1,p}$.

Remark When the support of g is not connected, we have a similar result, but on the condition that we can connect any two points of the support of g by a finite sequence of balls B_i with radii $\alpha_i < \alpha$, such that $\int_{B_i \cap B_{i+1}} g(y)dy > 0$ [77]. The proof is then achieved following the same approach as that of Theorem 15.11. This remark can also apply to the study of the irreducibility and aperiodicity of the Metropolis-Hastings algorithms.

The reversibility condition given in the theorem is quite cumbersome. To ensure the reversibility of the chain, it in fact suffices to update each component of y not cyclically, but according to a permutation of $\{1, \dots, p\}$ chosen randomly at the beginning of each loop of the algorithm [74].

Theorem 15.11 shows that we can simulate according to the distribution of a random vector by iteratively simulating each of its components conditionally to the others. We note that we do not find a similar property in deterministic optimisation problems in the sense that if we optimise a function of several variables iteratively with respect to each of the variables, convergence towards the optimum is not achieved in general. The property that we have here comes from the fact that knowledge of the conditional distributions g_k entirely characterises the distribution g, up to one factor. Indeed, we show (see Appendix Y) that for a fixed $y' \in E^p$

$$g(y_{1:p}) = \prod_{k=1,p} \frac{g_k(y_k | y_{1:k-1}, y'_{k+1:p})}{g_k(y'_k | y'_{1:k-1}, y_{k+1:p})} g(y'_{1:p}), \tag{15.64}$$

on condition that the ratios that appear in this relation are defined.

For practical use of a subchain of interest X of the simulated chain $Y = (X, Z)$, it may seem preferable to use $\mathbb{E}[X|Z]$ rather than X. Indeed, from Jensen's inequality (see Appendix A),

$$\text{var}[\mathbb{E}[h(X)|Z]] \leq \text{var}[h(X)], \tag{15.65}$$

which suggests that estimators of $\int_E h(x)f(x)dx$ of the form

$$\frac{1}{n} \sum_{k=1,n} \mathbb{E}[h(X)|Z_k] \tag{15.66}$$

could have a weaker variance than

$$\frac{1}{n} \sum_{k=1,n} h(X_k). \tag{15.67}$$

This is evident in the particular case where the random variables X_k are uncorrelated. The conditions under which using this technique, called *Rao-Blackwellisation*, is justified, can be found in [77].

Let us also note that it will be in our interest to jointly simulate strongly correlated variables, in order to ensure better performance of the algorithm.

Finally, to end this section, we note the existence of hybrid simulation techniques that involve replacing certain steps, of the type

- generate $y_{n+1,k} \sim g_k(y_k|y_{n+1,1:k-1}, y_{n,k+1:p})$

by Metropolis-Hastings simulation steps for conditional distributions q_k:

1. generate $\tilde{y}_{n+1,k} \sim q_k(y_k|y_{n+1,1:k-1}, y_{n,k+1:p})$,
2.

$$y_{n+1,k} = \begin{cases} \tilde{y}_{n+1,k} \text{ with the probability } \rho \\ y_{n,k} \text{ with the probability } 1 - \rho, \end{cases} \tag{15.68}$$

with $\rho = \min\{1, \dfrac{g_k(\tilde{y}_{n+1,k}|y_{n+1,1:k-1}, y_{n,k+1:p})}{g_k(y_{n,k}|y_{n+1,1:k-1}, y_{n,k+1:p})}$

$$\times \dfrac{q_k(y_{n,k}|y_{n+1,1:k-1}, y_{n,k+1:p})}{q_k(\tilde{y}_{n+1,k}|y_{n+1,1:k-1}, y_{n,k+1:p})}\}. \tag{15.69}$$

[77] can be referred to for certain properties of this algorithm, which is useful when some of the conditional distributions g_k are difficult to simulate. In certain situations, the variance of Monte Carlo estimators constructed with this hybrid method is weaker than that obtained from the corresponding Gibbs algorithm.

Remark MCMC algorithms generally provide samples that are distributed according to the target distribution only asymptotically. Therefore, in practice, the first generated samples should be discarded, only taking into account those that are obtained when the algorithm has roughly converged.

15.5 Application to Filtering

We consider a general state space model of the form

$$\begin{cases} X_k = F(X_{k-1}, V_k) \\ Y_k = G(X_k, W_k). \end{cases} \tag{15.70}$$

For such a model, the problems of filtering, prediction, or smoothing, that we presented in Chapter 9 in the context of the extensions of Kalman filtering, do not generally have any closed form solution, but only a solution in an integral form.

It is, however, possible to provide approximate numerical solutions to these problems by using the importance sampling technique that we presented in Section 15.3.1. In the case of the filtering problem to which we shall limit ourselves (see [70] for problems of prediction and smoothing), we shall proceed by generating an empirical estimator of the posterior PDF $p(x_{0:n}|y_{0:n})$. We shall then be able to compute estimators of integral expressions of the form

$$\int_{E^{n+1}} h_n(x_{0:n}) p(x_{0:n}|y_{0:n}) dx_{0:n}. \tag{15.71}$$

We shall assume here that the PDFs $p(x_k|x_{k-1})$ and $p(y_k|x_k)$ are available, and that it is possible to simulate according to the density $p(x_k|x_{k-1})$. As it is often impossible to simulate directly according to the density $p(x_{0:n}|y_{0:n})$, we usually use the importance sampling method to evaluate (15.71). But, unlike the presentation in Section 15.3.1, this method is here implemented for a sequence of random variables that are not independent, which is theoretically justified by additional convergence results that are presented in the next section.

15.5.1 Importance Sampling

Let $g(x_{0:n}|y_{0:n})$ be the so-called importance sampling PDF chosen to simulate $p(x_{0:n}|y_{0:n})$ by using the importance sampling method. To estimate the integral

$$\begin{aligned} I(h_n) &= \int_{E^{n+1}} h_n(x_{0:n}) p(x_{0:n}|y_{0:n}) dx_{0:n} \\ &= \int_{E^{n+1}} h_n(x_{0:n}) \frac{p(x_{0:n}|y_{0:n})}{g(x_{0:n}|y_{0:n})} g(x_{0:n}|y_{0:n}) dx_{0:n}, \end{aligned} \tag{15.72}$$

we simulate N realisations of $X_{0:n}$ according to the PDF $g(x_{0:n}|y_{0:n})$, denoted by $(x_{0:n}^{(i)})_{i=1,N}$.

The integral $I(h_n)$ is then estimated by

$$\hat{I}_N^*(h_n) = \frac{1}{N} \sum_{i=1,n} h_n(x_{0:n}^{(i)}) w_n^{*(i)}, \tag{15.73}$$

with

$$\begin{aligned} w_n^{*(i)} &= \frac{p(x_{0:n}^{(i)}|y_{0:n})}{g(x_{0:n}^{(i)}|y_{0:n})} \\ &= \frac{p(y_{0:n}|x_{0:n}^{(i)}) p(x_{0:n}^{(i)})}{p(y_{0:n}) g(x_{0:n}^{(i)}|y_{0:n})}. \end{aligned} \tag{15.74}$$

As is often the case, the computation of the constant factor $p(y_{0:n})^{-1}$ cannot be performed, so we compute

$$w_n^{(i)} = p(y_{0:n})w_n^{*(i)}$$

$$(15.75)$$

$$\text{and}\ \ \tilde{w}_n^{(i)} = w_n^{(i)}(\textstyle\sum_{j=1,N} w_n^{(j)})^{-1},$$

and we estimate $I(h_n)$ by

$$\tilde{I}_N(h_n) = \sum_{i=1,N} h_n(x_{0:n})\tilde{w}_n^{(i)}.$$

$$(15.76)$$

The coefficients $\tilde{w}_n^{(i)}$ are called the *importance weights*. The normalisation of the coefficients $w_n^{(i)}$ is made clear by noting that

$$\mathbb{E}[W_n] = \mathbb{E}[\frac{p(X_{0:n}|y_{0:n})}{g(X_{0:n}|y_{0:n})}],$$

$$(15.77)$$

and therefore $\mathbb{E}[W_n] = 1$, since $X_{0:n} \sim g(x_{0:n}|y_{0:n})$.

Under some conditions that are generally easy to satisfy [71], we show that $\tilde{I}_N(h_n)$ converges almost surely towards $I(h_n)$, and even that we have a central limit theorem when N tends towards $+\infty$:

Theorem 15.12 *We assume that* $(X_{0:n}^{(i)})_{i=1,N}$ *form a sequence of independent random vectors with the same distributions of density* $g(x_{0:n}|y_{0:n})$ *such that*

$$\forall x_{0:n} \in E^{n+1}\ \ \ p(x_{0:n}|y_{0:n}) > 0 \Rightarrow g(x_{0:n}|y_{0:n}) > 0.$$

$$(15.78)$$

We assume, moreover, that

$$\int_{E^{n+1}} h_n(x_{0:n})p(x_{0:n}|y_{0:n})dx_{0:n} \qquad < \infty,$$

$$\int_{E^{n+1}} w(x_{0:n})p(x_{0:n}|y_{0:n})dx_{0:n} \qquad < \infty,$$

$$(15.79)$$

and $\int_{E^{n+1}} h_n^2(x_{0:n})w(x_{0:n})p(x_{0:n}|y_{0:n})dx_{0:n} < \infty.$

Then,

$$\lim_{N \to \infty} \sqrt{N}[\tilde{I}_N(h_n) - I(h_n)] \sim \mathcal{N}(0, \sigma_{h_n}^2),$$

$$(15.80)$$

with

$$\sigma_{h_n}^2 = \int_{E^{n+1}} [h_n(x_{0:n}) - I(h_n(x_{0:n}))]^2 w(x_{0:n})p(x_{0:n}|y_{0:n})dx_{0:n}.$$

$$(15.81)$$

Proof See [71].

The corresponding empirical estimator of $p(x_{0:n}|y_{0:n})$ provides the estimate

$$d\hat{P}(x_{0:n}|y_{0:n}) = \sum_{i=1,N} \tilde{w}_n^{(i)}\delta_{x_{0:n}^{(i)}}.$$

$$(15.82)$$

15.5.2 Sequential Importance Sampling

For the filtering problem, in order to be able to obtain a recursive estimation scheme of $p(x_{0:n}|y_{0:n})$, we choose importance distribution functions $g(x_{0:n}|y_{0:n})$ such that

$$g(x_{0:n}|y_{0:n}) = g(x_{0:n-1}|y_{0:n-1})g(x_n|x_{0:n-1}, y_{0:n}), \qquad (15.83)$$

that is, we impose $g(x_{0:n-1}|y_{0:n}) = g(x_{0:n-1}|y_{0:n-1})$.

Thus, at instant n, $(x_{0:n}^{(i)})_{i=1,n}$ can be obtained simply by $x_{0:n}^{(i)} = (x_{0:n-1}^{(i)}, x_n^{(i)})$, where $x_n^{(i)} \sim g(x_n|x_{0:n-1}^{(i)}, y_{0:n})$, for $i = 1, N$. We can check that the non-normalised weights $w_n^{(i)}$ are given by

$$
\begin{aligned}
w_n^{(i)} &= \frac{p(y_{0:n}|x_{0:n}^{(i)})p(x_{0:n}^{(i)})}{g(x_{0:n}^{(i)}|y_{0:n})} \\
&= w_{n-1}^{(i)} \frac{p(y_n|x_n^{(i)})p(x_n^{(i)}|x_{n-1}^{(i)})}{g(x_n^{(i)}|x_{n-1}^{(i)}, y_{0:n})}.
\end{aligned}
\qquad (15.84)
$$

15.5.3 Degeneracy of the Algorithm

It can be shown that the variance of the non-normalised weights increases rapidly with n. This means that all the normalised weights, except one, converge towards zero. The variance of the non-normalised weights can be minimised by choosing [69]

$$g(x_n|x_{0:n-1}^{(i)}, y_{0:n}) = p(x_n|x_{0:n-1}^{(i)}, y_n). \qquad (15.85)$$

But, direct simulation under the posterior importance distribution function (15.85) is generally impossible. We can thus attempt to solve this problem in various ways: by simulating this distribution (which is very arduous) by an MCMC method, by taking an approximation of it (for example, a Gaussian approximation) or by linearising the equation of observation $Y_n = G(X_n, W_n)$ of the model. We can also use particular importance functions, such as the one with probability density function $p(x_n|x_{n-1})$, which is simple but provides a poor estimator.

In fact, it is with methods based on the analytical approximation of the optimal importance function or of the state space model that we generally obtain the best results, for we then strongly incorporate prior knowledge about the model in the choice of g.

To combat the degeneracy of the estimator of $p(x_{0:n}|y_{0:n})$, we also have to incorporate into the algorithm an additional step, called re-sampling. For this, we define

$$N_{eff} = (\sum_{i=1,N} (\tilde{w}_n^{(i)})^2)^{-1}, \qquad (15.86)$$

which varies between 1, when all the coefficients except one are equal to zero, and N, when all the coefficients are equal. When N_{eff} becomes lower than an experimentally fixed threshold, we then replace the sequences $(x_{0:n}^{(i)})_{i=1,N}$ by the sequences $(\tilde{x}_{0:n}^{(i)})_{i=1,N}$ generated according to the probabilities

$$P(\tilde{X}_{0:n}^{(i)} = x_{0:n}^{(l)}) = \tilde{w}_n^{(l)}, \qquad (15.87)$$

and we replace $\tilde{w}_n^{(i)}$ by N^{-1}. This kind of technique is often referred to as *bootstrap filter, particle filter* or *Monte Carlo filter* [72], [70].

15.6 Estimation of a Rational Transfer Function

In Chapter 14, we showed that identification of an unknown, not necessarily minimum-phase, scalar rational transfer function cannot be addressed from sole knowledge of the second order statistics of the observed process. Although higher order statistics allow us to solve this problem, they present, however, the drawback of only partly using the information relating to the data distribution.

We shall see that with very general parametric modelling of the distribution of the model input white noise by means of a mixture of Gaussian distributions, it is possible to provide a solution to this type of problem. The presentation below aims only to give the main outline of the method. [70] can be referred to for further details.

15.6.1 Principle of the Method

We consider a linear stationary state space model of the form

$$\begin{cases} X_n = AX_{n-1} + BU_n \\ Y_n = CX_n + \sigma_w W_n. \end{cases} \qquad (15.88)$$

The presence of the observation noise W shows that Y is an ARMA process corrupted by an additive noise. We shall assume that W is a white noise process, with $W_n \sim \mathcal{N}(0,1)$. The matrix A and the vector line C, as we saw in Chapter 9, are parameterised by the vectors $a = [a_1, \dots, a_p]$ and $b = [b_1, \dots, b_q]$ respectively of the coefficients of the filter to be estimated, the transfer function of this filter being

$$h(z) = \frac{1 + \sum_{l=1,q} b_l z^{-l}}{1 + \sum_{k=1,p} a_k z^{-k}}. \qquad (15.89)$$

We consider a general model for the distribution of the input white noise U by setting $U_n = m_n + \sigma_n V_n$, where $V_n \sim \mathcal{N}(0,1)$ and where (m_n, σ_n^2) has a prior distribution of the form

$$dP(m, \sigma^2 | \alpha) = \sum_{i=1,n_d} \pi_{l_i} \delta_{(m_{l_i}, \sigma_{l_i}^2)}(m, \sigma^2) + \pi_c \delta_{m_c}(m) f_c(\sigma^2) d\sigma^2, \quad (15.90)$$

where $\sum_{i=1,n_d} \pi_{l_i} + \pi_c = 1$. In this expression, the indices l_i are symbols, that is, not numerical values (we shall understand later the importance of this point) and α represents the values of the parameters of the distribution of (m_n, σ_n^2):

$$\alpha = ((\pi_{l_i}, m_{l_i}, \sigma_{l_i}^2)_{i=1,n_d}, \pi_c, m_c). \quad (15.91)$$

The model (15.90) presents the advantage of being fairly general without being too complex: it can be used to approximate multimodal distributions (when $n_d > 1$), but it can also be used to model heavy tail distributions, that is, distributions whose density $f(x)$ is equivalent to x^γ ($\gamma < -1$) when x tends towards infinity. Hence, if we assume, for example, that $n_d = 0$, $m_c = 0$ and that f_c is the PDF of an inverse gamma distribution $\mathcal{IG}(1/2, 1/2)$, that is,

$$f_c(\sigma^2) = \frac{1}{\sqrt{2\pi\sigma^3}} \exp(\frac{-1}{2\sigma^2}) \mathbb{I}_{\mathbb{R}_+}(\sigma^2), \quad (15.92)$$

then, the PDF $f(u)$ of U_n is given by

$$f(u) = \int_{\mathbb{R}_+} \frac{1}{\sqrt{2\pi}\sigma} \exp(\frac{-u^2}{2\sigma^2}) \frac{1}{\sqrt{2\pi\sigma^3}} \exp(\frac{-1}{2\sigma^2}) d\sigma^2$$

$$(15.93)$$

$$= \frac{1}{\pi(1 + u^2)}.$$

U_n is, therefore, distributed according to a Cauchy distribution.

We also consider prior distributions for a, b, and w. We can choose $(a, b)^T \sim \mathcal{N}(0, \varepsilon_1^{-1} I)$, and $\sigma_w \sim \mathcal{IG}(\varepsilon_2, \varepsilon_3)$. Choosing $\varepsilon_i << 1$ ($i = 1, 3$) ensures that the given priors for these parameters are not very informative, that is, the shape of their PDF is relatively flat.

15.6.2 Estimation of the Parameters

We shall use the Gibbs sampler to simulate the desired parameters according to the PDF $p(a, b, \alpha, \sigma_w^2 | y_{0:n})$. In order to be able to perform this operation simply, we complete the model with the random variables $\xi_{0:n}$ that have values in $\{l_1, \dots, l_{n_d}\} \cup \mathbb{R}_+$, and such that

$$\forall i = 1, n_d, \quad p(u_k|\xi_k = l_i, \alpha) \sim \mathcal{N}(m_{l_i}, \sigma_{l_i}^2)$$

$$\forall \lambda \in \mathbb{R}_+, \quad p(u_k|\xi_k = \lambda, \alpha) \sim \mathcal{N}(m_c, \lambda). \tag{15.94}$$

To simulate according to the PDF $p(a, b, \alpha, \sigma_w^2|y_{0:n})$, we may use the Gibbs algorithm described by the following steps

$$(\sigma_w^{2(i)}, \alpha^{(i)}) \sim g_1(\sigma_w^2, \alpha|X_{0:n}^{(i-1)}, \xi_{0:n}^{(i-1)}, a^{(i-1)}, b^{(i-1)}, y_{0:n})$$

$$\xi_{0:n}^{(i)} \quad \sim g_2(\xi_{0:n}|X_{0:n}^{(i-1)}, \sigma_w^{2(i)}, a^{(i-1)}, b^{(i-1)}, \alpha^{(i)}, y_{0:n})$$

$$(a^{(i)}, b^{(i)}) \sim g_3(a, b|X_{0:n}^{(i-1)}, \xi_{0:n}^{(i)}, \sigma_w^{2(i)}, \alpha^{(i)}, y_{0:n})$$

$$X_{0:n}^{(i)} \quad \sim g_4(X_{0:n}|\xi_{0:n}^{(i)}, \sigma_w^{2(i)}, \alpha^{(i)}, a^{(i)}, b^{(i)}, y_{0:n}). \tag{15.95}$$

Here, we make the following hypothesis of independence:

$$p(a, b, \alpha, \sigma_w^2) = p(a, b)p(\alpha)p(\sigma_w^2), \tag{15.96}$$

and we assume that the above-mentioned conditional distributions are easy to simulate, with the possible exception of $\xi_{0:n}$ conditional to $(X_{0:n}, \sigma_w^2, a, b, \alpha, y_{0:n})$, in which case we replace its direct simulation by one iteration of the Metropolis-Hastings algorithm. The instrumental distribution can then be that of $\xi_{0:n}$ conditional to $\alpha^{(i)}$.

We note that conditionally to $\xi_{0:n}$ the noise $U_{0:n}$ is Gaussian. In this case, a Kalman smoother can be implemented to compute the distribution of $X_{0:n}^{(i)}$ conditional to $(\xi_{0:n}^{(i)}, \sigma_w^{2(i)}, \alpha^{(i)}, a^{(i)}, b^{(i)}, y_{0:n})$. This is a main interest of modelling the noise as a mixture of Gaussian distributions.

In certain situations, the irreducibility of the simulated chain cannot be guaranteed. We may, however, solve this problem by removing $X_{0:n}$ from the simulation according to the conditional distributions of $\xi_{0:n}$ and (a, b) in (15.95). For a presentation of this type of approach, called partial conditioning, refer to [65].

Remark We can generalise the above study to the estimation of non-linear systems. In this case, the unknown non-linearities can be approximated as linear combinations of known non-linearities where the coefficients of the linear combinations have to be estimated.

Exercises

15.1 We want to simulate a random variable X that takes values $x_1, \ldots, x_N$ with respective probabilities $p_1, \ldots, p_N$ ($\sum_{n=1,N} p_n = 1$). Explain how X can be simulated from a random variable $U \sim \mathcal{U}_{[0,1]}$.

15.2 Explain how $X \sim \mathcal{E}(\lambda)$ can be simulated from a random variable $U \sim \mathcal{U}_{[0,1]}$.

15.3 (Box and Muller method) We want to generate a Gaussian vector X of size p.

a) Let $\mathbf{U} = [U_1, \ldots, U_p]^T$ denote a vector of p independent random variables with uniform distributions on $[0,1]$ ($U_k \sim \mathcal{U}_{[0,1]}$ for $k = 1, \ldots, p$). Show that the relations

$$
\begin{aligned}
X_1 &= \sqrt{-2\log(U_1)}\cos(2\pi U_2), \\
X_2 &= \sqrt{-2\log(U_1)}\sin(2\pi U_2)\cos(2\pi U_3), \\
&\cdots \\
X_k &= \sqrt{-2\log(U_1)}\sin(2\pi U_2)\ldots\sin(2\pi U_k)\cos(2\pi U_{k+1}), \\
&\cdots \\
X_p &= \sqrt{-2\log(U_1)}\prod_{n=2,p}\sin(2\pi U_n),
\end{aligned}
\tag{15.97}
$$

define a random vector $\mathbf{X} = [X_1, \ldots, X_p]^T$ with an $\mathcal{N}(0, I_p)$ distribution.

b) Explain how a vector with an $\mathcal{N}(m, R)$ distribution can be generated from $\mathbf{U}$.

15.4 What is the average number of iterations of the acceptance-rejection algorithm necessary for the simulated random variable to be accepted? Which instrumental distribution ensures the minimum of this acceptance rate?

15.5 Give an intuitive interpretation of the Metropolis-Hastings algorithm when $q(x|y) = q(y|x)$.

15.6 Show that if $P\left(f(x)q(y|x) = f(y)q(x|y)\right) = 1$, then the Markov chain generated by the Metropolis-Hastings algorithm is reversible with stationary distribution f.

15.7 Write the Metropolis-Hastings algorithm that simulates a Poisson distribution: $P(X = k) = e^{-\lambda}\lambda^k/k!$ ($k \in \mathbb{N}$), using the following random walk

$$
dQ(y|x) = \mathbb{I}_{\{x>0\}}\frac{1}{2}(\delta_{x-1}(y) + \delta_{x+1}(y)) + \mathbb{I}_{\{x=0\}}\delta_1(y)
\tag{15.98}
$$

as the instrumental distribution.

15.8 By means of a computer simulation, apply the simulated annealing algorithm to minimise the function $f(x) = x^2 + 3x\cos(x)$ on $[-8\pi, 8\pi]$ and compare its behaviour to the gradient method, when it is initialised at point $x_0 = 7.5$. For the gradient algorithm $x_{n+1} = x_n - \mu_n f'(x_n)$, choose $\mu_n = 0.025/n$. For the simulated annealing algorithm consider a uniform random walk on $[x_n - \pi/2, x_n + \pi/2]$ for data simulation and a temperature function of the form $T_n = T_0\alpha^n$, where $T_0 = 500$ and $\alpha = 0.999$.

15.9 We want to calculate the variance of the Monte Carlo estimator of $\int_{[0,1]} h(x)f(x)dx$ given by $N^{-1}\sum_{n=1,N} h(X_n)$, where $f(x) = \frac{1}{\sqrt{2\pi}}\exp(-x^2/2)$ and $X_n \sim f(x)$. Here, we assume that $X = (X_n)_{n\in\mathbb{N}^*}$ is a Gaussian process with $\mathbb{E}[X_m X_n] = R_X(m - n)$. Calculate the variance of $N^{-1}\sum_{n=1,N} h(X_n)$ in terms of the Fourier coefficients of h.

15.10 We want to simulate a random Gaussian vector $[X_1, X_2]^T \sim \mathcal{N}(m, R)$ by simulating X_1 and X_2 iteratively.

a) Propose a Gibbs algorithm that performs this task.

b) Check directly the convergence of the simulated vector distribution in this situation.

15.11 (EM algorithm for a mixture of Gaussian distributions) We want to estimate the parameters (weights, means and variances) of a mixture of Gaussian distributions. We observe the independent random variables $(X_k)_{k=1,n}$ with probability density distributions

$$\sum_{l=1,p} \pi_l f_l(x) = \sum_{l=1,p} \pi_l \frac{1}{\sqrt{2\pi}\sigma_l} \exp\left(\frac{-(x - m_l)^2}{2\sigma_l^2}\right). \tag{15.99}$$

We shall denote by $\theta = (\pi_l, m_l, \sigma_l^2)_{l=1,p}$ the parameter vector of interest. In order to apply an EM estimation procedure, we consider the random variables $(Y_k)_{k=1,n}$, where Y_k represents the index of the component of the mixture to which X_k belongs.

a) Check that

$$P(Y_k = i | X_k, \theta) = \left(\sum_{l=1,p} \pi_l \frac{1}{\sqrt{2\pi}\sigma_l} \exp\left(\frac{-1}{2\sigma_l^2}(X_k - m_l)^2\right)\right)^{-1}$$

$$\times \frac{\pi_i}{\sqrt{2\pi}\sigma_i} \exp\left(\frac{-1}{2\sigma_i^2}(X_k - m_i)^2\right). \tag{15.100}$$

b) We note $\theta^u = (\pi_l^u, m_l^u, (\sigma_l^2)^u)_{l=1,p}$ the parameters estimated by means of an EM algorithm at the u-th iteration, and $\alpha_k^u(i) = P(Y_k = i | X_k, \theta^u)$. Show that $Q(\theta, \theta^u) = \mathbb{E}[\log(X, Y | \theta) | Y, \theta^u]$ is given by

$$Q(\theta, \theta^u) = \sum_{i=1,p} \log \pi_i \sum_{k=1,n} \alpha_k^u(i) - \sum_{i=1,p} \log \sigma_i \sum_{k=1,n} \alpha_k^u(i)$$

$$- \sum_{i=1,p} \frac{1}{2\sigma_i^2} \sum_{k=1,n} (X_k - m_i)^2 \alpha_k^u(i), \tag{15.101}$$

and check that the EM algorithm can be written as follows:

- step E: calculate the posterior distributions

$$\alpha_k^u(i) \propto \pi_i^u \frac{1}{\sqrt{2\pi}\sigma_i^u} \exp\left(\frac{-1}{2(\sigma_i^u)^2}(X_k - m_i^u)^2\right) \tag{15.102}$$

yielding thus $Q(\theta, \theta^u)$ from (15.101).

- step M: maximise $Q(\theta, \theta_k)$:

$$\pi_i^{u+1} = \frac{\sum_{k=1,n} \alpha_k^u(i)}{\sum_{i=1,p} \sum_{k=1,n} \alpha_k^u(i)},$$

$$(\sigma_i^2)^{u+1} = \frac{\sum_{k=1,n} \alpha_k^u(i)(X_k - m_i^u)^2}{\sum_{k=1,n} \alpha_k^u(i)}, \qquad (15.103)$$

$$m_i^{u+1} = \frac{\sum_{k=1,n} \alpha_k^u(i) X_k}{\sum_{k=1,n} \alpha_k^u(i)}.$$

16. Adaptive Estimation

Purpose Descent algorithms, such as gradient or Newton methods, are commonly used for the optimisation of criteria given in the form of a deterministic function of a desired parameter. In signal processing, we often look for the value of a parameter θ that achieves the minimum of a function of the form $\mathbb{E}[V(\theta, X_n)]$, where $X = (X_n)_{n \in \mathbb{Z}}$ is a random process. Since, in practice, the calculation of $\mathbb{E}[V(\theta, X_n)]$ is not always possible, particularly in cases where X is not a stationary process, we thus modify conventional descent methods by trying to approach at each instant the optimal value of an instantaneous criterion such as $V(\theta, X_n)$. The aim of this chapter is to present some frequently used algorithms based on this principle, along with some results that enable us to specify contexts where their use is justified.

16.1 Classical Descent Techniques

Conventionally, the minimisation of a function $g(\theta)$ with $\theta \in \mathbb{R}^p$ can be performed by means of a descent algorithm when the solution is not available from any analytical formula. Such an algorithm provides a sequence of values $(\theta_n)_{n \in \mathbb{N}}$ such that the sequence $g(\theta_n)$ converges towards a local or global minimum of the function g, assuming that one exists. The existence of such a minimum and the possible convergence of various descent methods towards this minimum can be guaranteed as long as the function g satisfies certain properties, such as convexity.

Among descent methods, we distinguish two particularly important classes known as the *Newton* and *gradient* algorithms. The Newton algorithm involves defining the sequence

$$\theta_n = \theta_{n-1} - (\nabla^2 g(\theta_{n-1}))^{-1} \nabla g(\theta_{n-1}), \tag{16.1}$$

initialised by a value θ_0, where $\nabla g(\theta)$ and $\nabla^2 g(\theta)$ represent the gradient and the Hessian of g at point θ respectively. We can intuitively understand this method by noticing that θ_n cancels the gradient of the second order development of $g(\theta)$ in the neighbourhood of θ_{n-1}:

$$\nabla[g(\theta_{n-1}) + (\theta - \theta_{n-1})^T \nabla g(\theta_{n-1})$$

$$+ \frac{1}{2}(\theta - \theta_{n-1})^T \nabla^2 g(\theta_{n-1})(\theta - \theta_{n-1})]_{\theta = \theta_n} = 0, \tag{16.2}$$

and remembering that the condition $\nabla g(\theta) = 0$ is a necessary condition to ensure that $g(\theta)$ is an optimum of the function g when it is derivable. In formula (16.1), we often multiply the updating term $(\nabla^2 g(\theta_{n-1}))^{-1} \nabla g(\theta_{n-1})$ by a small factor μ_n, which makes it possible to ensure that $\theta_n - \theta_{n-1}$ remains small.

The gradient algorithm can be seen as a simplified version of Newton's algorithm where the matrix $(\nabla^2 g(\theta_{n-1}))^{-1}$ is replaced by $\mu_n I$, leading to the iteration

$$\theta_n = \theta_{n-1} - \mu_n \nabla g(\theta_{n-1}). \tag{16.3}$$

We can take a sequence of values μ_n that are constant, decreasing or even optimised at each iteration so that

$$\mu_n = \arg \min_{\mu} g(\theta_{n-1} - \mu \nabla g(\theta_{n-1})). \tag{16.4}$$

The gradient algorithm converges for elliptic functions, that is, for functions of class $\mathcal{C}^1$ such that

$$\forall u, v \in \mathbb{R}^p \, \exists \alpha > 0, \ (v - u)^T (\nabla g(v) - \nabla g(u)) \geq \alpha \| v - u \|^2 . \tag{16.5}$$

The quadratic functions of the form $g(v) = (v - b)^T A(v - b)$, with $A^T = A$ and $A > 0$ are examples of elliptic functions. For an elliptic function that, moreover, satisfies

$$\exists M > 0, \ \| \Delta g(v) - \Delta g(u) \| \leq M \| v - u \|, \tag{16.6}$$

any gradient method such as

$$\exists a, b > 0, \ \forall n > 0, 0 < a \leq \mu_n \leq b < \frac{2\alpha}{M^2} \tag{16.7}$$

converges towards a minimum of g.

In the case of a decreasing stepsize, we show that under the sole conditions $\lim_{n \to \infty} \mu_n = 0$ and $\sum_{n=0,\infty} \mu_n = +\infty$, the convergence of the algorithm

$$\theta_n = \theta_{n-1} - \mu_n \frac{\nabla g(\theta_{n-1})}{\| \nabla g(\theta_{n-1}) \|} \tag{16.8}$$

towards a local minimum of $g(\theta)$ is guaranteed. We can, for example, take $\mu_n = n^{-1}$.

The justification of these results, along with the presentation of other methods, can be found in [82], [86], [89].

16.2 Principle of Stochastic Optimisation

In many signal processing applications, we look for the minimum of functions of the form $\mathbb{E}[V_n(\theta, X)]$, where X is a process.

The linear prediction of a process X is an example of such a problem, where $\theta = (a_1, \ldots, a_p)$ and

$$V_n(\theta, X) = \left| X_n - \sum_{k=1,p} a_k X_{n-k} \right|^2. \tag{16.9}$$

In the same way, let us consider the problem of estimating a transfer function $h(z) = \sum_{k=1,p} h_k z^{-k}$ of a filter with an input process X, and for which we wish to obtain the output $(d_n)_{n \in \mathbb{N}}$. Here, we can take $\theta = (h_1, \ldots, h_p)$, and

$$V_n(\theta, X) = \left| \sum_{k=1,p} h_k X_{n-k} - d_n \right|^2. \tag{16.10}$$

In the next section, we shall describe a practical situation where we are faced with this problem.

Often, the statistics of the processes brought into play (the second order statistics for the two examples above) are time-varying, or else we want to be able to estimate θ before the data are completely recorded. In these situations, we do not always have exact knowledge of the functions of θ defined by $\mathbb{E}[V_n(\theta, X)]$, nor of reliable statistical estimators of these means. We are then led to replace at instant n the criterion $\mathbb{E}[V_n(\theta, X)]$ by a criterion $J_n(\theta, X)$ which no longer involves any mathematical expectation, and which can, for example, be $V_n(\theta, X)$, or $\sum_{k=0,n} \lambda^{n-k} V_k(\theta, X)$. In this latter expression, the factor λ $(0 < \lambda \le 1)$ represents a forgetting factor. For non-stationary systems the closer to 1 the chosen value of λ is, the slower the evolution of the statistical parameters of the process.

We can try to follow the evolution of the minimum of $J_n(\theta, X)$ by constructing a sequence of estimators of θ obtained from algorithms of the form

$$\theta_n = \theta_{n-1} - \mu_n \nabla_\theta J_n(\theta_{n-1}, X),$$

$$\text{or } \theta_n = \theta_{n-1} - \mu_n (\nabla_\theta^2 J_n(\theta_{n-1}, X))^{-1} \nabla_\theta J_n(\theta_{n-1}, X), \tag{16.11}$$

known as *stochastic gradient* and *stochastic Newton* algorithms respectively. In what follows, we shall consider algorithms of the general form

$$\theta_n = \theta_{n-1} + \mu_n H(\theta_{n-1}, \mathcal{X}_n). \tag{16.12}$$

We often have $\mathcal{X}_n = X_n$ or, as for the examples given above, $\mathcal{X}_n = [X_n, \ldots, X_{n-p}]^T$.

The sequence of gains μ_n will be assumed to be decreasing or constant. Concerning algorithms with decreasing stepsize, the step μ_n will be chosen in such a way that it decreases towards 0, and that $\sum_{n>0} \mu_n = +\infty$. For constant stepsize algorithms, it is clear that the sequence $(\theta_n)_{n \in \mathbb{N}}$ will not generally be convergent, in particular due to the stochastic nature of the term $H(\theta, \mathcal{X}_n)$. The choice of a constant stepsize is of interest for tracking the parameters of non-stationary phenomena.

16.3 LMS and RLS Algorithms

In signal processing, we often encounter two adaptive algorithms, known as the Least Mean Square (LMS) and Recursive Least Square (RLS) algorithms. We shall present these algorithms in the context of the classical example of the equalisation of digital communication signals.

Schematically, in digital transmission, we transmit a sequence of random variables that take their values in a finite set. During their transmission, these random variables undergo distortions that can often be modelled by the action of a filter, and by the addition of a noise often assumed to be white and Gaussian. This leads to the observation of a process X at the receiver side, from which we take a decision on the sequence of transmitted data.

It is often possible to obtain an estimation of the impulse response of the filter, called an equaliser, which compensates for the distortion of the transmitted signal by emitting an initial sequence of symbols that are known by the receiver. Subsequently, we have to take into account the possible evolutions of the transmission channel, which is done by means of an adaptive method.

We assume that the probability of making a wrong decision remains very small. The impulse response of the equaliser can then be chosen in such a way that the output of the filter is the closest possible to the sequence of the decisions upon transmitted data and which correspond to their true values up to the decision errors. We shall denote by A the process associated with the sequence of decisions.

For an equaliser filter of the form $\theta_n(z) = \sum_{k=0,p} \theta_{k,n}^* z^{-k}$, the iterative minimisation of the criterion $|\sum_{k=0,p} \theta_{k,n}^* X_{n-k} - A_n|^2 = |\theta_n^H \mathbf{X}_n - A_n|^2$, with $\mathbf{X}_n = [X_n, \ldots, X_{n-p}]^T$ and $\theta_n = [\theta_{0,n}, \ldots, \theta_{p,n}]^T$, leads to the stochastic gradient LMS algorithm:

$$\theta_n = \theta_{n-1} - \mu \nabla |\theta_{n-1}^H \mathbf{X}_n - A_n|^2$$

$$= \theta_{n-1} - \mu \mathbf{X}_n (\theta_{n-1}^H \mathbf{X}_n - A_n)^*. \tag{16.13}$$

Another approach involves minimising the criterion

$$J_n(\theta) = \sum_{l=1,n} \lambda^{n-l} |\theta_n^H \mathbf{X}_l - A_l|^2. \tag{16.14}$$

Iterative minimisation of $J_n(\theta)$ can be achieved by means of the algorithm described by the following theorem, known as the RLS algorithm. Unlike the LMS algorithm, the RLS algorithm at each instant leads to the exact value of the parameter θ, which achieves the minimum of the criterion at this instant.

Theorem 16.1 *The exact adaptive minimisation of the criterion*

$$\sum_{l=1,n} \lambda^{n-l} |\theta_n^H \mathbf{X}_l - A_l|^2 \qquad (16.15)$$

is given by the recurrence equations

$$R_n^{-1} = \lambda^{-1} R_{n-1}^{-1} - \frac{\lambda^{-1} R_{n-1}^{-1} \mathbf{X}_n \mathbf{X}_n^H R_{n-1}^{-1}}{\lambda + \mathbf{X}_n^H R_{n-1}^{-1} \mathbf{X}_n}$$

$$\theta_n = \theta_{n-1} - \frac{(\theta_{n-1}^H \mathbf{X}_n - A_n)^* R_{n-1}^{-1} \mathbf{X}_n}{\lambda + \mathbf{X}_n^H R_{n-1}^{-1} \mathbf{X}_n}. \qquad (16.16)$$

The proof of this result is based on the matrix inversion lemma, whose proof is straightforward:

Lemma 16.2 *Let $A, B, C, D,$ and E be matrices of compatible sizes such that*

$$A = B + CDE. \qquad (16.17)$$

Then,

$$A^{-1} = B^{-1} - B^{-1}C(D^{-1} + EB^{-1}C)^{-1}EB^{-1}. \qquad (16.18)$$

Proof (of the theorem) We first remark that

$$\nabla J_n(\theta) = \sum_{l=1,n} \lambda^{n-l} \mathbf{X}_l (\theta^H \mathbf{X}_l - A_l)^*, \qquad (16.19)$$

that is, since $\nabla J_n(\theta_n) = 0$, and by putting $R_n = \sum_{l=1,n} \lambda^{n-l} \mathbf{X}_l \mathbf{X}_l^H$,

$$R_n \theta_n = \sum_{l=1,n} \lambda^{n-l} \mathbf{X}_l A_l^*$$

$$= \lambda R_{n-1} \theta_{n-1} + \mathbf{X}_n A_n^*. \qquad (16.20)$$

R_n can be calculated recursively by

$$R_n = \lambda R_{n-1} + \mathbf{X}_n \mathbf{X}_n^H. \qquad (16.21)$$

We assume, moreover, that R_n is invertible. The matrix inversion lemma then leads to the relation

$$R_n^{-1} = \lambda^{-1} R_{n-1}^{-1} - \frac{\lambda^{-1} R_{n-1}^{-1} \mathbf{X}_n \mathbf{X}_n^H R_{n-1}^{-1}}{\lambda + \mathbf{X}_n^H R_{n-1}^{-1} \mathbf{X}_n}. \tag{16.22}$$

Relations (16.20) and (16.22) finally yield

$$\theta_n = (\lambda^{-1} R_{n-1}^{-1} - \frac{\lambda^{-1} R_{n-1}^{-1} \mathbf{X}_n \mathbf{X}_n^H R_{n-1}^{-1}}{\lambda + \mathbf{X}_n^H R_{n-1}^{-1} \mathbf{X}_n})(\lambda R_{n-1}\theta_{n-1} + \mathbf{X}_n A_n^*)$$

$$= \theta_{n-1} + \frac{-R_{n-1}^{-1} \mathbf{X}_n \mathbf{X}_n^H \theta_{n-1}}{\lambda + \mathbf{X}_n^H R_{n-1}^{-1} \mathbf{X}_n}$$

$$+ \frac{(R_{n-1}^{-1} + \lambda^{-1} \mathbf{X}_n^H R_{n-1}^{-1} \mathbf{X}_n R_{n-1}^{-1} - \lambda^{-1} R_{n-1}^{-1} \mathbf{X}_n \mathbf{X}_n^H R_{n-1}^{-1})\mathbf{X}_n A_n^*}{\lambda + \mathbf{X}_n^H R_{n-1}^{-1} \mathbf{X}_n}$$

$$= \theta_{n-1} - \frac{R_{n-1}^{-1} \mathbf{X}_n (\theta_{n-1}^H \mathbf{X}_n - A_n)^*}{\lambda + \mathbf{X}_n^H R_{n-1}^{-1} \mathbf{X}_n}. \square$$

$$\tag{16.23}$$

Usually, we initialise the RLS algorithm with a full rank matrix R_0, such that $R_0 = \alpha I$ where α is a small constant, and with $\theta_0 = 0$.

16.4 Convergence of LMS and RLS Algorithms

Studying the behaviour of stochastic algorithms is often complex. It can be undertaken fairly simply at the cost of crude simplifying hypotheses, which are unjustified in practice but which, however, give a good idea of the behaviour of the algorithms in many practical situations. On the other hand, a rigorous study based on more realistic models of the signal studied is often difficult to carry out well.

We notice that studying convergence is, in all cases, envisaged for stationary processes X, which often is not the usual context for applying stochastic methods. Moreover, the convergence properties that we can obtain are often fairly limited, concerning, for instance, the convergence in mean, that is, of $\mathbb{E}[\theta_n]$, at least for constant stepsize algorithms. In the case where X is not stationary, it is rather the ability of the algorithm to track the variations of the system that is interesting to consider.

Here, we propose a direct study of the convergence of the LMS algorithm, which requires the hypothesis of independence of the random variables $\mathbf{X}_n$. This very restrictive hypothesis may correspond to the case where X is an order p MA process and where the parameter θ is only updated periodically, by taking, for example, $\mathbf{X}_n = (X_{Kn}, \ldots, X_{Kn-p})$, with $K > p$. However, the results obtained give a good idea of the behaviour of the algorithm for more general situations.

Moreover, we assume that X is Gaussian complex and circular, which makes it possible to use simple expressions of the type

$$\mathbb{E}[X_a X_b^* X_c X_d^*] = \mathbb{E}[X_a X_b^*]\mathbb{E}[X_c X_d^*] + \mathbb{E}[X_a X_d^*]\mathbb{E}[X_c X_b^*] \qquad (16.24)$$

In the real Gaussian case, the term $\mathbb{E}[X_a X_c]\mathbb{E}[X_b X_d]$ would have to be added in the right-hand part of the equation. These relations are useful for calculating the asymptotic variance of the algorithms.

For the RLS algorithm, apart from the case where $\lambda = 1$, there exists no very satisfactory result concerning the behaviour of the algorithm, and we shall restrict ourselves to giving some partial indications.

16.4.1 Mean Convergence of the LMS Algorithm

In what follows, we shall note $T_p = \mathbb{E}[\mathbf{X}_n \mathbf{X}_n^H]$, and $r_{XA} = \mathbb{E}[\mathbf{X}_n A_n^*]$. It is clear that for the LMS algorithm,

$$\mathbb{E}[\theta_n] = (I - \mu T_p)\mathbb{E}[\theta_{n-1}] + \mu r_{XA}$$

$$= (I - \mu T_p)^n \mathbb{E}[\theta_0] + \mu(\sum_{k=0,n-1}(I - \mu T_p)^k) r_{XA}. \qquad (16.25)$$

Denoting by $U \Lambda U^H$ the eigenvalue decomposition of T_p, with

$$\Lambda = diag(\lambda_0, \ldots, \lambda_p), \qquad (16.26)$$

and $0 < \lambda_0 \leq \ldots \leq \lambda_p$, it is clear that

$$\mathbb{E}[\theta_n] = U \, diag((1 - \mu\lambda_0)^n, \ldots, (1 - \mu\lambda_p)^n)U^H \mathbb{E}[\theta_0]$$

$$+ U \, diag(\frac{1 - (1 - \mu\lambda_0)^n}{\lambda_0}, \ldots, \frac{1 - (1 - \mu\lambda_p)^n}{\lambda_p})U^H r_{XA}. \qquad (16.27)$$

The convergence can only be obtained for $-1 < 1 - \mu\lambda_k < 1$ ($k = 0, p$), that is, by putting $\lambda_{\max} = \lambda_p$, for

$$0 < \mu < \frac{2}{\lambda_{\max}}. \qquad (16.28)$$

We notice that $\lambda_{\max} < Tr(T_p) = (p + 1)R_X(0)$, which leads to the more restricting criterion often proposed

$$0 < \mu < \frac{2}{(p + 1)R_X(0)}. \qquad (16.29)$$

By considering relation (16.27), we then easily check that constraint (16.28) (or constraint (16.29)) means that $\lim_{n \to \infty} \mathbb{E}[\theta_n] = T_p^{-1} r_{XA}$. As the solution to the deterministic problem of minimisation of $\| \theta^H X_n - A_n \|^2$ is $\theta_* = T_p^{-1} r_{XA}$, it is satisfying to note that $\mathbb{E}[\theta_n]$ converges towards this same value.

16.4.2 Asymptotic Variance of the LMS Algorithm

Mean convergence is not sufficient to ensure the good practical behaviour of the algorithm. Studying the evolution of the variance of the error $E_n = \theta_n - \theta_*$ imposes additional conditions on the choice of the stepsize μ of the algorithm. We first note that

$$E_n = E_{n-1} - \mu \mathbf{X}_n (\theta_{n-1}^H \mathbf{X}_n - A_n)^*$$

$$= (I - \mu X_n X_n^H) E_{n-1} - \mu X_n (\theta_*^H X_n - A_n)^*. \tag{16.30}$$

Consequently,

$$E_n E_n^H = (I - \mu X_n X_n^H) E_{n-1} E_{n-1}^H (I - \mu X_n X_n^H)$$

$$- 2\mu \mathcal{R}e[(I - \mu X_n X_n^H) E_{n-1} X_n^H (\theta_*^H X_n - A_n)] \tag{16.31}$$

$$+ \mu^2 X_n X_n^H |\theta_*^H X_n - A_n|^2.$$

We shall use the following notations: $\Sigma_n = \mathbb{E}[E_n E_n^H]$, σ_A^2 is the variance of the random variable A_n (assumed to be zero mean) and $\sigma_{\min}^2 = \parallel \theta_*^H X_n - A_n \parallel^2$. We remark that

$$\sigma_{\min}^2 = \parallel r_{XA}^H T_p^{-1} X_n - A_n \parallel^2$$

$$= \sigma_A^2 - r_{XA}^H T_p^{-1} r_{XA}, \tag{16.32}$$

and $\sigma_{\min}^2$ is the minimal value that the criterion $J(\theta) = \parallel \theta^H X_n - A_n \parallel^2$ can reach. Σ_n satisfies the following recurrence:

$$\Sigma_n = \Sigma_{n-1} - \mu(T_p \Sigma_{n-1} + \Sigma_{n-1} T_p) + \mu^2 (T_p \Sigma_{n-1} T_p + T_p Tr(\Sigma_{n-1} T_p))$$

$$+ \mu^2 (T_p \sigma_{min}^2 + (T_p \theta_* - r_{XA})(T_p \theta_* - r_{XA})^H)$$

$$= \Sigma_{n-1} - \mu(T_p \Sigma_{n-1} + \Sigma_{n-1} T_p)$$

$$+ \mu^2 (T_p \Sigma_{n-1} T_p + T_p Tr(\Sigma_{n-1} T_p) + T_p \sigma_{min}^2)$$

$$= (I - \mu T_p) \Sigma_{n-1} (I - \mu T_p) + \mu^2 (T_p Tr(\Sigma_{n-1} T_p) + T_p \sigma_{min}^2). \tag{16.33}$$

The following result specifies the conditions of existence of an asymptotic limit for Σ_n and the value of this limit.

Theorem 16.3 *We denote by T_p the covariance matrix of $\mathbf{X}_n$, and $U \Lambda U^H$ its eigenvalue decomposition, with $\Lambda = diag(\lambda_0, \ldots, \lambda_p)$. We note, moreover, $\lambda_{\max} = \max\{\lambda_0, \ldots, \lambda_p\}$. When $0 < \mu < 2/\lambda_{\max}$, the error covariance $\theta_n - \theta_*$ ($\theta_* = T_p^{-1} r_{XA}$) converges if and only if*

$$\sum_{l=0,p} \frac{\mu\lambda_l}{2 - \mu\lambda_l} < 1. \tag{16.34}$$

The asymptotic covariance is then given by

$$\Sigma_\infty = \frac{\mu\sigma_{min}^2}{1 - \sum_{l=0,p} \mu\lambda_l(2 - \mu\lambda_l)^{-1}} \tag{16.35}$$

$$\times\, U \times diag((2 - \mu\lambda_0)^{-1}, \ldots, (2 - \mu\lambda_p)^{-1}) \times U^H.$$

Proof See Appendix Z.

We therefore see that mean convergence condition (16.28) of the LMS algorithm must necessarily be completed by condition (16.34) in order to ensure the convergence of the asymptotic covariance matrix of the error $\theta_n - \theta_*$. The total power error, defined by $\sigma_\infty^2 = Tr(\Sigma_\infty)$, is then given by

$$\sigma_\infty^2 = \sigma_{min}^2 \frac{\sum_{l=0,p} \mu(2 - \mu\lambda_l)^{-1}}{1 - \sum_{l=0,p} \mu\lambda_l(2 - \mu\lambda_l)^{-1}}. \tag{16.36}$$

We easily check that σ_∞^2 is an increasing function of μ.

16.4.3 Convergence of the RLS Algorithm

We shall not study the convergence of the algorithm RLS in detail. We remark, however, that for this algorithm

$$\theta_n = \Big(\sum_{k=1,n} \lambda^{n-k}\mathbf{X}_k\mathbf{X}_k^H\Big)^{-1}\Big(\sum_{k=1,n} \lambda^{n-k}\mathbf{X}_k A_k^*\Big). \tag{16.37}$$

For $\lambda = 1$, the strong law of large numbers makes it possible to ensure the almost sure convergence of θ_n towards $\theta_* = T_p^{-1} r_{XA}$ under fairly weak conditions.

For the case where $0 < \lambda < 1$, we simply indicate the following theorem, which represents a modified version of the strong law of large numbers:

Theorem 16.4 (Kolmogorov) *If a sequence $(Y_n)_{n\in\mathbb{N}^*}$ of random variables is such that the sequence $\sum_{n\in\mathbb{N}^*} \mathbb{E}[Y_n]$ and $\sum_{n\in\mathbb{N}^*} \| Y_n - \mathbb{E}[Y_n] \|^2$ converge, then the sequence of random variables $S_n = \sum_{k=1,n} Y_n$ converges almost surely towards a random variable, of mean $\sum_{n\in\mathbb{N}^*} \mathbb{E}[Y_n]$ and of variance $\sum_{n\in\mathbb{N}^*} \| Y_n - \mathbb{E}[Y_n] \|^2$.*

Proof See Appendix U.

Theorem 16.4 shows that $\sum_{k=1,n} \lambda^{n-k}X_kX_k^H$ and $\sum_{k=1,n} \lambda^{n-k}X_kA_k^*$ converge almost surely towards random variables with respective means $(1 - \lambda)^{-1}T_p$ and $(1 - \lambda)^{-1}r_{XA}$ and with bounded variance. Studying the convergence of expression (16.37) is made difficult by the presence of the inverse of $\sum_{k=0,n} \lambda^{n-k}X_kX_k^H$. For more results concerning the study of RLS algorithm convergence we can refer to [85].

16.5 The ODE Method

We have presented a direct study of the convergence of the LMS algorithm. We shall now more generally look into the behaviour of stochastic algorithms of the form (16.12):

$$\theta_n = \theta_{n-1} + \mu_n H(\theta_{n-1}, X_n). \tag{16.38}$$

Convergence is, of course, studied in the case where the observed process X is stationary. We shall indicate how we can characterise the points that such an algorithm is likely to converge towards. We shall also give a fairly general formula to use in order to evaluate the asymptotic variance of the sequence of estimators of the parameter θ provided by the algorithm. We shall restrict ourselves to indicating briefly those results presented in greater detail in [81].

16.5.1 Hypotheses

We assume that the process X has a representation of the form

$$P(\xi_n \in A | \xi_{n-1}, \xi_{n-2}, \dots ; \theta_{n-1}, \theta_{n-2}, \dots) = P(\xi_n \in A | \xi_{n-1}, \theta_{n-1})$$

$$X_n = f(\xi_n), \tag{16.39}$$

where $f(\xi_n)$ represents a random variable whose distribution is parameterised by ξ_n. In practice, X can therefore depend, or not, on the parameter θ. Among the particular important cases, we can cite the case where X is a Markov process independent of θ, or where X is described by a difference equation whose coefficients at the instant n are functions of θ_n.

We shall assume, moreover, that there exists a function $h(\theta)$ such that

$$h(\theta) = \lim_{n \to \infty} \mathbb{E}[H(\theta, X_n) | \theta]. \tag{16.40}$$

We will then define the differential equation, commonly called the ODE (Ordinary Differential Equation), by

$$\frac{d\theta}{dt} = h(\theta). \tag{16.41}$$

With these hypotheses and further hypotheses concerning the regularity of the function H, we can show that the behaviour of the stochastic algorithm (16.38) is linked to that of the ODE. An intuitive justification for this is given by the following approximations:

$$\theta_{n+N} = \theta_n + \sum_{k=0,N-1} \mu_{n+k+1} H(\theta_{n+k}, X_{n+k+1})$$

$$\approx \theta_n + \sum_{k=0,N-1} \mu_{n+k+1} H(\theta_n, X_{n+k+1}) \qquad (16.42)$$

$$\approx \theta_n + (\sum_{k=0,N-1} \mu_{n+k+1}) h(\theta_n).$$

The first approximation comes from the fact that the function H is assumed to be sufficiently regular with respect to the variable θ and that θ_n has slow evolution. The second approximation is justified for large N, by the possibility of applying the law of large numbers to the sequence of random variables $(H(\theta_n, X_{n+k+1}))_{k \geq 0}$.

By putting $t_n = \sum_{k=1,n} \mu_k$, the relation

$$\theta_{n+N} \approx \theta_n + (\sum_{k=0,N-1} \mu_{n+k+1}) h(\theta_n)$$

$$\approx \theta_n + (t_{N+n} - t_{n+1}) h(\theta_n) \qquad (16.43)$$

shows that the sequence $(\theta_n)_{n \in \mathbb{N}}$ can be seen as an approximation of a discretised version of the trajectory $\theta(t)$ of the ODE starting at $\theta(0) = \theta_0$, and considered at points $(t_n)_{n \in \mathbb{N}}$. It is then reasonable to think that the sequence $(\theta_n)_{n \in \mathbb{N}}$ and the solution $\theta(t)$ of the ODE starting at $\theta(0) = \theta_0$ and discretised at instants t_n have close trajectories. Consequently, it seems that the set of points towards which the sequence $(\theta_n)_{n \in \mathbb{N}}$ is likely to converge in mean belongs to the set of limit values of the trajectories of the ODE.

We shall already give a result valid for algorithms with both constant or decreasing stepsize, which justifies the remarks above. Noting $t_n = \sum_{k=1,n} \mu_k$, and $\mu = \max\{\mu_n; t_n < T\}$, it can be shown that ([81] p. 46)

$$\forall \varepsilon > 0, \forall T > 0, \lim_{\mu \to 0} P[\max_{0 \leq t_n \leq T} |\theta_n - \theta(n\mu)| > \varepsilon] = 0, \qquad (16.44)$$

which means that on the interval $[0, T]$ the values of θ_n remain confined in a tube centred on the trajectory of the ODE with a probability that increases when μ becomes smaller. We shall see below that we can, in fact, obtain stronger results of convergence for decreasing stepsize algorithms than for constant stepsize algorithms.

16.5.2 Convergence of Stochastic Algorithms

Definitions In what follows, $\theta(t)$ will denote the solution of the ODE starting from $\theta(0) = \theta_0$. In order to specify the asymptotic behaviour of the trajectories of the ODE, we begin by recalling the definitions below:

Definition 16.1 *A point θ_* is a stationary point of the ODE if $h(\theta_*) = 0$. A stationary point θ_* is said to be:*

- *stable if $\forall \varepsilon > 0, \exists \nu > 0, |\theta_0 - \theta_*| < \nu \Rightarrow \forall t \in \mathbb{R}_+, |\theta(t) - \theta_*| < \varepsilon$;*

- *asymptotically stable if $\exists \nu > 0$, $|\theta_0 - \theta_*| < \nu \Rightarrow \lim_{t \to \infty} \theta(t) = \theta_*$;*
- *globally asymptotically stable if $\mathcal{A}(\theta_*) = \{\theta_0; \lim_{t \to \infty} \theta(t) = \theta_*\}$ coincides with the set of all possible values of θ_0. $\mathcal{A}(\theta_*)$ is called the domain of attraction of θ_*.*

Moreover, we shall say that a set E of stationary points is globally asymptotically stable if all the trajectories of the ODE converge towards E.

Algorithms with Decreasing Stepsize We consider algorithms with decreasing stepsize such that $\exists \alpha > 1$, $\sum_{n=1,\infty} \mu_n^\alpha < \infty$, and $\sum_{n=1,\infty} \mu_n = \infty$. Then, we have the following result, which justifies considering the ODE for studying stochastic algorithms ([81] p. 51):

Theorem 16.5 *The sequence of values $(\theta_n)_{n \in \mathbb{N}}$ of the algorithm $\theta_n = \theta_{n-1} + \mu_n H(\theta_{n-1}, X_n)$ can only converge to a stable stationary point of the ODE $\mathbb{E}[H(\theta, X_n)] = d\theta/dt$.*

Moreover, it can be shown that:

Theorem 16.6 *If the ODE has a finite number of stable stationary points and if each trajectory of the ODE converges towards one of these points, then the sequence $(\theta_n)_{n \in \mathbb{N}}$ converges almost surely towards one of them.*

Algorithms with Constant Stepsize For constant stepsize algorithms, it can be shown that if the ODE has a unique globally asymptotically stable point θ_*, then under a certain number of hypotheses (such as $H(\theta, X_n) - h(\theta)$ has at most polynomial growth in X_n, ... see [81]), we obtain a central limit theorem: if we denote by θ_n^μ the values of θ_n associated with the algorithm of stepsize μ and

$$\tilde{\theta}_n^\mu = \frac{\theta_n^\mu - \theta_*}{\sqrt{\mu}}, \tag{16.45}$$

then when μ tends towards 0 and n tends towards infinity, $\tilde{\theta}_n^\mu$ tends towards a zero mean random variable with Gaussian distribution.

16.5.3 Asymptotic Variance

Here, we consider the problem of calculating the asymptotic variance of $\tilde{\theta}_n^\mu$.

In practice, μ is small and fixed, but we shall assume that we may still approximate the asymptotic distribution of $\mu^{-1/2}(\theta_n - \theta_*)$ when n tends towards infinity by a Gaussian distribution. The asymptotic covariance matrix of the residual error $\theta_n - \theta_*$, defined by $\Sigma_\infty = E[(\theta_n - \theta_*)(\theta_n - \theta_*)^T]$, can then be approximated by the solution to the following matrix equation, called Lyapunov's equation, ([81] p.103)

$$G\Sigma_\infty + \Sigma_\infty G^T + \mu R = 0, \tag{16.46}$$

where

$$G = \frac{dh(\theta_*)}{d\theta},$$

and $R = \sum_{n \in \mathbb{Z}} \mathbb{E}[\{H(\theta, X_n) - h(\theta)\}\{H(\theta, X_0) - h(\theta)\}^T | \theta = \theta^*],$

$$(16.47)$$

and where $dh(\theta)/d\theta$ is the matrix with general term

$$[\frac{dh(\theta)}{d\theta}]_{ij} = \frac{\partial[h(\theta)]_i}{\partial\theta_j}. \tag{16.48}$$

We note that if the random vectors X_n are independent and zero mean, R is simply equal to the covariance matrix of $H(\theta_*, X_0)$. Unless we have sufficient information about X, which very often is not the case, in practice we consider this simplifying hypothesis.

16.5.4 Stationary Stable Points of the ODE

We have seen the importance of the stable stationary points of the ODE for studying the convergence of stochastic algorithms. Therefore, we shall now indicate some methods that can be useful for characterising them. We begin by a simple and often useful result:

Theorem 16.7 *If $h(\theta) = -\nabla J(\theta)$, where $J(\theta)$ is a positive scalar function, all the trajectories of the ODE converge towards the set of stable stationary points of the ODE.*

This result shows that when $h(\theta)$ is the opposite of the gradient of a positive function, the stable stationary points of the ODE are globally asymptotically stable.

Proof For $h(\theta) = -\nabla J(\theta)$, with $J(\theta) \geq 0$, and a solution $\theta(t)$ of the ODE,

$$\frac{dJ(\theta(t))}{dt} = (\frac{d\theta(t)}{dt})^T \nabla J(\theta)$$

$$= h(\theta)^T \nabla J(\theta)$$

$$(16.49)$$

$$= - \| h(\theta) \|^2$$

$$\leq 0.$$

Therefore, $J(\theta(t))$ is a positive and decreasing function of t and has a finite limit. Consequently,

$$\lim_{t \to \infty} \| h(\theta) \|^2 = -\lim_{t \to \infty} \frac{dJ(\theta(t))}{dt}$$

$$(16.50)$$

$$= 0.$$

Therefore, $\theta(t)$ converges towards the set of stationary points of the ODE when t tends towards $+\infty$. $\square$

Therefore, when $h(\theta) = -\nabla J(\theta)$, and $J(\theta) \geq 0$, it appears that if the set of stable stationary points of the ODE is discrete, then each trajectory of the ODE converges towards such a point. If this is not the case, the convergence of each trajectory towards a precise point of this set is not guaranteed.

The eigenvalues of $dh(\theta_*)/d\theta$ also provide information about the stability of the point θ_* ([35] p.203):

Theorem 16.8 *Let θ_* be a stationary point of the ODE, and $\lambda_1, \ldots, \lambda_p$ the eigenvalues of $dh(\theta_*)/d\theta$.*

- *If $\forall i = 1, p$ $\mathcal{Re}(\lambda_i) < 0$, θ_* is asymptotically stable;*
- *if $\exists i \in \{1, \ldots, p\}$, $\mathcal{Re}(\lambda_i) > 0$, θ_* is unstable;*
- *if $\forall i = 1, p$ $\mathcal{Re}(\lambda_i) \leq 0$, and $\mathcal{Re}(\lambda_{i_0}) = 0$ for $i_0 \in \{1, \ldots, p\}$, we cannot conclude.*

16.5.5 Application to the LMS Algorithm

The ODE associated with the LMS algorithm is $d\theta/dt = h(\theta) = T_p\theta - r_{XA}$. The only stable stationary point of the ODE is $\theta_* = T_p^{-1}r_{XA}$, and since $h(\theta) = -\nabla \parallel \theta^H X_n - A_n \parallel^2$, Theorem 16.7 shows that θ_* is globally asymptotically stable. The only point towards which the LMS algorithm is likely to converge in mean is therefore $\theta_* = T_p^{-1}r_{XA}$.

Formula (16.46) leads to an expression of the asymptotic variance of $\theta_n - \theta_*$ up to the second order in μ. Formula (16.46) applied to the LMS algorithm leads to the equation

$$T_p\Sigma_\infty + \Sigma_\infty T_p = \mu T_p(Tr(\Sigma_\infty T_p) + \sigma_{\min}^2), \tag{16.51}$$

whose solution is

$$\Sigma_\infty = \frac{\mu\sigma_{\min}^2}{2 - \mu Tr(T_p)}I. \tag{16.52}$$

This solution coincides with the first order development in μ of formula (16.35) of the direct calculation of the asymptotic covariance matrix. We also note that the positivity condition $\Sigma_\infty > 0$ is expressed here by $\mu < 2(Tr(T_p))^{-1}$, which is a slightly weaker condition than the condition

$$\sum_{l=0,p} \mu\lambda_l(2 - \mu\lambda_l)^{-1} < 1 \tag{16.53}$$

that we had obtained before.

Exercises

16.1 Express the LMS and the RLS adaptive algorithms applied to the linear prediction problem $\min_a \parallel X_n - \sum_{k=1,p} a_k X_{n-k} \parallel^2$, where $a = [a_1, \ldots, a_p]^T$.

16.2 (Relation between RLS and Kalman algorithms) Show that minimising

$$\sum_{n=0,N} \lambda^{N-n} |u_n^H \mathbf{X}_n - Y_n|^2 \tag{16.54}$$

with respect to $\mathbf{X}_n$ amounts to performing the Kalman filtering for the model

$$\begin{cases} \mathbf{X}_{n+1} = (\lambda)^{-1/2} \mathbf{X}_n \\ Y_n = u_n^H \mathbf{X}_n + V_n, \end{cases} \tag{16.55}$$

where $\sigma_V^2 = 1$.

16.3 (Rectangular windowing) We consider the stochastic adaptive minimisation of $\theta^H \mathbf{X}_n - Y_n$ using the following update of θ:

$$theta_n = \arg\min_\theta \sum_{l=n-N,n} |\theta^H \mathbf{X}_l - Y_l|. \tag{16.56}$$

a) Express θ_n in terms of θ_{n-1}.
(Hint: consider the discussion to be found in the study of the RLS algorithm.)
b) Compare this approach with the RLS approach.

16.4 (Instrumental variable method) We consider a process $Y = (Y_n)_{n \in \mathbb{Z}}$ of the form $Y_n = \theta^H \mathbf{X}_n + W_n$, and we are looking for the vector θ. Y_n and $\mathbf{X}_n$ are observed but not W_n. In order to deal with this problem, we introduce random vectors $(\mathbf{Z}_n)_{n \in \mathbb{Z}}$ such that $\mathbb{E}[W_n \mathbf{Z}_n^H] = 0$. The RLS estimator sequence of θ that minimises $\sum_{l=1,n} \lambda^{n-l} |Y_l - \theta^H \mathbf{X}_l - W_l|^2$, given by

$$\theta_n = \left(\sum_{l=1,n} \lambda^{n-l} \mathbf{X}_l \mathbf{X}_l^H \right)^{-1} \left(\sum_{l=1,n} \lambda^{n-l} \mathbf{X}_l (Y_l - W_l)^* \right), \tag{16.57}$$

is replaced by

$$\theta_n = \left(\sum_{l=1,n} \lambda^{n-l} \mathbf{Z}_l \mathbf{X}_l^H \right)^{-1} \left(\sum_{l=1,n} \lambda^{n-l} \mathbf{Z}_l Y_l^* \right). \tag{16.58}$$

a) Rewrite θ_n given by Equation (16.58) as a function of θ_{n-1}.
b) We assume here that $\mathbb{E}[W_n \mathbf{Z}_n^H] = 0$ and that $\mathbb{E}[\mathbf{Z}_n \mathbf{X}_n^H]$ has an inverse. In addition, we choose $\lambda = 1$. Show that θ_n converges almost surely to θ.
c) Application. We consider an ARMA process $X: X_n + \sum_{k=1,p} a_k X_{n-k} = \sum_{l=0,q} b_l V_{n-l}$. Check that using the instrumental variable method enables $(a_k)_{k=1,p}$ to be updated without considering $(b_l)_{l=0,q}$. Give a possible choice for $\mathbf{Z}_n$.

16.5 (Alternative voltage component removal) A process of interest, denoted by $X = (X_n)_{n \in \mathbb{Z}}$, is corrupted by an alternative voltage component that we want to remove. The observed process is $Y_n = X_n + \alpha e^{2i\pi f_0 n}$, where f_0 is known.

a) α being unknown, express the LMS and the RLS updates that can be used to estimate it adaptively.
b) Study the convergence and the asymptotic variance of these algorithms.

16.6 (Adaptive cumulant estimation) We consider the following updating scheme for the second and fourth order cumulants of a zero mean random variable X from independent observations $X_1, X_2, \ldots$:

$$\begin{pmatrix} \hat{C}_{X,2}(n) \\ \hat{C}_{X,4}(n) \end{pmatrix}$$

$$= \begin{pmatrix} \hat{C}_{X,2}(n-1) \\ \hat{C}_{X,4}(n-1) \end{pmatrix} + \mu \begin{pmatrix} X_n^2 - \hat{C}_{X,2}(n-1) \\ X_n^4 - 3X_n^2 \hat{C}_{X,2}(n-1) - \hat{C}_{X,4}(n) \end{pmatrix}.$$
(16.59)

Study the convergence of this algorithm.

16.7 (Robbins-Monro and Kiefer-Wolfowitz procedures) A signal $(S_n)_{n \in \mathbb{N}}$ is used to control a system that generates an output signal $Y_{n+1} = g(S_n, V_{n+1})$, where g is unknown. The random variables $(V_n)_{n \in \mathbb{N}}$ are independent and V_n is independent of $S_{n-1}, S_{n-2}, \ldots$

a) Give an intuitive justification for the following algorithm, known as the Robbins-Monro procedure, that is designed to obtain outputs Y_n as close as possible to a fixed constant value c:

$$S_{n+1} = S_n - \mu_n (Y_{n+1} - c).$$
(16.60)

b) Now, we consider a given sequence $(c_n)_{n \in \mathbb{N}}$ that decreases to 0. We note $Y_{n+1}^+ = g(S_n + c_n, V_{n+1})$ and $Y_{n+1}^- = g(S_n - c_n, V_{n+1})$. The Kiefer-Wolfowitz procedure is defined by

$$S_{n+1} = S_n + (\mu_n/c_n)(Y_{n+1}^+ - Y_{n+1}^-).$$
(16.61)

What is this the aim of this procedure ?

16.8 (Convergence of the gradient and Newton algorithms) First, let us recall the following result (see ,for instance, [84] p.11):

Theorem 16.9 *Let g be a real continuous function, with $g(x^*) = c$ and*

$$\forall x \in \mathbb{R}, \, (g(x) - c)(x - x^*) < 0,$$
$$|g(x)| \qquad\qquad\qquad \leq C(1 + |x|),$$
(16.62)

where C is a constant. Let $(\mu_n)_{n \in \mathbb{N}}$ denote a sequence of positive stepsizes, with $\lim_{n \to \infty} \mu_n = 0$ and $\sum_{n \in \mathbb{N}} \mu_n = +\infty$, and $(V_n)_{n \in \mathbb{N}}$ a sequence of random variables such that $\sum_{n \in \mathbb{N}} \mu_n V_n$ converges almost surely. Then, the sequence defined iteratively by

$$Y_{n+1} = Y_n + \mu_n (g(Y_n) - c + V_{n+1})$$
(16.63)

converges almost surely to c for any initial value Y_0.

Now, we consider a function h of class C^2 having a minimum at point x^*.

a) Prove that the gradient algorithm $x_{n+1} = x_n - \mu_n h'(x_n)$ converges to x^* when $|h'(x)| \leq K(1 + |x|)$ on $\mathbb{R}$, for some constant value K.

b) Prove that the Newton algorithm $x_{n+1} = x_n - \mu_n (h''(x_n))^{-1} h'(x_n)$ converges to x^* when $|(h''(x_n))^{-1} h'(x)| \leq K(1 + |x|)$ on $\mathbb{R}$, for some constant value K.

16.9 (Phase lock loop) We consider a digital communication signal of the form

$$X_t = \exp\left(i(\frac{\pi}{4} + \frac{\pi}{2} \sum_{k \in \mathbb{Z}} A_k \mathbb{1}_{[0,T[}(t - kT) + \phi) \right), \tag{16.64}$$

where $A_k \in \{0, 1, 2, 3\}$ are independent random variables and ϕ is an unknown phase error term.

a) Check that X_t can be rewritten as $X_t = \sum_{k \in \mathbb{Z}} D_k \mathbb{1}_{[0,T[}(t - kT) e^{i\phi}$, and relate A_k to D_k.

b) We observe $Y_t = h(t) \star X_t$, where $h(t) = \mathbb{1}_{[0,T[}(t)$, at instant kT: $Y_{kT} = \int_{\mathbb{R}} h(t) X_{kT-t} dt$. Check that $Y_{kT} = D_{n-1} T e^{i\phi}$. What does filtering $h(t)$ correspond to when X is observed in the presence of white additive noise ?

c) In order to recover D_{n-1} from Y_{nT}, ϕ is estimated by means of a stochastic gradient algorithm applied to the criterion $J(\phi) = \mathbb{E}[|Y_{nT}^4 e^{-4i\phi} + T|^2]$. Justify the use of this criterion and check that the corresponding stochastic gradient updating scheme is given by

$$\phi_n = \phi_{n-1} - \mu_n \mathcal{I}m(Y_{nT}^4 e^{-4i\phi_{n-1}}). \tag{16.65}$$

This technique is known as *Costa's loop*.

d) We take $\mu_n = n^{-1}$. Check that ϕ_n converges to ϕ^*, where $\phi^* = \phi + k(\pi/2)$ and $k \in \mathbb{Z}$, when the algorithm stepsize decreases.

e) Check that taking $\hat{D}_{n-1} = T^{-1} Y_{nT} e^{-i\phi^*}$ may yield a biased estimator of D_{n-1}. In order to overcome this problem, in (16.64) data A_k are replaced by data $B_k = A_k + B_{k-1} mod[4]$. Explain how this strategy enables bias suppression.

16.10 (Blind equalisation)[87] Let us consider a sequence $D_n \in \{-1, +1\}$ of independent random variables transmitted over a propagative channel with unknown transfer function $F(z)$. $F(z)$ may be seen as an MA filter that might not be minimum-phase. The received signal is denoted by $X_n = [F(z)]D_n + V_n$, where $V = (V_n)_{n \in \mathbb{Z}}$ is a white noise with variance σ_V^2. Let $C(z)$ be the transfer function of the filter, called the equaliser, used to recover the transmitted symbols: D_n is estimated by $\hat{D}_n = sign([C(z)]X_n)$. In practice, adaptive implementations of $C(z)$ must be considered in order to account for time variable propagation environments.

a) Show that the optimal choice for $C(z)$ in the mean square sense, that is, $C(z)$ such that $\| [C(z)]X_n - D_n \|^2$ is minimum, is

$$C(z) = \frac{F^*(z^{-1})}{F(z)F^*(z^{-1}) + \sigma_V^2}. \tag{16.66}$$

b) Let $\alpha G(z)G^*(z^{-1})$ represent the minimum-phase factorisation of the denominator of $C(z)$: $G(z) = \sqrt{\alpha} \prod_{k=1,K}(1 - z_k z^{-1})$ with $|z_k| < 1$. We note $C(z) = [G(z)]^{-1}T(z)$. Check that $[G(z)]^{-1}X_n$ is a white noise and that $T(z)$ is an all-pass transfer function.

c) Show that $[G(z)]^{-1}$ can be implemented approximately by a filter with input X_n and output Y_n defined by $Y_n = -[A(z)]Y_n + X_n$, where $A(z) = \sum_{k=1,p} a_k z^{-k}$ minimises the criterion $\| X_n - [A(z)]Y_n \|$. Give the expression of the gradient algorithm that realises an adaptive implementation of the criterion.

d) $T(z)$ is approximated by an MA filter with input Y_n and output $Z_n = \sum_{k=0,q} b_k Y_{n-k}$. The coefficients $(b_k)_{k=0,q}$ are chosen so as to minimise the following criterion, known as the *Godard criterion*: $\| |Z_n|^2 - 1 \|^2$. Explain why it is necessary to consider a criterion involving higher order statistics of the data to estimate $T(z)$. Show that a transfer function $H(z)$, such that $\| |[H(z)]Y_n|^2 - 1 \|^2$ is minimum, is equal to $T(z)$ up to one phase error factor: $H(z) = e^{i\theta}T(z)$. Express the stochastic gradient algorithm that updates $(b_k)_{k=0,q}$.

e) Finally, in order to suppress the phase error factor $e^{i\theta}$, the following algorithm is used

$$\begin{aligned} S_n &= Z_n e^{-i\theta_{n-1}}, \\ \theta_n &= \theta_{n-1} + \mu \mathcal{I}m[S_n(\hat{D}_n - S_n)^*], \end{aligned} \tag{16.67}$$

where $\hat{D}_n = sign[S_n]$ are the decisions at the output of the equaliser. Find the criterion that this stochastic gradient algorithm implements and explain it.

A. Elements of Measure Theory

We recall here, without proofs, some important results of measure theory (see for instance [1, 29]). Unless otherwise stated, the functions under consideration map $\mathbb{R}$ onto $\mathbb{C}$, and μ represents a measure on the σ-algebra of Borel sets, denoted by $\mathcal{B}(\mathbb{R})$.

Monotone convergence theorem

Theorem A.1 *If $(f_n)_{n \in \mathbb{N}^*}$ represents an increasing sequence of positive measurable functions that converge simply towards a function f, then f is measurable and*

$$\lim_{n \to \infty} \int_{\mathbb{R}} f_n d\mu = \int_{\mathbb{R}} f d\mu. \tag{A.1}$$

Fatou's theorem

Theorem A.2 (Fatou) *If μ represents a positive measure, and if $(f_n)_{n \in \mathbb{N}^*}$ represents a sequence of measurable positive functions,*

$$\int_{\mathbb{R}} (\liminf_{n \to \infty} f_n) d\mu \leq \liminf_{n \to \infty} \int_{\mathbb{R}} f_n d\mu. \tag{A.2}$$

Lebesgue's dominated convergence theorem

Theorem A.3 (Lebesgue) *If $(f_n)_{n \in \mathbb{N}^*}$ represents a sequence of measurable functions, such that*

$$\lim_{n \to \infty} f_n(x) = f(x) \ \mu\text{-almost everywhere}$$

and $|f_n(x)| \leq g(x)$ μ-almost everywhere, where $g \in L^1(\mathbb{R}, \mathcal{B}(\mathbb{R}), d\mu)$,
$$\tag{A.3}$$

then, $f \in L^1(\mathbb{R}, \mathcal{B}(\mathbb{R}), d\mu)$ and

$$\lim_{n \to \infty} \int_{\mathbb{R}} f_n d\mu = \int_{\mathbb{R}} f d\mu. \tag{A.4}$$

Remark A function, integrable in the Riemann sense on a bounded interval, is also integrable in the Lebesgue sense and both integrals are equal. More generally, a function integrable in the Riemann sense is also integrable in the Lebesgue sense if and only if it is absolutely integrable.

Hölder's inequality Hölder's inequality generalises Cauchy-Schwarz's inequality (obtained for $p = 1/2$): we consider the norms $\| f \|_p = (\int |f|^p d\mu)^{1/p}$. Then, for $1 \leq p < \infty$, and $p^{-1} + q^{-1} = 1$,

$$\| fg \|_1 \leq \| f \|_p \| g \|_q .$$
(A.5)

Jensen's inequality

Theorem A.4 (Jensen) *If φ is a measurable convex function and μ a probability measure, then for any function $f \in L^1(\mathbb{R}, \mathcal{B}(\mathbb{R}), d\mu)$,*

$$\varphi(\int_{\mathbb{R}} f(x)d\mu(x)) \leq \int_{\mathbb{R}} \varphi(f(x))d\mu(x).$$
(A.6)

Product of measures and Fubini's theorem A measure μ on a measurable space (E, B) is said to be σ-finite if there is a partition $(A_n)_{n \in \mathbb{N}^*}$ of E such that $\forall n \in \mathbb{N}^*, \mu(A_n) < \infty$.

Let (E_1, B_1, μ_1) and (E_2, B_2, μ_2) be two measured spaces, where μ_1 and μ_2 are σ-finite. We denote by $B_1 \otimes B_2$ the σ-algebra generated by the elements $A_1 \times A_2 \in B_1 \times B_2$.

Then, there is a unique measure on $(E_1 \times E_2, B_1 \otimes B_2)$, denoted by $\mu_1 \otimes \mu_2$, such that

$$\forall A_1 \times A_2 \in B_1 \times B_2, \quad \mu_1 \otimes \mu_2(A_1 \times A_2) = \mu_1(A_1)\mu_2(A_2).$$
(A.7)

Theorem A.5 (Fubini) *If $f \in L^1(E_1 \times E_2, B_1 \otimes B_2, d(\mu_1 \otimes \mu_2)(x, y))$, then for almost every x the function $y \to f(x, y)$ belongs to $L^1(E_2, B_2, d\mu_2(y))$ and for almost every y the function $x \to f(x, y)$ belongs to $L^1(E_1, B_1, d\mu_1(y))$. Moreover,*

$$\int_{E_1 \times E_2} f(x, y)d(\mu_1 \otimes \mu_2)(x, y) = \int_{E_1} d\mu_1(x)[\int_{E_2} f(x, y)d\mu_2(y)]$$

$$= \int_{E_2} d\mu_2(y)[\int_{E_1} f(x, y)d\mu_1(x)].$$
(A.8)

Tonelli's theorem In order to be able to apply Fubini's theorem, we are often led to test that the function $f(x, y)$ belongs to the set $L^1(E_1 \times E_2, B_1 \otimes B_2, d(\mu_1 \otimes \mu_2)(x, y))$ by means of Tonelli's theorem.

Theorem A.6 (Tonelli) *If for almost every x the function $y \to f(x, y)$ belongs to $L^1(E_2, B_2, d\mu_2(y))$ and $\int_{E_2} f(x, y)d\mu_2(y)$ belongs to $L^1(E_1, B_1, d\mu_1(x))$, then $f \in L^1(E_1 \times E_2, B_1 \otimes B_2, d(\mu_1 \otimes \mu_2)(x, y))$.*

Change of variables Let ϕ be a derivable application, mapping an open set O of $\mathbb{R}^n$ onto $\mathbb{R}^n$. We denote by $y = \phi(x)$ and $J_\phi(x)$ the matrix with general term $(\frac{\partial y_i}{\partial x_j})$ at the point x.

We assume that ϕ is a bijective mapping from $A_x \subset O$ onto A_y, where A_x and A_y belong to $\mathcal{B}(\mathbb{R}^n)$, and that $|J_\phi(x)| \neq 0$ at any point of A_x. Then,

$$f(y) \in L^1(A_y, \mathcal{B}(\mathbb{R}^n) \cap A_y, dy)$$
$$\Leftrightarrow f(\phi(x)) \times |J_\phi(x)| \in L^1(A_x, \mathcal{B}(\mathbb{R}^n) \cap A_x, dx),$$

(A.9)

and

$$\int_{A_y} f(y)dy = \int_{A_x} f(\phi(x)) |(|J_\phi(x)|)| dx. \tag{A.10}$$

Continuity and differentiation of integrals depending on the parameter Let $f(x,t)$ be μ-integrable with respect to x for any value of t.

If f is continuous at point t_0 for μ-almost every x and if there is a neighbourhood V_{t_0} of t_0 and $g(x) \in L^1(\mathbb{R}, \mathcal{B}(\mathbb{R}), d\mu)$ such that

$$\forall t \in V_{t_0}, \quad |f(x,t)| \leq g(x) \text{ for } \mu\text{-almost every } x, \tag{A.11}$$

then $\int f(x,t)d\mu(x)$ is continuous at point t_0.

If $\frac{\partial f}{\partial t}(x,t)$ exists at $t = t_0$ for μ-almost every x and if there is a neighbourhood V_{t_0} of t_0 and $g(x) \in L^1(\mathbb{R}, \mathcal{B}(\mathbb{R}), d\mu)$ such that

$$\forall t \in V_{t_0}, \quad |f(x,t) - f(x,t_0)| \leq |t - t_0| g(x) \text{ for } \mu\text{-almost every } x, \tag{A.12}$$

then $\int f(x,t)d\mu(x)$ is derivable at point t_0 and

$$\left(\frac{\partial}{\partial t} \int f(x,t)d\mu(x) \right)_{t=t_0} = \int \left(\frac{\partial f(x,t)}{\partial t} \right)_{t=t_0} d\mu(x). \tag{A.13}$$

Helly's selection theorem Let $(\mu_n)_{n\geq 1}$ be a set of positive measures such that $\mu_n(\mathbb{R}) < c$ for some positive constant c, and denote by $(F_n)_{n\geq 1}$ their distribution functions: $F_n(x) = \mu_n(] - \infty, x])$. Then,

Theorem A.7 (Helly) *There exists a subsequence $(F_{n_k})_{k\geq 1}$ of $(F_n)_{n\geq 1}$ that converges to a distribution function F, at any point of continuity of F.*

Proof See for instance [1] p.289 or [29] p.158.

Letting μ denote the measure defined by $\mu(] - \infty, x]) = F(x)$, it comes that the sequence of measures $(\mu_{n_k})_{k\geq 1}$ converges to μ in the sense that $\forall A \in \mathcal{B}(\mathbb{R})$, $\lim_{k\to\infty} \mu_{n_k}(A) = \mu(A)$.

B. $L^2(\Omega, \mathcal{A}, dP)$ is a Complete Space

Theorem B.1 $L^2(\Omega, \mathcal{A}, P)$ *is a Complete Space.*

Proof To show that $L^2(\Omega, \mathcal{A}, dP)$ is a complete space, we consider a Cauchy sequence $(X_n)_{n \in \mathbb{N}}$ of $L^2(\Omega, \mathcal{A}, dP)$. It is clear that

$$\exists n_1 < n_2 < \ldots, \quad \| X_{n_{k+1}} - X_{n_k} \| < 2^{-k}. \tag{B.1}$$

We write $n_0 = 0$, and $X_0 = 0$. We show the almost sure convergence of X_{n_k}. This amounts to showing that the series whose partial sums are of the form $\sum_{l=0, k-1}(X_{n_{l+1}} - X_{n_l}) = X_{n_k}$ converges almost surely. From the theorem of monotone convergence and Cauchy-Schwarz's inequality,

$$\mathbb{E}[\sum_{l=0,\infty} |X_{n_{l+1}} - X_{n_l}|] = \sum_{l=0,\infty} \mathbb{E}[|X_{n_{l+1}} - X_{n_l}|]$$

$$\leq \sum_{l=0,\infty} \| X_{n_{l+1}} - X_{n_l} \| \tag{B.2}$$

$$\leq \| X_{n_1} \| + \sum_{l=1,\infty} 2^{-k} < \infty.$$

Therefore, almost surely the series $\sum_{l \geq 0}(X_{n_{l+1}} - X_{n_l})$ converges absolutely towards a random variable and thus converges. Consequently, $(X_{n_k})_{k \geq 0}$ converges almost surely towards a random variable denoted by X. To complete the proof, it is sufficient to notice that

$$\| X_n - X \| \leq \| X_n - X_{n_k} \| + \| X_{n_k} - X \|, \tag{B.3}$$

and that the two terms of the right-hand side of the inequality tend towards 0 when n and n_k tend towards $+\infty$. $\square$

C. Continuous Extension of a Linear Operator

Theorem C.1 *Let $T : D \to B$ be a bounded linear operator, where D is a dense subset of a space A equipped with a norm, and B a Banach space (a complete normed vector space). Then, there exists a bounded linear operator $\tilde{T}$ defined on A such that $\forall x \in D$, $\tilde{T}x = Tx$, and $\| \tilde{T} \| = \| T \|$.*

Proof For $x \in D$, we let $\tilde{T}x = Tx$. For $x \in A$, but $x \notin D$, we put $\tilde{T}x = \lim_n Tx_n$, where $(x_n)_{n \in \mathbb{N}}$ is a sequence of elements of D that converges towards x.

The limit of the sequence $(Tx_n)_{n \in \mathbb{N}}$ does exist for $\| Tx_n - Tx_m \| \leq \| T \| \times \| x_n - x_m \|$, and $(x_n)_{n \in \mathbb{N}}$ is a Cauchy sequence. Therefore, the sequence $(Tx_n)_{n \in \mathbb{N}}$ is also a Cauchy sequence, and as B is complete, the sequence $(Tx_n)_{n \in \mathbb{N}}$ converges.

Moreover, the limit of the sequence $(Tx_n)_{n \in \mathbb{N}}$ does not depend on the choice of any particular sequence $(x_n)_{n \in \mathbb{N}}$ converging towards x. Indeed, if $(x_n)_{n \in \mathbb{N}}$ and $(x'_n)_{n \in \mathbb{N}}$ converge towards x, the limits a and a' of Tx_n and Tx'_n satisfy:

$$\| a - a' \| \leq \| a - Tx_n \| + \| T \| \times (\| x_n - x \| + \| x - x'_n \|)$$
$$+ \| Tx'_n - a' \|, \tag{C.1}$$

and the right-hand terms converge towards 0 when $n \to \infty$. Therefore, $a = a'$.

To show that $\| T \| = \| \tilde{T} \|$, we remark that $\| Tx_n \| \| x_n \|^{-1} \leq \| T \|$, and therefore $\| \tilde{T} \| \leq \| T \|$. We have, moreover,

$$\| \tilde{T} \| = \sup_{x \in A, \|x\|=1} \| \tilde{T}(x) \| \geq \sup_{x \in D, \|x\|=1} \| \tilde{T}(x) \| = \| T \| . \tag{C.2}$$

Therefore, $\| T \| = \| \tilde{T} \|$. $\square$

We note, moreover, that if T preserves the norm, the same is true for $\tilde{T}$. Indeed, $\forall x \in A$, x is the limit of a sequence $(x_n)_{n \in \mathbb{N}}$ of D and from the continuity of the norm operator,

$$\| \tilde{T}(x) \| = \lim_{n \to \infty} \| \tilde{T}(x_n) \| = \lim_{n \to \infty} \| T(x_n) \| = \lim_{n \to \infty} \| x_n \| = \| x \| . \tag{C.3}$$

D. Kolmogorov's Isomorphism and Spectral Representation

Theorem *The function*

$$T_X : \sum_n \alpha_n e^{2i\pi t_n f} \to \sum_n \alpha_n X_{t_n}, \tag{D.1}$$

where the summation is finite, can be extended to an isomorphism that maps $L^2(\mathbb{R}, \mathcal{B}(\mathbb{R}), d\mu_X(f))$ onto H_X, and that maps $L^2(\mathcal{I}, \mathcal{B}(\mathcal{I}), d\mu_X(f))$ onto H_X for the discrete case.

Theorem *For a zero mean, mean square continuous, WSS process X indexed by $\mathbb{R}$, there is a unique stochastic measure $\hat{X}(f)$ such that $X_t = \int_{\mathbb{R}} e^{2i\pi f t} d\hat{X}(f)$.*
For a discrete time, zero mean, WSS process, there is a single stochastic measure $\hat{X}(f)$ such that $X_n = \int_{\mathcal{I}} e^{2i\pi n f} d\hat{X}(f)$.

Proof We consider the application

$$T_X : \sum_{k=1,n} \alpha_k e^{2i\pi f t_k} \to \sum_{k=1,n} \alpha_k X_{t_k}. \tag{D.2}$$

It defines a normed vector space homomorphic transform, mapping the space of finite linear combinations $\sum_{k=1,n} \alpha_k e^{2i\pi f t_k}$ into H_X, since, from Bochner's theorem,

$$\int_{\mathbb{R}} e^{2i\pi f t} d\mu_X(f) = R_X(t) = \mathbb{E}[X_{t+\tau} X_\tau^*]. \tag{D.3}$$

Therefore, T_X can be extended to a homomorphism $\tilde{T}_X$ mapping $L^2(\mathbb{R}, \mathcal{B}(\mathbb{R}), d\mu_X)$ onto H_X since the set of the functions of the form $\sum_{k=1,n} \alpha_k e^{2i\pi f t_k}$ is a dense subset of $L^2(\mathbb{R}, \mathcal{B}(\mathbb{R}), d\mu_X)$.

We show that $\tilde{T}_X$ is bijective. $\tilde{T}_X$ is injective since $\forall Y \in H_X$, if $\tilde{T}_X(\phi_1) = \tilde{T}_X(\phi_2) = Y$, then

$$\| \phi_1 - \phi_2 \| = \| \tilde{T}_X(\phi_1) - \tilde{T}_X(\phi_2) \| = 0, \tag{D.4}$$

and, therefore, $\phi_1 = \phi_2$.

To show that $\tilde{T}_X$ is surjective, we consider any element Y of H_X. Y is the limit of a sequence of random variables $Y_n = \sum_{k=1,n} \alpha_{k,n} X_{t_{k,n}}$, and we have $Y_n = \tilde{T}_X(\phi_n)$, with $\phi_n = \sum_{k=1,n} \alpha_{k,n} e^{2i\pi f t_{k,n}}$. The sequence $(\phi_n)_{n \in \mathbb{N}}$

is convergent since it is a Cauchy sequence of $L^2(\mathbb{R}, \mathcal{B}(\mathbb{R}), d\mu_X)$. Indeed, $\| \phi_n - \phi_m \| = \| Y_n - Y_m \|$, which tends towards 0 when $(m, n) \to \infty$, since the sequence $(Y_n)_{n \in \mathbb{N}}$ converges. Let ϕ be the limit of $(\phi_n)_{n \in \mathbb{N}}$. It is clear that $\tilde{T}_X(\phi) = Y$, for

$$\| \tilde{T}_X(\phi) - Y \| \le \| \tilde{T}_X(\phi) - \tilde{T}_X(\phi_n) \| + \| Y_n - Y \|$$
$$\le \| \phi - \phi_n \| + \| Y_n - Y \|,$$
(D.5)

and the two right-hand terms tend towards 0 when n tends towards $+\infty$. Therefore, $\tilde{T}_X$ is surjective, and finally $\tilde{T}_X$ is an isomorphism.

We easily check that the function $\hat{X}$ defined by $\hat{X}(\Delta) = \tilde{T}_X(\mathbb{1}_\Delta)$, for any Borel set Δ, is a stochastic measure. To show that X_t, the image of $e^{2i\pi ft}$, is equal to $\int_\mathbb{R} e^{2i\pi ft} d\hat{X}(f)$, we shall use the fact that the functions $\mathbb{1}_\Delta$, where $\Delta \in \mathcal{B}(\mathbb{R})$, constitute a generating family of $L^2(\mathbb{R}, \mathcal{B}(\mathbb{R}), d\mu_X)$, and therefore that $\chi_t(f) = e^{2i\pi ft}$ is the limit of a sequence of finite linear combinations of index functions $\phi_n(f) = \sum_{k=1,N_n} \alpha_{k,n} \mathbb{1}_{\Delta_{k,n}}(f)$. As $\tilde{T}_X$ is an isomorphism,

$$\tilde{T}_X(e^{2i\pi ft}) = \tilde{T}_X \left(\lim_{n \to \infty} \sum_{k=1,N_n} \alpha_{k,n} \mathbb{1}_{\Delta_{k,n}}(f) \right)$$

$$= \lim_{n \to \infty} \sum_{k=1,N_n} \alpha_{k,n} \hat{X}(\Delta_{k,n})(f)$$
(D.6)

$$= \lim_{n \to \infty} \int_\mathbb{R} \phi_n(f) d\hat{X}(f).$$

It thus results from the very definition of the stochastic integral that

$$X_t = \tilde{T}_X(e^{2i\pi ft}) = \int_\mathbb{R} e^{2i\pi ft} d\hat{X}(f).$$
(D.7)

For the discrete case, a similar approach can be used, defining

$$T_X : \sum_{k=1,n} \alpha_k e^{2i\pi kf} \to \sum_{k=1,n} \alpha_k X_k$$
(D.8)

and from the fact that the functions $\mathbb{1}_\Delta(f)$, where $\Delta \in \mathcal{B}(\mathcal{I})$, constitute a generating family of the space $L^2(\mathcal{I}, \mathcal{B}(\mathcal{I}), d\mu_X)$. $\square$

E. Wold's Decomposition

Theorem *Wold's decomposition leads to a representation of X in the form $X_n = Y_n + Z_n$, with $Y_n = X_n/H_{\nu,n}$ and $Z_n = X_n/H_{X,-\infty}$, where Y_n and Z_n are regular and singular respectively.*

If Lebesgue's decomposition of μ_X is given by $d\mu_X(f) = S_X(f)df + d\mu_X^s(f)$, where μ_X^s is carried by a set of measure zero, the innovation I_n of X satisfies

$$\| I_n \|^2 = \exp\left(\int_{\mathcal{I}} \log S_X(f)df\right). \tag{E.1}$$

Singular processes are characterised by the fact that $I_n = 0$. When $I_n \neq 0$, the spectral measures of Y and of Z then satisfy $d\mu_Y(f) = S_X(f)df$, and $d\mu_Z(f) = d\mu_X^s(f)$.

Proof Let us show first that Z is singular. To do this, we remark that $\forall n$ $H_{X,-\infty} = H_{X,n}/H_{X,-\infty}$, since $H_{X,-\infty} \subset H_{X,n}$. $H_{X,-\infty}$ is, therefore, generated by the random variables $Z_k = X_k/H_{X,-\infty}$, for $k \leq n$. Consequently, $H_{Z,n} = \overline{\operatorname{span}}\{Z_k; k \leq n\} = H_{X,-\infty}$ and $H_{Z,-\infty} = H_{X,-\infty} = H_{Z,\infty}$. Hence the singularity of Z.

We show now that Y is regular. $Y_n = Y_n/H_{\nu,n}$, therefore $Y_n \in H_{\nu,n}$. Consequently, $H_{Y,n} \subset H_{\nu,n}$ and $H_{Y,-\infty} \subset H_{\nu,-\infty}$. Since $H_{\nu,-\infty} = \{0\}$ (a white noise is a regular process), $H_{Y,-\infty} = \{0\}$, and Y is regular.

The relation $\| I_n \|^2 = \exp(\int_{\mathcal{I}} \log S_X(f)df)$ is a consequence of one of Szegö's theorems that states that for a positive, measurable function S

$$\min_{(a_k)_{k\geq 1}} \int_{\mathcal{I}} |1 - \sum_{k\geq 1} a_k e^{-2i\pi kf}|^2 S(f)df = \exp\left(\int_{\mathcal{I}} \log S(f)df\right). \tag{E.2}$$

For the proof of this result, refer, for example, to [41] p.189.

We now remark that

$$\| I_n \|^2 = \min_{(a_k)_{k\geq 1}} \| X_n - \sum_{k\geq 1} a_k X_{n-k} \|^2$$
$$= \min_{(a_k)_{k\geq 1}} \int_{\mathcal{I}} |1 - \sum_{k\geq 1} a_k e^{-2i\pi kf}|^2 d\mu_X(f)df, \tag{E.3}$$

and we denote by α_k the values of the coefficients a_k that achieve the minimum. It suffices, therefore, to show that the function $1 - \sum_{k\geq 1} \alpha_k e^{-2i\pi kf}$, which ensures the minimum of the above integral, is equal to zero on the set E of the mass points of μ_X^S, to ensure that

$$\| I_n \|^2 = \min_{(a_k)_{k\geq 1}} \int_{\mathcal{I}} |1 - \sum_{k\geq 1} a_k e^{-2i\pi k f}|^2 S_X(f) df$$

(E.4)

$$= \exp\left(\int_{\mathcal{I}} \log S_X(f) df\right).$$

It is clear that E is a finite set or a denumerable set, otherwise we would have $\mu_X^s(\mathbb{R}) = +\infty$, and consequently $\| X \|^2 = +\infty$ (check this as an exercise). We assume that $1 - \sum_{k\geq 1} \alpha_k e^{-2i\pi k f_l}$ is not equal to zero at a certain point f_l of E. The sequence of functions

$$g_N(f) = 1 - \frac{1}{N} \sum_{n=1,N} e^{-2i\pi n(f-f_l)}$$

(E.5)

$$= 1 - \frac{\sin(\pi N(f - f_l))}{N \sin(\pi(f - f_l))} e^{-i\pi(N+1)(f-f_l)}$$

is equal to 0 in f_l and converges uniformly towards 1 on any closed interval not containing f_l. The limit of $g_N(f)$ in $L^2(\mathcal{I}, \mathcal{B}(\mathcal{I}), d\mu_X)$ is therefore the function $\mathbb{1}_{\mathcal{I}-\{f_l\}}(f)$, and the functions $1 - \sum_{k>0} \beta_k^N e^{-2i\pi k f} = g_N(f)(1 - \sum_{k>0} \alpha_k e^{-2i\pi k f})$ are such that

$$\| X_n - X_n/H_{X,n-1} \| = \| X_n - \sum_{k\geq 1} \alpha_k X_{n-k} \|$$

(E.6)

$$> \lim_{N\to+\infty} \| X_n - \sum_{k\geq 1} \beta_k^N X_{n-k} \|,$$

which is contradictory. Therefore, $1 - \sum_{k\geq 1} \alpha_k e^{-2i\pi k f_l}$ is equal to zero on E, which completes the proof of this part of the theorem.

We show that $I_n = 0$ if and only if X is singular. If $I_n = 0$, $Y_n = 0$, and $X_n = Z_n$. Therefore, X_n is singular. Conversely, if X is singular, $H_{X,\infty} = H_{X,-\infty}$. In particular, $X_n \in H_{X,n-1}$, and $I_n = X_n - X_n/H_{X,n-1} = 0$. The condition $I_n = 0$ therefore characterises singular processes.

We now assume that $I_n \neq 0$, and we show that $d\mu_Y(f) = S_X(f)df$, and that $d\mu_Z(f) = d\mu_X^s(f)$. We shall begin by establishing that $S_X(f)df$ and $d\mu_X^s$ are the spectral measures of a regular process and of a singular process respectively. Then we shall show that these processes coincide with Y and Z.

We first consider the process S defined by

$$S_n = \int_{\mathcal{I}} \mathbb{1}_E(f) e^{2i\pi n f} d\hat{X}(f)$$

(E.7)

$$= \sum_{f_k \in E} e^{2i\pi n f_k} \hat{X}(\{f_k\}).$$

The spectral measure of S is μ_X^s. To show that S is singular, it is sufficient to consider the singular processes S^k defined by $S_n^k = e^{2i\pi n f_k} \hat{X}(\{f_k\})$. The processes S^k generate orthogonal spaces; indeed, $\forall m, n, k \neq l, H_{S^k,m} \perp H_{S^l,n}$ since the variables $(\hat{X}(\{f_k\}))_{f_k \in E}$ are uncorrelated. As $H_{S,n} = \oplus_k H_{S^k,n}$, we have finally

$$H_{S,+\infty} = \oplus_k H_{S^k,+\infty} = \oplus_k H_{S^k,-\infty} = H_{S,-\infty}. \tag{E.8}$$

S is therefore a singular process.

Similarly, we define the process R by

$$R_n = \int_{\mathcal{I}} (1 - \mathbb{I}_E(f)) e^{2i\pi n f} d\hat{X}(f). \tag{E.9}$$

Thus, $X = R + S$, and $\hat{X} = \hat{X}_R + \hat{X}_S$, with $d\hat{X}_S(f) = \mathbb{I}_E(f) d\hat{X}(f)$ and $d\hat{X}_R(f) = \mathbb{I}_{\mathcal{I}-E}(f) d\hat{X}(f)$. $d\hat{X}_S(f)$ represents the stochastic measure of S and $d\mu_X^S(f)$ its spectral measure. The process R, of stochastic measure $d\hat{X}_R(f)$, is orthogonal to S, since

$$\forall m, n \; \mathbb{E}[R_n S_m^*] = \int_{\mathcal{I}} e^{2i\pi(n-m)f} \mathbb{I}_{\mathcal{I}-E}(f) \mathbb{I}_E(f) d\mu_X(f) = 0. \tag{E.10}$$

Moreover, the spectral measure of R is $S_X(f) df$ (or more exactly, $S_X(f)$ matches the power spectral density of R, except on the zero measure set E).

We show that R is a regular process. If this is not the case, there exists a non-zero random variable $U \in H_{R,-\infty}$. Consequently, since $\forall n, U \in H_{R,n}$,

$$\exists n, \quad U = \sum_{k \geq 0} a_k R_{n-k}, \text{ and } a_0 \neq 0. \tag{E.11}$$

Let there be $V = a_0^{-1} U$. $V \in H_{R,-\infty}$, therefore, $V \in H_{n-1}$, and $V = \sum_{k \geq 0} b_k R_{n-k-1}$. We write

$$1 - \sum_{k \geq 1} c_k e^{-2i\pi k f} = a_0^{-1} \sum_{k \geq 0} a_k e^{-2i\pi k f} - \sum_{k \geq 1} b_{k-1} e^{-2i\pi k f}. \tag{E.12}$$

It is clear that

$$\| V - V \|^2 = \| \int_{\mathcal{I}} e^{2i\pi n f} (1 - \sum_{k \geq 1} c_k e^{-2i\pi k f}) d\hat{R}(f) \|^2$$

$$= \int_{\mathcal{I}} |1 - \sum_{k \geq 1} c_k e^{-2i\pi k f}|^2 S_X(f) df \tag{E.13}$$

$$\geq \min_{(\alpha_k)_{k \geq 1}} \int_{\mathcal{I}} |1 - \sum_{k \geq 1} \alpha_k e^{-2i\pi k f}|^2 S_X(f) df.$$

The left-hand term of the last inequality is equal to 0 and the right-hand term is $\| I_n \|^2$, which is impossible, from the hypothesis $I_n \neq 0$. Therefore, R is a regular process.

Consequently,

$$H_{X,-\infty} = H_{R,-\infty} \oplus H_{S,-\infty} = H_{S,-\infty}, \tag{E.14}$$

and since $S \perp R$, and S is singular,

$$Z_n = X_n / H_{X,-\infty}$$

$$= R_n / H_{S,-\infty} + S_n / H_{S,-\infty} = S_n. \tag{E.15}$$

As $X_n = R_n + S_n = Y_n + Z_n$, we also have $Y_n = R_n$, which completes the proof. $\square$

F. Dirichlet's Criterion

Lemma F.1 (Riemann-Lébesgue) *If $S(f)$ is a continuous function, periodic with period 1,*

$$\lim_{t \to \infty} \int_{\mathcal{I}} e^{2i\pi tf} S(f) df = 0. \tag{F.1}$$

Theorem F.2 *If a function S, periodic with period 1, has limits $S(f^-)$ and $S(f^+)$ on the left and on the right at point f, and if the function*

$$u \to u^{-1}[S(f+u) + S(f-u) - S(f^+) - S(f^-)] \tag{F.2}$$

is bounded in the neighbourhood of 0, then the Fourier series associated with S converges towards $\frac{1}{2}[S(f^+) + S(f^-)]$ at point f.

Corollary F.3 *If S, periodic with period 1, is continuous, piecewise derivable, and with a bounded derivative, the Fourier series associated with S converges towards $S(f)$ at any point f in $\mathcal{I}$.*

Proof We easily check that the lemma is satisfied for the step functions. Since any continuous function $S(f)$ is the limit, in the sense of uniform convergence, of a sequence of step functions $(\phi_n(f))_{f \geq 0}$, we obtain

$$\forall \varepsilon > 0, \exists N \in \mathbb{N}, \forall n > N, \exists T_n \in \mathbb{R}, \forall t > T_n,$$

$$\left| \int_{\mathcal{I}} e^{2i\pi tf} S(f) df \right| \leq \left| \int_{\mathcal{I}} e^{2i\pi tf} \phi_n(f) df \right| + \int_{\mathcal{I}} |\phi_n(f) - S(f)| df \tag{F.3}$$

$$\leq \varepsilon.$$

To prove the theorem, we note that

$$\sum_{k=-n,n} R(k) e^{-2i\pi kf} = \int_{\mathcal{I}} \sum_{k=-n,n} e^{2i\pi k(u-f)} S(u) du$$

$$= \int_{\mathcal{I}} \frac{\sin(\pi(2n+1)(u-f))}{\sin(\pi(u-f))} S(u) du, \tag{F.4}$$

$$\sum_{k=-n,n} R(k)e^{-2i\pi kf} = \int_{\mathcal{I}} \frac{\sin(\pi(2n+1)u)}{\sin(\pi u)}S(u+f)du$$

$$= \frac{1}{2}\int_{\mathcal{I}} \frac{\sin(\pi(2n+1)u)}{\sin(\pi u)}[S(f+u)+S(f-u)]du.$$

$$(F.5)$$

For $S(f) = 1$ on $\mathcal{I}$, we have $R(k) = \delta_{0,k}$, and therefore $\int_{\mathcal{I}} \frac{\sin(\pi(2n+1)u)}{\sin(\pi u)}du = 1$.
Thus,

$$|\sum_{k=-n,n} R(k)e^{-2i\pi kf} - \frac{1}{2}[S(f^+)+S(f^-)]|$$

$$\leq \frac{1}{2}|\int_{\mathcal{I}} (u\frac{\sin(\pi(2n+1)u)}{\sin(\pi u)})(\frac{1}{u}[S(f+u)+S(f-u)-S(f^+)-S(f^-)]du|$$

$$\leq C|\int_{\mathcal{I}} \frac{u}{\sin(\pi u)}\sin(\pi(2n+1)u)du|,$$

$$(F.6)$$

Since $\frac{u}{\sin(\pi u)}$ is continuous on $\mathcal{I} - \{0\}$, and has a continuous extension on $\mathcal{I}$, Riemann-Lebesgue's lemma allows us to conclude.

The corollary is a direct consequence of the theorem. $\square$

Remark The hypothesis $\sum_{k\in\mathbb{Z}} |R_X(k)| < \infty$ often made in this book is also found to be satisfied for processes whose spectral measure is absolutely continuous (with respect to Lebesgue's measure) and whose PSD is continuous, piecewise derivable, and with bounded derivative.

Indeed, for such a process X, with PSD $S_X(f)$, by using the partial integration formula it results that

$$R_X(n) = \int_{\mathcal{I}} e^{2i\pi nf}S_X(f)df = [S_X(f)\frac{e^{2i\pi nf}}{2i\pi n}]_{-1/2}^{1/2} - \int_{\mathcal{I}} \frac{e^{2i\pi nf}}{2i\pi n}S_X'(f)df$$

$$= -\int_{\mathcal{I}} \frac{e^{2i\pi nf}}{2i\pi n}S_X'(f)df,$$

$$(F.7)$$

with $S_X'(f) = \frac{d}{df}S_X(f)$. Denoting by c_n the Fourier coefficients of $S_X'(f)$, we see that $R_X(n) = -(2i\pi n)^{-1}c_n$. Therefore,

$$(|c_n| - n^{-1})^2 = |c_n|^2 - 4\pi|R_X(n)| + n^{-2}$$

$$(F.8)$$

$$\geq 0,$$

and from Perseval's inequality,

$$\sum_{n \in \mathbb{Z}} |R_X(n)| \leq R_X(0) + \frac{1}{4\pi} \sum_{n \in \mathbb{Z}^*} (|c_n|^2 + n^{-2})$$

$$\leq R_X(0) + \frac{1}{4\pi} \sum_{n \in \mathbb{N}^*} n^{-2} + \frac{1}{4\pi} \int_{\mathcal{I}} |S_X'(f)|^2 df,$$

(F.9)

which justifies the remark.

G. Viterbi Algorithm

From the sole observation of $y_{0:n}$, we want to find the value of the sequence $x_{0:n}$ that maximises the probability $P(x_{0:n}|y_{0:n})$. We assume here that X is a Markov chain with L discrete states. To clarify our ideas, we assume that Y has continuous values (we can use the same approach when Y takes discrete values). We note that

$$P(x_{0:n}|y_{0:n}) = \frac{p(y_{0:n}|x_{0:n})P(x_{0:n})}{p(y_{0:n})} \tag{G.1}$$

and that as $y_{0:n}$ is known, the problem amounts to the maximisation of $p(y_{0:n}|x_{0:n})P(x_{0:n})$. In order to maximise this expression, we note that

$$\delta_k(e_i) = \max_{x_0,\dots,x_{k-1}} p(y_{0:k}|x_{0:k-1}, x_k = e_i)P(x_{0:k-1}, x_k = e_i). \tag{G.2}$$

From the relation

$$p(y_{0:k+1}|x_{0:k}, x_{k+1} = e_j)P(x_{0:k}, x_{k+1} = e_j) =$$
$$p(y_{0:k}|x_{0:k})p(y_{k+1}|x_{k+1} = e_j)P(x_{k+1} = e_j|x_k)P(x_{0:k}), \tag{G.3}$$

we see that δ_k satisfies a simple recurrence relation:

$$\delta_{k+1}(e_j)$$
$$= p(y_{k+1}|x_{k+1} = e_j) \max_{e_i} [P(x_{k+1} = e_j|x_k = e_i)$$
$$\max_{x_0,\dots,x_{k-1}} [p(y_{0:k}|x_{0:k-1}, x_k = e_i)P(x_{0:k})]]$$
$$= p(y_{k+1}|x_{k+1} = e_j) \max_{e_i} [\delta_k(e_i)P(x_{k+1} = e_j|x_k = e_i)]. \tag{G.4}$$

We propagate the recurrence from $\delta_0(e_i) = p(y_0|x_0 = e_i)P(x_0 = e_i)$, and at each step we memorise the following values:

$$c(e_j, k) = \arg\max_{e_i}[\delta_k(e_i)P(x_{k+1} = e_j|x_k = e_i)], \quad (j = 1, L). \tag{G.5}$$

We next estimate x_n, then, going backwards, $x_{n-1}, x_{n-2}, \dots, x_0$:

$$\hat{x}_n = \arg\max_{e_i} \delta_n(e_i)$$

$$\hat{x}_k = \arg\max_{e_i} [\delta_k(e_i)P(x_{k+1} = \hat{x}_{k+1}|x_k = e_i)] \tag{G.6}$$

$$= c(\hat{x}_{k+1}, k), \quad k = n-1, n-2, \ldots, 0.$$

The computational cost of the algorithm is about nL^2 operations instead of n^L operations that would be required for an exhaustive direct search.

H. Minimum-phase Spectral Factorisation of Rational Spectra

Theorem *If X is a rational spectral process, $S_X(f)$ can be factorised in the form*

$$S_X(f) = G(e^{2i\pi f}) = \left| \frac{b(e^{2i\pi f})}{a(e^{2i\pi f})} \right|^2, \tag{H.1}$$

where $a(z)$ and $b(z)$ have no common zeroes, and $a(z) \neq 1$, for $|z| = 1$.

In particular, there is a unique factorisation up to a modulus 1 factor, for which $b(z) \neq 0$ for $|z| > 1$, $a(z) \neq 0$ for $|z| \geq 1$, and the numerator and the denominator of $G(z)$ have degrees that are twice the degree of $b(z)$ and of $a(z)$ respectively.

$$degree(b(z)a^{-1}(z)) = (1/2)\,degree(G(z)). \tag{H.2}$$

This factorisation coincides with the minimum-phase causal factorisation of $S_X(f)$.

Proof $a(z) \neq 0$ for $|z| = 1$, otherwise we would have $\| X_n \|^2 = \int_{\mathcal{I}} S_X(f)df = +\infty$. $G(z)$ can be written in the form

$$G(z) = \alpha z^{-r_0} \prod_k (1 - z^{-1} z_k)^{r_k}. \tag{H.3}$$

Since $S_X(f)$ is a real-valued function, $G(z) = [G(z)]^*$ for $|z| = 1$, and

$$\alpha z^{-r_0} \prod_k (1 - z^{-1}z_k)^{r_k} = \alpha^* z^{r_0} \prod_k (1 - zz_k^*)^{r_k}, \quad \text{for } |z| = 1. \tag{H.4}$$

This equality property of two rational functions on the unit circle must clearly be satisfied for any complex number z. Consequently, it appears that if z_k is a zero, or a pole, of order r_k of $G(z)$ such that $|z_k| \neq 1$, it is the same for $(z_k^*)^{-1}$. For a zero $z_k = e^{2i\pi f_k}$ located on the unit circle, the order of multiplicity will be even, since $S_X(f)$ is positive and in the neighbourhood of f_k, $S_X(f) = \lambda(f - f_k)^{r_k} + o((f - f_k)^{r_k})$. $G(z)$ can, therefore, be written in the form

$$\alpha z^{-r_0} \prod_{|z_k| \leq 1} (1 - z^{-1} z_k)^{r_k} (1 - zz_k^*)^{r_k}. \tag{H.5}$$

Moreover, the positivity of $G(z)$ on the unit circle means that $\alpha > 0$, and $r_0 = 0$. Among the possible factorisations, the particular factorisation satisfying the terms of the theorem is thus obtained by taking

$$b(z)a^{-1}(z) = \alpha^{1/2} \prod_{|z_k| \leq 1} (1 - z^{-1}z_k)^{r_k}. \tag{H.6}$$

This is the only factorisation that verifies $b(z) \neq 0$ for $|z| > 1$, $a(z) \neq 0$ for $|z| \geq 1$ and that satisfies the degree conditions stated by the theorem.

To show that $b(z)a^{-1}(z)$ is the minimum-phase causal factorisation of $S_X(f)$, we begin by assuming that $b(z) \neq 0$ for $|z| = 1$, and we consider the process W of stochastic measure

$$d\hat{W}(f) = \frac{a(e^{2i\pi f})}{b(e^{2i\pi f})} d\hat{X}(f). \tag{H.7}$$

W is a white noise process with variance equal to 1. The filter $a(z)b^{-1}(z)$ is causal since $b(z) \neq 0$ for $|z| \geq 1$. Therefore, $H_{W,n} \subset H_{X,n}$. In addition, since $b(z)a^{-1}(z)$ is causal and

$$d\hat{X}(f) = \frac{b(e^{2i\pi f})}{a(e^{2i\pi f})} d\hat{W}(f), \tag{H.8}$$

it is clear that $H_{W,n} = H_{X,n}$. W is, therefore, the normalised innovation of X (up to one modulus 1 factor). $b(z)a^{-1}(z)$ is, therefore, the minimum-phase causal factorisation of $S_X(f)$. In the case where the finite set $E = \{f; b(e^{2i\pi f}) = 0\}$ is not empty, we proceed similarly by defining $d\hat{W}$ by (H.7) for any f of $\mathcal{I} - E$. $\square$

I. Compatibility of a Given Data Set with an Autocovariance Set

Following [43], we shall adopt a geometrical approach that requires recalling a few basics.

I.1 Elements of Convex Analysis

Given a closed convex cone K of $\mathbb{R}^p$, we shall note

$$K^\tau = \{x \in \mathbb{R}^p; \forall y \in K, y^T x \geq 0\}. \tag{I.1}$$

K^τ is called the dual set of K.

Theorem I.1 $K = (K^\tau)^\tau$.

Proof

$$(K^\tau)^\tau = \{b; a \in K^\tau \Rightarrow b^T a \geq 0\}, \tag{I.2}$$

and any element b of K verifies the implication $a \in K^\tau \Rightarrow b^T a \geq 0$, therefore $K \subset (K^\tau)^\tau$. Conversely, if $b \notin K$, there exists $a \in \mathbb{R}^p$ such that the hyperplane of equation $x^T a = 0$ separates b and K, with $K \subset \{x; x^T a \geq 0\}$. Therefore, $a \in K^\tau$. Since $b^T a < 0$, $b \notin (K^\tau)^\tau$, which shows that $(K^\tau)^\tau \subset K$ and completes the proof. $\square$

Let $(u_k(t))_{k=1,p}$ be a set of continuous, linearly independent functions, defined on an interval $[a, b]$. We note $u(t) = [u_1(t), ..., u_p(t)]^T$ and

$$K = \{c \in \mathbb{R}^p; \exists \mu \geq 0, c = \int_{[a,b]} u(t) d\mu(t)\} \tag{I.3}$$

Theorem I.2 K is a closed convex cone.

Proof K is clearly a convex cone. To establish that K is closed, it is sufficient to show that, for any sequence of points $(c_n)_{n \in \mathbb{N}}$ of K that converges towards a point c, c belongs to K. We eliminate the trivial case where $c = 0$ ($0 \in K$). We may then assume without loss of generality that the points c_n are also non-zero. The points c_n are written in the form $c_n = \int_{[a,b]} u(t) d\mu_n(t)$, where the measures μ_n are positive measures. We now write $c'_n = \int_{[a,b]} u(t) d\mu'_n(t)$, where $d\mu'_n(t) = (\mu_n([a, b]))^{-1}d\mu_n(t)$. Since K is a cone, and the vectors c'_n are equal to the vectors c_n up to positive factors, it is sufficient to show that the limit c' of the sequence c'_n exists and belongs to K to complete the proof, since point c will then necessarily be located on the half straight line of K passing through c' and originating at 0. For this, we use Helly's selection theorem (see Appendix A). Thus, there exists a positive measure μ', carried by $[a, b]$, which is the weak limit of a sub-sequence $(\mu'_{n_k})_{k \in \mathbb{N}}$ of the sequence of measures $(\mu'_n)_{n \in \mathbb{N}}$ carried by $[a, b]$, since the measures μ'_n are bounded $(\forall n \in \mathbb{N}, \mu'_n([a, b]) = 1)$. The sequence of vectors c_{n_k} converges, therefore, towards $\int_{[a,b]} u(t)d\mu'(t)$. But, this limit is also equal to c'. Therefore $c' \in K$, which completes the proof. $\square$

The following important result provides a necessary and sufficient condition to ensure the existence of solutions to problems referred to as *moment problems*.

Theorem I.3 $\forall c \in \mathbb{R}^N, \exists \mu \geq 0, c = \int_{[a,b]} u(t)d\mu(t)$, *if and only if*

$$\forall \alpha \in \mathbb{R}^p, \left([\forall t \in [a, b], \alpha^T u(t) \geq 0] \Rightarrow [\alpha^T c \geq 0]\right). \qquad (I.4)$$

Proof The necessary condition is obvious since $\alpha^T c = \int_{[a,b]} \alpha^T u(t)d\mu(t)$. Conversely, let there be c such that (I.4) is satisfied. We write

$$E = \{x; \exists \mu \geq 0, x = \int_{[a,b]} u(t)d\mu(t)\}. \qquad (I.5)$$

We note that

$$E^\tau = \{\alpha; \forall \mu \geq 0, \int_{[a,b]} \alpha^T u(t)d\mu(t) \geq 0\}$$
$$= \{\alpha; \forall t \in [a, b], \alpha^T u(t) \geq 0\}. \qquad (I.6)$$

Indeed, to show that

$$\forall \mu \geq 0, \int_{[a,b]} \alpha^T u(t)d\mu(t) \geq 0 \Rightarrow \forall t \in [a, b], \alpha^T u(t) \geq 0, \qquad (I.7)$$

it is sufficient to take $d\mu(t) = \delta_t$, and the converse of this implication is obvious. Therefore, the implication (I.4) can be rewritten as $\alpha \in E^\tau \Rightarrow \alpha^T c \geq 0$, that is $c \in (E^\tau)^\tau$, or $c \in E$, which was to be proved. $\square$

I.2 A Necessary and Sufficient Condition

Theorem I.3 makes it possible to prove the following result simply.

Theorem *A sequence of coefficients* $(R(k))_{k=0,N}$ *represents the* $N + 1$ *first*

autocovariance coefficients of a certain WSS process if and only if the Toeplitz matrix T_N of size $N + 1$ and with general term $[T_N]_{i,j} = R(i - j)$ is positive.
Proof It suffices to consider the linearly independent functions $u_0(f) = 1$, $u_{2n}(f) = \cos(2\pi n f)$, $u_{2n+1}(f) = \sin(2\pi n f)$, $(n = 1, N)$, defined on $\mathcal{I}$ and to check that any polynomial of the functions $(u_k(f))_{k=0,2N+1}$, positive on $\mathcal{I}$, can be written in the form

$$P(f) = \Big| \sum_{k=0,N} \beta_k e^{2i\pi k f} \Big|^2. \tag{I.8}$$

Indeed, rewriting the functions $\cos(2\pi n f)$ and $\sin(2\pi n f)$ as functions of $e^{2i\pi n f}$, this result can be seen as a particular case of the spectral factorisation of rational spectra (see Appendix H), for which the denominator of the PSD is constant.

Then, the condition

$$\forall \alpha \in \mathbb{R}^p, \ [\forall t \in [a, b], \alpha^T u(t) \geq 0] \Rightarrow [\alpha^T c \geq 0], \tag{I.9}$$

of Theorem I.3 is simply put in the form

$$\forall \beta \in \mathbb{C}^{N+1}, \ \beta^H T_N \beta \geq 0, \tag{I.10}$$

with $\beta = [\beta_0, \ldots, \beta_N]^T$. It is clear that this condition is always satisfied if and only if the matrix T_N is positive, which completes the proof. $\square$
Remarks
1) Bochner's theorem, for the case of discrete processes, stems directly from the result above.
2) In the proof above, the closed interval $[a, b]$ of the previous paragraph is, in fact, replaced by the left open interval $\mathcal{I} =] - 1/2, 1/2]$. This point does not present a problem, since we can begin by reasoning on the interval $\mathcal{I}' = [-1/2, 1/2]$, then noting that here the vector function $u(f)$ verifies $u(-1/2) = u(1/2)$, which for any positive measure ν makes it possible to write that

$$\int_{\mathcal{I}'} u(f) d\nu(f) = \int_{\mathcal{I}} u(f) d\mu(f), \tag{I.11}$$

with $d\mu(f) = d\nu(f)$ on $] - 1/2, 1/2[$, and $\mu(\{1/2\}) = \nu(\{-1/2\}) + \nu(\{+1/2\})$.

The case where the matrix T_N is singular is addressed in the following result:
Theorem (Caratheodory) *The matrix T_N is positive singular, with rank $p < N + 1$, if and only if there is a unique positive measure μ, carried by p mass points, whose coefficients $(R(k))_{k=0,N}$ are the first Fourier coefficients.*
Proof If the matrix T_N is positive singular, with rank $p < N + 1$, there exists a vector $u^0 = [u_0^0, \ldots, u_N^0] \neq 0$ such that $u^0 T_N (u^0)^H = 0$. A positive measure μ, whose coefficients of T_N are the Fourier coefficients, will therefore satisfy

$$\int_{\mathcal{I}} | \sum_{n=0,N} u_n^0 e^{2i\pi nf} |^2 d\mu(f) = (u^0)^H T_N u^0$$

(I.12)

$$= 0.$$

μ is therefore a discrete measure carried by the points f of $\mathcal{I}$, for which $u^0(e^{2i\pi f}) = -\sum_{n=0,N} u_n^0 e^{2i\pi nf}$ is equal to zero. We note that there exists a basis of $N + 1 - p$ vectors of the kernel of T_N, and consequently $N + 1 - p$ corresponding polynomials $(u^k(z))_{k=0,N-p}$, which form an independent family. Let l be the number of mass points of μ. The polynomials $u^k(z)$, with degree smaller than or equal to N, have common zeroes on the unit circles corresponding to these l points. Since $(u^k(z))_{k=0,N-p}$ form an independent family ($l \leq p$). Moreover, if we had $l < p$, we would easily check that the rank of T_N would be lower than p, by constructing $N + 1 - l$ independent polynomials, which are equal to zero on the support of μ (the $N + 1 - l$ corresponding independent vectors would then be in the kernel of T_N).

To characterise the parameters of such a measure μ and show that it is unique, we consider a process X whose first autocovariance coefficients are the coefficients of T_N. The prediction error of X at the order $p-1$ is non-zero and it is zero at the order p. Since

$$\min_{\{a_k\}_{k=1,p}} \| X_n - \sum_{k=1,p} a_k X_{n-k} \|^2 = [1 - \alpha^T] T_p [1 - \alpha^T]^H$$

(I.13)

$$= 0,$$

where α is the vector of the coefficients a_k that achieves the minimum, the mass points of μ are given by the zeroes $(f_k)_{k=1,p}$ of the transfer function $\alpha(z) = 1 - \sum_{k=1,p} \alpha_k z^{-k}$: $\alpha(e^{2i\pi f_k}) = 0$.

The values ρ_k of the measure $d\mu(f) = \sum_{k=1,p} \rho_k \delta_{f_k}$ are then obtained by solving the system of linear equations

$$\sum_{k=1,p} \rho_k e^{2i\pi nf_k} = R(n), \quad n = 0, p - 1,$$

(I.14)

which proves that μ is unique.

Conversely, if μ is carried by p distinct points of $\mathcal{I}$, with $p < N + 1$ then, by noting $d\mu(f) = \sum_{k=1,p} \rho_k \delta_{f_k}$, it is clear that

$$T_N = \sum_{k=1,p} \rho_k d(f_k) d(f_k)^H,$$

(I.15)

with $d(f) = [1, e^{2i\pi f}, \ldots, e^{2i\pi Nf}]^T$. Therefore, T_N is positive and of rank p, since the vectors $(d(f_k))_{k=1,p}$ constitute an independent family, as established by the theorem below. $\square$

Finally, we indicate the following result, useful for characterising the values of the frequencies of a line spectrum.

Theorem *For p distinct values of $\mathcal{I}$ $(p \leq N+1)$, $f_1, \ldots, f_p$, the family of vectors $\{d(f_1), \ldots, d(f_p)\}$ is independent.*

Proof It suffices to show that for $N+1$ distinct frequencies $f_1, \ldots, f_{N+1}$, $\{d(f_1), \ldots, d(f_{N+1})\}$ constitutes a basis of $\mathbb{C}^{N+1}$. For this, note that $M = (d(f_1), \ldots, d(f_{N+1}))$ is a Vandermonde matrix, whose determinant is

$$|M| = \prod_{1 \leq k < l \leq N+1} \left(e^{2i\pi f_k} - e^{2i\pi f_l} \right)$$

$$\neq 0.$$

(I.16)

Therefore, the matrix M is full rank and $\{d(f_1), \ldots, d(f_{N+1})\}$ constitutes a basis of $\mathbb{C}^{N+1}$. $\square$

J. Levinson's Algorithm

Theorem (Levinson's algorithm) *The polynomials $Q_n(z)$ can be obtained by means of the following recurrence relations:*

$$Q_0(z) \qquad = 1,$$

$$\sigma_0^2 \qquad = R(0),$$

for $n = 0, N-1$,

$$Q_{n+1}(z) \qquad = zQ_n(z) - k_{n+1}\tilde{Q}_n(z), \qquad (J.1)$$

$$\tilde{Q}_{n+1}(z) \qquad = \tilde{Q}_n(z) - k_{n+1}^* zQ_n(z),$$

$$k_{n+1} \qquad = \left(\sum_{k=0,n} q_{k,n} R(n+1-k)\right)/\sigma_n^2,$$

$$\sigma_{n+1}^2 \qquad = \sigma_n^2(1 - |k_{n+1}|^2),$$

where $\tilde{Q}_n(z) = z^n Q_n^*(z^{-1}) = \sum_{k=0,n} q_{k,n}^* z^k$.

Proof Here, we shall use index n for the time, and index p for the prediction order. We remark that

$$X_n - X_n/H_{X,n-p,n-1} \qquad = [Q_p(z)]X_{n-p},$$

$$X_{n-p} - X_{n-p}/H_{X,n-p+1,n} = [\tilde{Q}_p(z)]X_{n-p}, \qquad (J.2)$$

where $[h(z)]X_n$ represents the output at the instant n of the filter with input X and with transfer function $h(z)$. The second relation comes from the fact that $[\tilde{Q}_p(z)]X_{n-p}$ is of the form $X_{n-p} + \sum_{k=1,p} q_{k,p}^* X_{n-p+k}$ and that $[\tilde{Q}_p(z)]X_{n-p} \perp X_{n-p+k}$ for $k = 1, p$. Indeed, for $k = 1, p$ we clearly have

$$< \tilde{Q}_p(z), z^k > = R(-k) + \sum_{l=1,p} q_{l,p}^* R(l-k)$$

$$= [R(k) + \sum_{l=1,p} q_{l,p} R(k-l)]^* \qquad (J.3)$$

$$= 0.$$

Now, we note that

$$[Q_{p+1}(z)]X_{n-p} = X_{n+1} - X_{n+1}/H_{X,n-p,n}$$

$$= X_{n+1} - X_{n+1}/H_{X,n+1-p,n}$$

$$-(X_{n+1}/H_{X,n-p,n} - X_{n+1}/H_{X,n-p+1,n})$$

$$= [zQ_p(z)]X_{n-p} - X_{n+1}/\{H_{X,n-p,n} \ominus H_{X,n-p+1,n}\}.$$

(J.4)

The random variable $X_{n+1}/\{H_{X,n-p,n} \ominus H_{X,n-p+1,n}\}$ belongs to $H_{X,n-p,n}$ and is orthogonal to $H_{X,n-p+1,n}$. It is the same for $[\tilde{Q}_p(z)]X_{n-p}$. Since $H_{X,n-p,n} \ominus H_{X,n-p+1,n}$ is a vector space of dimension 1, we obtain

$$X_{n+1}/\{H_{X,n-p,n} \ominus H_{X,n-p+1,n}\} = k_{p+1}[\tilde{Q}_p(z)]X_{n-p}.$$

(J.5)

The independence of k_{p+1} with respect to n is a direct consequence of the stationarity of X. The above results finally lead to

$$Q_{p+1}(z) = zQ_p(z) - k_{p+1}\tilde{Q}_p(z).$$

(J.6)

The calculation of k_{p+1} can be done by noting that $< Q_{p+1}(z), 1 > \ = 0$ and that $< Q_{p+1}(z), 1 > \ = \ < zQ_p(z), 1 > -k_{p+1} < \tilde{Q}_p(z), 1 >$. A direct calculation leads to $< zQ_p(z), 1 > \ = \sum_{k=0,p} q_{k,p}R(p+1-k)$. Moreover,

$$< \tilde{Q}_p(z), 1 > \ = \ < z^p Q_p^*(1/z), 1 > \ = \ < z^p, Q_p(z) >$$

$$= \ < Q_p(z), Q_p(z) > \ = \sigma_p^2,$$

(J.7)

and, therefore,

$$k_{p+1} = \sigma_p^{-2}[\sum_{k=0,p} q_{k,p}R(p+1-k)].$$

(J.8)

In addition,

$$\sigma_{p+1}^2 = \ < zQ_p(z) - k_{p+1}\tilde{Q}_p(z), zQ_p(z) - k_{p+1}\tilde{Q}_p(z) >$$

$$= \ < Q_p(z), Q_p(z) > -2\mathcal{R}e[k_{p+1}^* < zQ_p(z), \tilde{Q}(z) >]$$

(J.9)

$$+ |k_{p+1}|^2 < Q_p(z), Q_p(z) >$$

and

$$< zQ_p(z), \tilde{Q}(z) > \ = \ < zQ_p(z), 1 > \ = \ < Q_{p+1}(z) + k_{p+1}\tilde{Q}_p(z), 1 >$$

$$= k_{p+1} < \tilde{Q}_p(z), 1 > \ = k_{p+1}\sigma_p^2.$$

(J.10)

Therefore, we finally obtain $\sigma_{p+1}^2 = \sigma_p^2(1 - |k_p|^2).$ $\square$

K. Maximum Principle

Theorem K.1 *A non-constant function $f(z)$, holomorphic in a domain, that is, in an open connected set $\mathcal{D}$, does not have a maximum in $\mathcal{D}$.*

In particular, for a holomorphic function $f(z)$ in a domain containing the unit disk $\mathbb{D}$, it is clear that either $f(z)$ is constant on the closed unit disk $\overline{\mathbb{D}}$, or its maximum $\mathbb{D}$ is reached on the unit circle.

Proof To show the theorem, we shall establish that if $f(z)$ has a maximum in $\mathcal{D}$, then $f(z)$ is constant in $\mathcal{D}$. In fact, if $f(z)$ is maximum in $z_0 \in \mathcal{D}$, it suffices to show that $f(z)$ is constant on a whole neighbourhood of z_0.

From the Cauchy integral formula, for any point z_0 of $\mathcal{D}$, and any positively oriented circle $C(z_0, \rho)$ with centre z_0 and radius ρ,

$$f(z_0) = \frac{1}{2i\pi} \int_{C(z_0,\rho)} \frac{f(z)}{z - z_0} dz$$

$$= \int_{\mathcal{I}} f(z_0 + \rho e^{2i\pi u}) du. \tag{K.1}$$

Therefore,

$$|f(z_0)| \leq \int_{\mathcal{I}} |f(z_0 + \rho e^{2i\pi u})| du. \tag{K.2}$$

Furthermore, if the maximum of $f(z)$ on $\mathcal{D}$ is obtained for $z = z_0$, it is clear that

$$\int_{\mathcal{I}} |f(z_0 + \rho e^{2i\pi u})| du \leq |f(z_0)|. \tag{K.3}$$

It then results from relations (K.2) and (K.3) that

$$\int_{\mathcal{I}} (|f(z_0 + \rho e^{2i\pi u})| - |f(z_0)|) du = \int_{\mathcal{I}} |f(z_0 + \rho e^{2i\pi u})| du - |f(z_0)| = 0,$$

$$\tag{K.4}$$

and since the term under the integral is negative on $\mathcal{I}$, it results that necessarily $|f(z_0 + \rho e^{2i\pi u})| = |f(z_0)|$, $\forall u \in \mathcal{I}$. $f(z)$, therefore, has a constant modulus on the circle $C(z_0, \rho)$. As the relation $|f(z_0 + \rho e^{2i\pi u})| = |f(z_0)|$ does not depend on ρ, for $C(z_0, \rho) \subset \mathcal{D}$, it is clear that $f(z)$ has a constant modulus on a whole disk B of $\mathcal{D}$.

We now note $z = x + iy$, and $f(z) = re^{i\phi} = u + iv$. The Cauchy-Riemann equations are given by

$$\frac{\partial u}{\partial x} = \frac{\partial v}{\partial y},$$

$$\text{and} \quad \frac{\partial v}{\partial x} = -\frac{\partial u}{\partial y},$$

(K.5)

which, by taking into account the fact that $r = |f(z)|$ is constant in B, leads to the relations

$$-r\left(\frac{\partial \phi}{\partial x} \sin \phi + \frac{\partial \phi}{\partial y} \cos \phi\right) = 0$$

$$r\left(\frac{\partial \phi}{\partial x} \cos \phi - \frac{\partial \phi}{\partial y} \sin \phi\right) = 0.$$

(K.6)

The determinant of this system in $(\frac{\partial \phi}{\partial x}, \frac{\partial \phi}{\partial y})$ is equal to $-r^2$, and it appears that at any point where $r \neq 0$, $(\frac{\partial \phi}{\partial x}, \frac{\partial \phi}{\partial y}) = (0, 0)$.

Therefore, $f(z)$ is constant in B, and consequently in $\mathcal{D}$. $\square$

L. One Step Extension of an Autocovariance Sequence

Theorem *Let $(R(n))_{n=0,N}$ denote a sequence of autocovariances. The set of coefficients $R(N+1)$, such that $(R(n))_{n=0,N+1}$ is a sequence of autocovariances, is the closed disk $D(C_N, \sigma_N^2)$ with centre*

$$C_N = -\sum_{k=1,N} q_{k,N} R(N+1-k), \tag{L.1}$$

and radius σ_N^2.

Proof Let $(R(n))_{n=0,N}$ be a sequence of autocovariances. If $(R(n))_{n=0,N+1}$ is a sequence of autocovariances, the corresponding reflection coefficient k_{N+1} is given by

$$k_{N+1} = \sigma_N^{-2}[R(N+1) + \sum_{k=1,N} q_{k,N} R(N+1-k)]. \tag{L.2}$$

Let there be $R(N+1) = \sigma_N^2 k_{N+1} + C_N$. As $|k_{N+1}| \leq 1$, it is clear that $R(N+1)$ belongs to the disk $D(C_N, \sigma_N^2)$ with centre C_N and radius σ_N^2.

Conversely, for any $R_{N+1} \in D(C_N, \sigma_N^2)$, we write $R(N+1) = \sigma_N^2 k_{N+1} + C_N$, where $|k_{N+1}| \leq 1$, and we define

$$Q_{N+1}(z) = zQ_N(z) - k_{N+1}\tilde{Q}_N(z)$$

$$= \sum_{k=0,N+1} q_{k,N+1} z^{N+1-k}. \tag{L.3}$$

We notice that for $k = 1, N$,

$$< zQ_N(z), z^k > \, = \, < Q_N(z), z^{k-1} >$$

$$= 0,$$

and $< \tilde{Q}_N(z), z^k > \, = \, < z^N Q_N^*(z^{-1}), z^k > \tag{L.4}$

$$= \, < z^{N-k}, Q_N(z) >$$

$$= 0.$$

Moreover, it is clear that

$$< Q_{N+1}(z), 1 > = R_{N+1} - C_{N+1} - k_{N+1}\sigma_N^2 = 0. \tag{L.5}$$

Therefore, $< Q_{N+1}(z), z^k > = 0$, $k = 0, N$. Consequently,

$$q = \arg(\min_\alpha \alpha T_{N+1} \alpha^H), \tag{L.6}$$

where $\alpha = [1, \alpha_1, \ldots, \alpha_{N+1}]$, and $q = [1, q_{1,N+1}, \ldots, q_{N+1,N+1}]$. But,

$$q T_{N+1} q^H = \sigma_{N+1}^2 = \sigma_N^2(1 - |k_{N+1}^2|) \geq 0, \tag{L.7}$$

which ensures that any vector $u = [u_0, .., u_{N+1}]$ of $\mathbb{C}^{N+2}$ verifies $u T_{N+1} u^H \geq 0$. Indeed, for $u_0 \neq 0$, we obtain

$$u T_{N+1} u^H \geq |u_0|^2 q T_{N+1} q^H = |u_0|^2 \sigma_{N+1}^2 \geq 0, \tag{L.8}$$

and for $u_0 = 0$, $u T_{N+1} u^H = [u_1, \ldots, u_{N+1}] T_N [u_1, \ldots, u_{N+1}]^H \geq 0$ for $T_N \geq 0$. Consequently, $T_{N+1} \geq 0$, which completes the proof. $\square$

M. Recurrence Relation upon $P_n(z)$

Theorem

$$P_n(z) = [(R(0) + 2R(1)z^{-1} + ... + 2R(n)z^{-n})Q_n(z)]_+, \quad (n = 0, N),$$
(M.1)

where $[.]_+$ represents the polynomial part for the variable z, and the polynomials $P_n(z)$ satisfy the following recurrence:

$$P_0(z) \quad = R_0,$$

(M.2)

$$\text{and } P_{n+1}(z) = zP_n(z) + k_{n+1}\tilde{P}_n(z), \quad n = 0, N-1.$$

Proof It is clear that for $|z| < 1$:

$$P_n(z)$$

$$= \int_I \frac{e^{2i\pi f} + z}{e^{2i\pi f} - z}[Q_n(e^{2i\pi f}) - Q_n(z)]d\mu(f)$$

$$= \int_I (1 + 2\sum_{k=1,\infty} z^k e^{-2i\pi k f})(\sum_{m=0,n} q_{m,n}[e^{2i\pi(n-m)f} - z^{n-m}])d\mu(f)$$

$$= \int_I \sum_{m=0,n} q_{m,n} z^{n-m} \times [e^{2i\pi(n-m)f} z^{m-n}$$

$$+ 2\sum_{k=1,\infty} z^{-n+k+m} e^{2i\pi(n-k-m)f} - 1 - 2\sum_{k=1,\infty} z^k e^{-2i\pi k f}]d\mu(f)$$

$$= \int_I \sum_{m=0,n} q_{m,n} z^{n-m}$$

$$\times [e^{2i\pi(n-m)f} z^{m-n} - 1 + 2\sum_{k=m-n+1,0} z^k e^{-2i\pi k f}]d\mu(f)$$

(M.3)

$$P_n(z)$$

$$= \int_I \sum_{m=0,n} q_{m,n} z^{n-m}$$

$$\times [-e^{2i\pi(n-m)f} z^{m-n} + 1 + 2\sum_{k=1,n-m} z^{-k} e^{2i\pi kf}] d\mu(f)$$

(M.4)

$$= -\sum_{m=0,n} q_{m,n} R(n-m)$$

$$+ [Q_n(z)(R(0) + 2R(1)z^{-1} + \ldots + 2R(n)z^{-n})]_+.$$

Moreover, $\sum_{m=0,n} q_{m,n} R(n-m) = <Q_n(z), 1> = 0$, which establishes relation (M.1) for $|z| < 1$, and consequently for any z, since $P_n(z)$ is a polynomial function.

In addition,

$$zP_n(z) + k_{n+1}\tilde{P}_n(z)$$

$$= \int_I \frac{e^{2i\pi f} + z}{e^{2i\pi f} - z}$$

$$\times [z(Q_n(z) - Q_n(e^{2i\pi f})) - k_{n+1}(\tilde{Q}_n(z) - z^n\{Q_n(e^{2i\pi f})\}^*)] d\mu(f)$$

$$= P_{n+1}(z) + \int_I \frac{e^{2i\pi f} + z}{e^{2i\pi f} - z}$$

$$\times [Q_{n+1}(e^{2i\pi f}) - zQ_n(e^{2i\pi f}) + k_{n+1}z^n\{Q_n(e^{2i\pi f})\}^*] d\mu(f)$$

$$= P_{n+1}(z) + \int_I \frac{e^{2i\pi f} + z}{e^{2i\pi f} - z}$$

$$\times [(e^{2i\pi f} - z)Q_n(e^{2i\pi f}) - k_{n+1}(e^{2i\pi nf} - z^n)\{Q_n(e^{2i\pi f})\}^*] d\mu(f)$$

$$= P_{n+1}(z) + \int_I (e^{2i\pi f} + z)$$

$$\times [Q_n(e^{2i\pi f}) - k_{n+1}(\sum_{l=0,n-1} z^{n-1-l} e^{2i\pi lf})\{Q_n(e^{2i\pi f})\}^*] d\mu(f)$$

$$= P_{n+1}(z) + \int_I [e^{2i\pi f} Q_n(e^{2i\pi f}) - k_{n+1}\tilde{Q}_n(e^{2i\pi f})] d\mu(f)$$

(M.5)

$$zP_n(z) + k_{n+1}\tilde{P}_n(z) = P_{n+1}(z) + \int_{\mathcal{I}} Q_{n+1}(e^{2i\pi f})d\mu(f)$$

(M.6)

$$= P_{n+1}(z).$$

Therefore,

$$zP_n(z) + k_{n+1}\tilde{P}_n(z) = P_{n+1}(z), \quad n = 0, N - 1. \square$$

(M.7)

N. General Solution to the Trigonometric Moment Problem

Theorem *The set of positive measures μ, whose first Fourier coefficients are the coefficients $(R(n))_{n=0,N}$, correspond to the Caratheodory functions of the form*

$$F(z) = \int_I \frac{e^{2i\pi f} + z}{e^{2i\pi f} - z} d\mu(f) = \frac{\tilde{P}_N(z) + S(z)zP_N(z)}{\tilde{Q}_N(z) - S(z)zQ_N(z)}. \tag{N.1}$$

where $S(z)$ is any Schur function. The Schur functions therefore parameterise the set of solutions to the trigonometric moment problem. When μ is an absolutely continuous measure with respect to Lebesgue's measure, almost everywhere its density $g(f)$ can be written in the form

$$g(f) = \frac{\sigma_N^2(1 - |S(e^{2i\pi f})|^2)}{|\tilde{Q}_N(e^{2i\pi f}) - S(e^{2i\pi f})e^{2i\pi f}Q_N(e^{2i\pi f})|^2}. \tag{N.2}$$

Proof Let μ be any positive measure, and $(Q_n(z))_{n \geq 0}$ and $(P_n(z))_{n \geq 0}$ the corresponding orthogonal Szegö polynomials of the first and second kind. To begin with, we show that for $|z| < 1$, the Caratheodory function

$$F_\mu(z) = \int_I \frac{e^{2i\pi f} + z}{e^{2i\pi f} - z} d\mu(f) \tag{N.3}$$

has a series expansion, which up to the order n coincides with that of $\tilde{P}_n(z)\tilde{Q}_n^{-1}(z)$. Indeed,

$$\tilde{P}_n(z) = -\int_I \frac{e^{2i\pi f} + z}{e^{2i\pi f} - z}[z^n(Q_n(e^{2i\pi f}))^* - \tilde{Q}_n(z)]d\mu(f)$$

$$= -z^n \int_I (1 + 2\sum_{k=1,\infty} z^k e^{2i\pi kf})[Q_n(e^{2i\pi f})]^* d\mu(f)$$

$$+ \tilde{Q}_n(z)F_\mu(z), \tag{N.4}$$

$$= -2z^{n+1} \int_I (\sum_{k=1,\infty} z^{k-1}e^{2i\pi kf})[Q_n(e^{2i\pi f})]^* d\mu(f)$$

$$+ \tilde{Q}_n(z)F_\mu(z),$$

and $\tilde{Q}_n^{-1}(z)$ is holomorphic in the unit disk, therefore

$$\frac{\tilde{P}_n(z)}{\tilde{Q}_n(z)} - F_\mu(z) = O(z^{n+1}). \tag{N.5}$$

Now, we define the transforms

$$\omega_a : S(z) \to \frac{a + zS(z)}{1 + a^* zS(z)}. \tag{N.6}$$

For $|a| \leq 1$, we can check that $S(z)$ is a Schur function if and only if $\omega_a[S(z)]$ is also a Schur function, and that the set of Schur functions that are equal to a in 0 are written in the form $\omega_a[S(z)]$. The proof of these results can be shown by using the maximum principle (see Appendix K) and is left up to the reader.

We now prove the theorem by induction. We consider the following hypothesis:

H_n : the set of Caratheodory functions of the form

$$F(z) = R(0) + 2 \sum_{k=0,n} R(-k)z^k + O(z^{n+1}) \tag{N.7}$$

is given by

$$F(z) = \frac{\tilde{P}_n(z) + S_{n+1}(z)zP_n(z)}{\tilde{Q}_n(z) - S_{n+1}(z)zQ_n(z)}, \tag{N.8}$$

where $S_{n+1}(z)$ is any Schur function.

We show H_0. For that, let us remark that $F(z)$ is a Caratheodory function if and only if $F(z) = (1+S_0(z))(1-S_0(z))^{-1}$, where $S_0(z)$ is a Schur function (the proof is straightforward). We then have $S_0(z) = (F(z)-1)(F(z)+1)^{-1}$, and the condition $F(0) = R(0)$ is expressed by $S_0(0) = (R(0)-1)(R(0)+1)^{-1}$. This relation is satisfied if and only if $S_0(z) = \omega_{\frac{R(0)-1}{R(0)+1}}[S_1(z)]$, where $S_1(z)$ is a Schur function. It thus results that

$$F(z) = \frac{1 + \omega_{\frac{R(0)-1}{R(0)+1}}[S_1(z)]}{1 - \omega_{\frac{R(0)-1}{R(0)+1}}[S_1(z)]}$$

$$= \frac{\left(1 + \dfrac{(R(0) - 1) + (R(0) + 1)zS_1(z)}{(R(0) + 1) + (R(0) - 1)zS_1(z)}\right)}{\left(1 - \dfrac{(R(0) - 1) + (R(0) + 1)zS_1(z)}{(R(0) + 1) + (R(0) - 1)zS_1(z)}\right)} \tag{N.9}$$

$$= \frac{R(0) + zS_1(z)R(0)}{1 - zS_1(z)}$$

$$= \frac{\tilde{P}_0 + S_1(z)zP_0(z)}{\tilde{Q}_0 - S_1(z)zQ_0(z)}.$$

We now assume that $F(z) = R(0) + 2\sum_{k=1,n+1} R(-k)z^k + O(z^{n+2})$. According to H_n, we obtain

$$F(z) = \frac{\tilde{P}_n(z) + S_{n+1}(z)zP_n(z)}{\tilde{Q}_n(z) - S_{n+1}(z)zQ_n(z)}. \qquad (N.10)$$

We show that we must have $S_{n+1}(0) = k_{n+1}^*$. We know that

$$\frac{\tilde{P}_{n+1}(z)}{\tilde{Q}_{n+1}(z)} - F(z) = O(z^{n+2}). \qquad (N.11)$$

Therefore, setting $A(z) = \tilde{Q}_{n+1}(z)[\tilde{Q}_n(z) - zS_{n+1}(z)Q_n(z)]$ yields:

$$O(z^{n+2})A(z) = [\tilde{P}_n(z) + k_{n+1}^* zP_n(z)][\tilde{Q}_n(z) - zS_{n+1}(z)Q_n(z)]$$

$$-[\tilde{P}_n(z) + zS_{n+1}(z)P_n(z)][\tilde{Q}_n(z) - k_{n+1}^* zQ_n(z)]$$

$$= z[k_{n+1}^* P_n(z)\tilde{Q}_n(z) - S_{n+1}(z)Q_n(z)\tilde{P}_n(z)]$$

$$+ z[k_{n+1}^* Q_n(z)\tilde{P}_n(z) - S_{n+1}(z)P_n(z)\tilde{Q}_n(z)]$$

$$= z[k_{n+1}^* - S_{n+1}(z)][Q_n(z)\tilde{P}_n(z) + P_n(z)\tilde{Q}_n(z)]$$

$$= 2\sigma_n^2 z^{n+1}[k_{n+1}^* - S_{n+1}(z)], \qquad (N.12)$$

taking into account the relation

$$Q_n(z)\tilde{P}_n(z) + P_n(z)\tilde{Q}_n(z) = 2\sigma_n^2 z^n, \qquad (N.13)$$

whose proof (by induction) we leave up to the reader.

The relation $2\sigma_n^2 z^{n+1}[S_{n+1}(z) - k_{n+1}^*] = O(z^{n+2})A(z)$, where $A(z)$ is holomorphic in the unit disk, clearly implies that $S_{n+1}(0) = k_{n+1}^*$. $S_{n+1}(z)$ is therefore of the form $S_{n+1}(z) = \omega_{k_{n+1}^*}[S_{n+2}(z)]$, where $S_{n+2}(z)$ is a Schur function. Then,

$$F(z) = \frac{\tilde{P}_n(z) + S_{n+1}(z)zP_n(z)}{\tilde{Q}_n(z) - S_{n+1}(z)zQ_n(z)}$$

$$= \frac{[1 + k_{n+1}zS_{n+2}(z)]\tilde{P}_n(z) + [k_{n+1}^* + zS_{n+2}(z)]zP_n(z)}{[1 + k_{n+1}zS_{n+2}(z)]\tilde{Q}_n(z) - [k_{n+1}^* + zS_{n+2}(z)]zQ_n(z)} \qquad (N.14)$$

$$= \frac{\tilde{P}_{n+1}(z) + S_{n+2}(z)zP_{n+1}(z)}{\tilde{Q}_{n+1}(z) - S_{n+2}(z)zQ_{n+1}(z)},$$

which completes this part of the proof.

The relation

$$g(f) = \frac{\sigma_n^2(1 - |S(e^{2i\pi f})|^2)}{|\tilde{Q}_n(e^{2i\pi f}) - S(e^{2i\pi f})e^{2i\pi f}Q_n(e^{2i\pi f})|^2} \tag{N.15}$$

is a direct consequence of the relation $Q_n(z)\tilde{P}_n(z) + P_n(z)\tilde{Q}_n(z) = 2\sigma_n^2 z^n$ and of the fact that when the measure μ is absolutely continuous on the unit circle, its density $g(f)$ is equal almost everywhere (see [37] Chapter 11) to the limit

$$\lim_{r \to 1^-} \mathcal{R}e[F(re^{2i\pi f})]. \square \tag{N.16}$$

O. A Central Limit Theorem for the Empirical Mean

Theorem *If X_n is of the form $X_n = m_X + \sum_{k \in \mathbb{Z}} h_k V_{n-k}$, where V is a white noise process with variance σ^2, $\sum_{k \in \mathbb{Z}} |h_k| < \infty$ and $\sum_{k \in \mathbb{Z}} h_k \neq 0$, then,*

$$\lim_{n \to \infty} \sqrt{n}[\hat{m}_{X,n} - m_X] \sim \mathcal{N}(0, \sigma^2 |\sum_{k \in \mathbb{Z}} h_k|^2). \tag{0.1}$$

Proof We note $X_{n,p} = m_X + \sum_{|k| \leq p} h_k V_{n-k}$, and $Y_{l,p} = l^{-1} \sum_{k=1,l} X_{k,p}$. For $l > 2p$,

$$Y_{l,p} = m_X + l^{-1} \sum_{k=l-p,1+p} V_k (\sum_{|u| \leq p} h_u)$$

$$+ l^{-1} \sum_{k=1,2p} [V_{k-p} (\sum_{u=p-k+1}^{p} h_u) + V_{l-p+k} (\sum_{u=-p}^{p-k} h_u)]. \tag{0.2}$$

The last right-hand terms of the equality converge towards 0 in the mean square sense when l tends towards $+\infty$. Therefore, it is clear that

$$\lim_{l \to \infty} \sqrt{l}[Y_{l,p} - m_X] \sim \mathcal{N}(0, \sigma^2 |\sum_{|k| \leq p} h_k|^2). \tag{0.3}$$

We denote by Y_p the limit of $\sqrt{l}[Y_{l,p} - m_X]$. When p tends towards $+\infty$, Y_p converges in distribution towards a variable $Y \sim \mathcal{N}(0, \sigma^2 |\sum_{k \in \mathbb{Z}} h_k|^2)$.

We now note $\tilde{m}_{X,n} = \sqrt{n}[\hat{m}_{X,n} - m_X]$, and $\tilde{Y}_{n,p} = \sqrt{n}[Y_{n,p} - m_X]$, and we consider the inequality

$$|\phi_{\tilde{m}_{X,n}}(u) - \phi_Y(u)| \leq |\phi_{\tilde{m}_{X,n}}(u) - \phi_{\tilde{Y}_{n,p}}(u)| + |\phi_{\tilde{Y}_{n,p}}(u) - \phi_{Y_p}(u)|$$

$$+ |\phi_{Y_p}(u) - \phi_Y(u)|. \tag{0.4}$$

To complete the proof, it suffices to show that the three right-hand terms of this equality tend towards 0. When n tends towards $+\infty$, the second right-hand term tends towards 0, and when p tends towards $+\infty$, the third right-hand term tends towards 0.

To show the convergence towards 0 of the first right-hand term when m and n tend towards infinity, we note that

$$\lim_{n\to\infty} \| (\tilde{m}_{X,n} - \tilde{Y}_{n,p}) \| = \lim_{n\to\infty} \| \sum_{|k|>p} h_k (n^{-1/2} \sum_{l=1,n} V_{l-k}) \|$$

$$\leq \sigma \sum_{|k|>p} |h_k|,$$

(O.5)

and therefore

$$\lim_{p\to\infty} (\lim_{n\to\infty} \| \tilde{m}_{X,n} - \tilde{Y}_{n,p} \|) = 0.$$

(O.6)

Moreover,

$$|\phi_{\tilde{m}_{X,n}}(u) - \phi_{\tilde{Y}_{n,p}}(u)| \leq \mathbb{E}[|1 - e^{iu(\tilde{m}_{X,n} - \tilde{Y}_{n,p})}|]$$

$$\leq \mathbb{E}[|1 - e^{iu(\tilde{m}_{X,n} - \tilde{Y}_{n,p})}| \mathbb{1}_{|\tilde{m}_{X,n} - \tilde{Y}_{n,p}| \leq \delta}]$$

(O.7)

$$+ \mathbb{E}[|1 - e^{iu(\tilde{m}_{X,n} - \tilde{Y}_{n,p})}| \mathbb{1}_{|\tilde{m}_{X,n} - \tilde{Y}_{n,p}| > \delta}].$$

Choosing a sufficiently small δ, the first right-hand term of (O.7) can be made arbitrarily small since $\lim_{x\to 0}(1 - e^{iux}) = 0$. The second right-hand term of (O.7) is smaller than $2P(|\tilde{m}_{X,n} - \tilde{Y}_{n,p}| > \delta)$, which can be made arbitrarily small according to relation (O.6). Thus, the third right-hand term of (O.4) tends towards 0 when n and p tend towards $+\infty$, which completes the proof.

□

P. Covariance of the Empirical Autocovariance Coefficients

Theorem *Let X_n be a linear process of the form $X_n = \sum_{k \in \mathbb{Z}} h_k V_{n-k}$, where $\mathbb{E}[|V_n|^4] = \nu \sigma^4 < \infty$. Thus, if X is real-valued,*

$$\lim_{n \to \infty} n.\mathrm{cov}[\hat{R}_{X,n}(k), \hat{R}_{X,n}(l)]$$

$$= (\nu - 3) R_X(k) R_X(l) \tag{P.1}$$

$$+ \sum_{p \in \mathbb{Z}} [R_X(p) R_X(p - k + l) + R_X(p + l) R_X(p - k)].$$

If X is a complex-valued circular process, that is, if $\| \mathcal{R}e[X_n] \| = \| \mathcal{I}m[X_n] \|$ and $\mathbb{E}[X_k X_l^*] = 0$,

$$\lim_{n \to \infty} n.\mathrm{cov}[\hat{R}_{X,n}(k), \hat{R}_{X,n}(l)] = (\nu - 2) R_X(k) R_X(l)$$

$$+ \sum_{p \in \mathbb{Z}} R_X(p + l) R_X(k - p). \tag{P.2}$$

Proof We first remark that $R_X(k) = \sigma^2 \sum_{u \in \mathbb{Z}} h_{u+k} h_u^*$. Moreover, it is clear that

$$\mathbb{E}[\hat{R}_{X,n}(k) \hat{R}_{X,n}^*(l)]$$

$$= n^{-2} \sum_{a=1}^{n-k} \sum_{b=1}^{n-l} \mathbb{E}[X_{a+k} X_a^* X_{b+l} X_b^*]$$

$$= n^{-2} \sum_{a=1}^{n-k} \sum_{b=1}^{n-l} \left(\sum_{p,q,r,s \in \mathbb{Z}} h_{p+k} h_q^* h_{r+l} h_s^* \mathbb{E}[V_{a-p} V_{a-q}^* V_{b-r} V_{b-s}^*] \right). \tag{P.3}$$

$\mathbb{E}[V_\alpha V_\beta^* V_\gamma V_\delta^*]$ can only take the values $\nu \sigma^4$, σ^4, or 0, according to whether the four indices are identical, equal in pairs, or whether one of them is distinct from the others. But, to specify these values, we must distinguish the case where V is real-valued from the case where V is complex-valued. Indeed, in the real case, $\mathbb{E}[V_\alpha V_\beta^* V_\gamma V_\delta^*]$ is equal to σ^4 if the coefficients are equal in pairs, and the pairs are distinct (for example, if $\alpha = \beta \neq \gamma = \delta$). The complex case generally requires the terms of the form $\mathbb{E}[V_\alpha V_\beta^* V_\alpha V_\beta^*]$ to be split into real and imaginary parts. However, for the case where V, and consequently X, are complex circular processes, the calculations are simplified due to the relations $\mathbb{E}[V_n V_n] = 0$. In what follows, we shall give a general formulation of $\mathbb{E}[\hat{R}_{X,n}(k)\hat{R}_{X,n}^*(l)]$, and we will only separate the real case and the complex case at the end of the calculations.

$$\mathbb{E}[\hat{R}_{X,n}(k)\hat{R}_{X,n}^*(l)]$$

$$= n^{-2}\sum_{a=1}^{n}\sum_{b=1}^{n}$$

$$\sum_{p,q,r,s\in\mathbb{Z}}h_{p+k}h_q^*h_{r+l+(b-a)}h_{s+(b-a)}^*\mathbb{E}[V_{a-p}V_{a-q}^*V_{a-r}V_{a-s}^*]$$

$$+ O(\frac{kl}{n^2})$$

$$= n^{-2}\sum_{a=1}^{n}\sum_{b=1}^{n}(\sigma^4\sum_{p,r\in\mathbb{Z}}h_{p+k}h_p^*h_{r+l+(b-a)}h_{r+(b-a)}^*\mathbb{I}_{p\neq r}$$

$$+ \sum_{p,q\in\mathbb{Z}}h_{p+k}h_q^*h_{p+l+(b-a)}h_{q+(b-a)}^*\mathbb{E}[V_{a-p}V_{a-q}^*V_{a-p}V_{a-q}^*]\mathbb{I}_{p\neq q}$$

$$+ \sigma^4\sum_{p,q\in\mathbb{Z}}h_{p+k}h_q^*h_{q+l+(b-a)}h_{p+(b-a)}^*\mathbb{I}_{p\neq q}$$

$$+ \nu\sigma^4\sum_{p\in\mathbb{Z}}h_{p+k}h_p^*h_{p+l+(b-a)}h_{p+(b-a)}^*) + O(\frac{kl}{n^2})$$

$$= R_X(k)R_X(l) + n^{-2}\sum_{a=1}^{n}\sum_{b=1}^{n}$$

$$(\sum_{p,q\in\mathbb{Z}}h_{p+k}h_q^*h_{p+l+(b-a)}h_{q+(b-a)}^*\mathbb{E}[V_{a-p}V_{a-q}^*V_{a-p}V_{a-q}^*]\mathbb{I}_{p\neq q}$$

$$+ \sigma^4\sum_{p,q\in\mathbb{Z}}h_{p+k}h_q^*h_{q+l+(b-a)}h_{p+(b-a)}^*\mathbb{I}_{p\neq q}$$

$$+ (\nu-1)\sigma^4\sum_{p\in\mathbb{Z}}h_{p+k}h_p^*h_{p+l+(b-a)}h_{p+(b-a)}^*) + O(\frac{kl}{n^2}).$$

$$\text{(P.4)}$$

In the real case, by setting $u = b - a$, we shall get

$$\text{cov}[\hat{R}_{X,n}(k), \hat{R}_{X,n}^*(l)]$$

$$= \mathbb{E}[\hat{R}_{X,n}(k)\hat{R}_{X,n}(l)^*] - R_X(k)R_X(l)$$

$$= n^{-2}\sum_{a=1}^{n}\sum_{b=1}^{n}(\sigma^4\sum_{p,q\in\mathbb{Z}}h_{p+k}h_q^*h_{p+l+(b-a)}h_{q+(b-a)}^*\mathbb{I}_{p\neq q}$$

$$+ \sigma^4\sum_{p,q\in\mathbb{Z}}h_{p+k}h_q^*h_{q+l+(b-a)}h_{p+(b-a)}^*\mathbb{I}_{p\neq q}$$

$$+ (\nu-1)\sigma^4\sum_{p\in\mathbb{Z}}h_{p+k}h_p^*h_{p+l+(b-a)}h_{p+(b-a)}^*) + O(\frac{kl}{n^2}),$$

$$\text{(P.5)}$$

$$\text{cov}[\hat{R}_{X,n}(k), \hat{R}^*_{X,n}(l)]$$

$$= n^{-2} \sum_{|u|<n} (n - |u|)(\sigma^4 \sum_{p,q\in\mathbb{Z}} h_{p+k} h^*_q h_{p+l+u} h^*_{q+u} \, \mathbb{I}_{p\neq q}$$

$$+ \sigma^4 \sum_{p,q\in\mathbb{Z}} h_{p+k} h^*_q h_{q+l+u} h^*_{p+u} \, \mathbb{I}_{p\neq q}$$

$$+ (\nu - 1)\sigma^4 \sum_{p\in\mathbb{Z}} h_{p+k} h^*_p h_{p+l+u} h^*_{p+u}) + O(\frac{kl}{n^2})$$

$$= n^{-1} \sum_{|u|<n} (1 - n^{-1}|u|)$$

$$\times [R_X(k - l - u) R_X(-u) + R_X(k - u) R_X(l + u)$$

$$+ \sum_{|u|<n}(\nu - 3)\sigma^4 \sum_{p\in\mathbb{Z}} h_{p+k} h^*_p h_{p+l+u} h^*_{p+u}] + O(\frac{kl}{n^2}).$$

$$\tag{P.6}$$

As $\sum_{k\in\mathbb{Z}} |h_k| < \infty$, the sequences above are absolutely summable. From Lebesgue's dominated convergence theorem and Fubini's theorem, we then have

$$\lim_{n\to\infty} n.\text{cov}[\hat{R}_{X,n}(k), \hat{R}_{X,n}(l)]$$

$$= (\nu - 3) R_X(k) R_X(l) \tag{P.7}$$

$$+ \sum_{p\in\mathbb{Z}} [R_X(p) R_X(p - k + l) + R_X(p + l) R_X(p - k)].$$

In the complex circular case, the terms $R_X(l - k + u) R_X(u)$ disappear and the factor $\nu - 3$ becomes $\nu - 2$. We therefore obtain

$$\lim_{n\to\infty} n.\text{cov}[\hat{R}_{X,n}(k), \hat{R}_{X,n}(l)] = (\nu - 2) R_X(k) R_X(l)$$

$$+ \sum_{p\in\mathbb{Z}} R_X(p + l) R_X(k - p). \,\Box$$

$$\tag{P.8}$$

Q. A Central Limit Theorem for Empirical Autocovariances

Theorem *If X is a linear process, with $X_n = \sum_{k \in \mathbb{Z}} h_k V_{n-k}$, $\sum_{k \in \mathbb{Z}} h_k \neq 0$, $\mathbb{E}[|V_n|^2] = \sigma^2$, and $\mathbb{E}[|V_n|^4] = \nu \sigma^4 < \infty$,*

$$\lim_{n \to \infty} \sqrt{n}\left(\begin{bmatrix} \hat{R}_{X,n}(0) \\ \cdot \\ \cdot \\ \hat{R}_{X,n}(N) \end{bmatrix} - \begin{bmatrix} R_X(0) \\ \cdot \\ \cdot \\ R_X(N) \end{bmatrix} \right) \sim \mathcal{N}(0, \Gamma_{X,N}), \qquad (Q.1)$$

where

$$[\Gamma_{X,N}]_{kl} = (\nu - 3) R_X(k) R_X(l)$$
$$+ \sum_{p \in \mathbb{Z}} [R_X(p) R_X(p - k + l) + R_X(p + l) R_X(k - p)] \qquad (Q.2)$$

if X is a real-valued process and

$$[\Gamma_{X,N}]_{kl} = (\nu - 2) R_X(k) R_X(l) + \sum_{p \in \mathbb{Z}} R_X(p + l) R_X(k - p) \qquad (Q.3)$$

if X is a complex second order circular process.

Proof We shall present the proof in the complex case, the proof being similar in the real case. We note $Y_m = [X_m X_m^*, \ldots, X_m X_{m+N}^*]^T$. Let u be any element of $\mathbb{C}^{N+1}$, and define the random variables

$$Z_n = u^H [n^{-1} \sum_{k=1,n} Y_k]$$
$$= u^H [\hat{R}_{X,n}(0), \ldots, \hat{R}_{X,n}(N)]^T + O(\frac{N}{n}). \qquad (Q.4)$$

It is easy to check that the term in $O(n^{-1} N)$ converges in the mean square sense towards 0 when n tends towards $+\infty$. Therefore, to prove the theorem, we must simply check that

$$\lim_{n \to \infty} \sqrt{n}(Z_n - u^H [R_X(0), \ldots, R_X(N)]^T) \sim \mathcal{N}(0, u^H \Gamma_{X,N}(n) u). \qquad (Q.5)$$

We already note that

$$\lim_{n\to\infty} \mathbb{E}[Z_n] = u^H[R_X(0),\dots,R_X(N)]^T,$$

$$\text{and } \lim_{n\to\infty} \text{var}[Z_n] = u^H \Gamma_{X,N}(n)u.$$

(Q.6)

Now, we must prove that Z_n converges in distribution towards a Gaussian random variable.

$$Z_n = n^{-1}\sum_{l=0,N} u_l^*\left(\sum_{k=1,n} X_k X_{k+l}^*\right)$$

$$= n^{-1}\sum_{l=0,N} u_l^*\left(\sum_{k=1,n}\sum_{a,b\in\mathbb{Z}} h_a h_b^* V_{k-a} V_{k+l-b}^*\right)$$

$$= n^{-1}\sum_{l=0,N} u_l^*\left(\sum_{a\in\mathbb{Z}} h_a h_{a+l}^* \sigma^2\right)$$

(Q.7)

$$+ n^{-1}\sum_{l=0,N} u_l^*\left(\sum_{k=1,n}\sum_{a\in\mathbb{Z}} h_a h_{a+l}^* (|V_{k-a}|^2 - \sigma^2)\right)$$

$$+ n^{-1}\sum_{l=0,N} u_l^*\left(\sum_{k=1,n}\sum_{a\neq b-l} h_a h_b^* V_{k-a} V_{k+l-b}^*\right).$$

Let us consider the third term of the sum. For fixed k and l,

$$\left|\sum_{a\neq b-l} h_a h_b^* V_{k-a} V_{k+l-b}^*\right|$$

$$= \left|\sum_{i\neq j} h_{k-i} h_{k+l-j}^* V_i V_j^*\right|$$

(Q.8)

$$\leq \sum_{m\in\mathbb{N}}\sum_{v=1,m-1}\left|\sum_{u\in\mathbb{Z}} h_{k-mu} h_{k+l-mu-v}^* V_{mu} V_{mu+v}^*\right|.$$

The series $\sum_{u\in\mathbb{Z}} h_{k-mu} h_{k+l-mu-v}^* V_{mu} V_{mu+v}^*$ are made up of independent, zero mean, random variables, and

$$\sum_{u\in\mathbb{Z}}\| h_{k-mu} h_{k+l-mu-v}^* V_{mu} V_{mu+v}^* \|^2 \leq \sum_{a\in\mathbb{Z}} |h_{a+l-v}|^4 \sigma^4$$

(Q.9)

$$< \infty.$$

From the version of the strong law of large numbers presented in Appendix V, it thus appears that

$$\sum_{u\in\mathbb{Z}} h_{k-mu} h_{k+l-mu-v}^* V_{mu} V_{mu+v}^* \overset{a.s.}{=} 0.$$

(Q.10)

We now consider the linear process defined by

$$T_k = \sum_{a\in\mathbb{Z}}\left(\sum_{l=0,N} u_l^* h_a h_{a+l}^*\right)(|V_{k-a}|^2 - \sigma^2).$$

(Q.11)

Since

$$\lim_{n\to\infty} Z_n = \sum_{l=0,N} u_l^*\left(\sum_{a\in\mathbb{Z}} h_a h_{a+l}^* \sigma^2\right) + \lim_{n\to\infty} n^{-1}\sum_{k=1,n} T_k,$$

(Q.12)

by applying Theorem 12.4, we see that the limit of the right-hand term of relation (Q.12) converges towards a Gaussian distribution, which completes the proof. $\square$

R. Distribution of the Periodogram for a White Noise

Theorem *If X is a white noise process of variance σ^2, $\hat{S}_{X,n}(f)$ converges towards a $\mathcal{E}(\sigma^{-2})$ distribution if $f \in \mathcal{I} - \{0, 1/2\}$ and a $\chi^2(1)$ distribution if $f \in \{0, 1/2\}$, of mean σ^2.*
Moreover, for $0 < f_1 < f_2 < 1/2$, the vector

$$[c_n(f_1)^T \mathbf{X}_n, s_n(f_1)^T \mathbf{X}_n, c_n(f_2)^T \mathbf{X}_n, s_n(f_2)^T \mathbf{X}_n]^T \tag{R.1}$$

converges in distribution towards a Gaussian random vector with covariance matrix $(\sigma^2/2)I_4$, and $\hat{S}_{X,n}(f_1)$ and $\hat{S}_{X,n}(f_2)$ are asymptotically independent.
Proof We denote by $\varphi_{c,n}(u)$ the characteristic function of $c_n(f)^T \mathbf{X}_n$.

$$\varphi_{c,n}(u) = \prod_{k=1,n} \mathbb{E}[e^{iun^{-1/2} X_k \cos(2\pi k f)}]$$

$$= \prod_{k=1,n} \left(1 - \sigma^2 \frac{u^2}{2n} \cos^2(2\pi k f) + \frac{u^2}{n} \varepsilon_k\left(\frac{u}{\sqrt{n}}\right) \right), \tag{R.2}$$

with $\lim_{u \to 0} \varepsilon_k(u) = 0$ $(k = 1, n)$. Let there be $\varepsilon > 0$; for $n^{-1/2}u$ sufficiently close to 0,

$$\left| \log[\varphi_{c,n}(u)] + \sigma^2 \frac{\sum_{k=1,n} \cos^2(2\pi k f)}{2n} u^2 \right| < \frac{\varepsilon}{2}. \tag{R.3}$$

We now remark that

$$\frac{1}{n} \sum_{k=1,n} \cos^2(2\pi k f) = \frac{1}{2} + \frac{1}{2n} \sum_{k=1,n} \cos(4\pi k f)$$

$$= \frac{1}{2} + \frac{\cos(2(n+1)\pi f) \sin(2n\pi f)}{2n \sin(2\pi f)} \tag{R.4}$$

converges towards $l_c = 1/2$ if $f \in \mathcal{I} - \{0, 1/2\}$, and towards $l_c = 1$ if $f \in \{0, 1/2\}$. Therefore, for $n^{-1/2}u$ sufficiently close to 0,

$$\left| \sigma^2 \frac{\sum_{k=1,n} \cos^2(2\pi k f)}{2n} u^2 - \frac{\sigma^2 l_c u^2}{2} \right| < \frac{\varepsilon}{2}. \tag{R.5}$$

Consequently, for fixed u and sufficiently large n, (R.3) and (R.5) lead to

$$\left| \log[\varphi_{c,n}(u)] + \frac{\sigma^2 l_c u^2}{2} \right| < \varepsilon. \tag{R.6}$$

When n tends towards infinity, it therefore appears that $\varphi_{c,n}(u)$ converges towards the function $e^{-\sigma^2 l_c u^2/2}$, which is continuous for $u = 0$. Therefore, from Levy's theorem, $c_n(f)^T \mathbf{X}_n$ converges in distribution towards a Gaussian variable, with variance $\sigma^2/2$ if $f \in \mathcal{I} - \{0, 1/2\}$ and with variance σ^2 if $f \in \{0, 1/2\}$.

Noting that

$$\frac{1}{n} \sum_{k=1,n} \sin^2(2\pi k f) = \frac{1}{2} - \frac{\cos(2(n+1)\pi f)\sin(2n\pi f)}{2n\sin(2\pi f)} \tag{R.7}$$

converges towards $l_s = 1/2$ if $f \in \mathcal{I} - \{0, 1/2\}$ and towards $l_s = 0$ if $f \in \{0, 1/2\}$, we could similarly show that $s_n(f)^T \mathbf{X}_n$ converges in distribution towards a Gaussian variable with variance $\sigma^2/2$ if $f \in \mathcal{I} - \{0, 1/2\}$ and towards 0 if $f \in \{0, 1/2\}$.

To show that the limits of $c_n(f)^T \mathbf{X}_n$ and $s_n(f)^T \mathbf{X}_n$ are independent, it suffices to note that the vectors $c_n(f)$ and $s_n(f)$ are asymptotically orthonormal:

$$c_n(f)^T s_n(f) = \frac{\sin(2(n+1)\pi f)\sin(2n\pi f)}{2n} = O(\frac{1}{n}). \tag{R.8}$$

Consequently,

$$\mathbb{E}[(c_n(f)^T \mathbf{X}_n)(s_n(f)^T \mathbf{X}_n)] = \sigma^2 c_n(f)^T s_n(f) = O(\frac{1}{n}) \tag{R.9}$$

and the two variables $c_n(f)^T \mathbf{X}_n$ and $s_n(f)^T \mathbf{X}_n$ are asymptotically uncorrelated.

Finally, if $f \in \mathcal{I} - \{0, 1/2\}$, $\hat{S}_{X,n}(f) = |c_n(f)^T \mathbf{X}_n|^2 + |s_n(f)^T \mathbf{X}_n|^2$ converges in distribution towards a $\chi^2(2)$ distribution, that is, towards an exponential distribution with parameter σ^{-2}. If $f \in \{0, 1/2\}$, $\hat{S}_{X,n}(f)$ converges in distribution towards a $\chi^2(1)$ distribution. In both cases, the mean of the limit is equal to σ^2.

Accounting for the fact that, for $0 < f_1 < f_2 < 1/2$, the vectors $\{c_n(f_i), s_n(f_i)\}_{i=1,2}$ form an asymptotically orthonormal family (scalar product in $O(n^{-1})$), it appears that the vector $(c_n(f_1)^T \mathbf{X}_n, s_n(f_1)^T \mathbf{X}_n, c_n(f_2)^T \mathbf{X}_n, s_n(f_2)^T \mathbf{X}_n)$, is asymptotically Gaussian with covariance matrix $(\sigma^2/2)I_4$. The four variables are therefore asymptotically independent, and $\hat{S}_{X,n}(f_1)$ and $\hat{S}_{X,n}(f_2)$ are therefore also asymptotically independent. $\square$

Remark A shorter proof could have been derived by using a version of the central limit theorem established for independent random variables with different distributions (Lindeberg's condition, see for example [1], Section 27).

S. Periodogram of a Linear Process

Theorem

$$\hat{S}_{X,n}(f) = |h(e^{2i\pi f})|^2 \hat{S}_{V,n}(f) + R_n(f)$$

$$= S_X(f)\hat{S}_{V,n}(f) + R_n(f), \tag{S.1}$$

with

$$\lim_{n\to\infty} \left(\sup_{f\in\mathcal{I}} \mathbb{E}[|R_n(f)|] \right) = 0. \tag{S.2}$$

Proof We denote by $F_{X,n}(f)$ the discrete Fourier transform of $\mathbf{X}_n = [X_1, \dots, X_n]^T$.

$$F_{X,n}(f) = \frac{1}{\sqrt{n}} \sum_{k=1,n} X_k e^{-2i\pi k f}$$

$$= \frac{1}{\sqrt{n}} \sum_{l\in\mathbb{Z}} h_t h e^{-2i\pi l f} \left(\sum_{k=1,n} V_{k-l} e^{-2i\pi(k-l)f} \right)$$

$$= \frac{1}{\sqrt{n}} \sum_{l\in\mathbb{Z}} h_t h e^{-2i\pi l f} \left(\sum_{k=1,n} V_k e^{-2i\pi k f} + U_{n,l}(f) \right) \tag{S.3}$$

$$= h(e^{2i\pi f}) F_{V,n}(f) + Y_n(f),$$

with

$$U_{n,l}(f) = \sum_{k=1-l,n-l} V_k e^{-2i\pi k f} - \sum_{k=1,n} V_k e^{-2i\pi k f}, \tag{S.4}$$

and

$$Y_n(f) = \frac{1}{\sqrt{n}} \sum_{l\in\mathbb{Z}} h_t h e^{-2i\pi l f} U_{n,l}(f). \tag{S.5}$$

Noting that

$$\| U_{n,l}(f) \|^2 \le 2\min(|l|, n), \tag{S.6}$$

we show that $Y_n(f)$ converges towards 0 in the mean square sense.

$$\| Y_n(f) \|^2 \le \frac{1}{n} \sum_{k,l \in \mathbb{Z}} |h_k||h_l||\mathbb{E}[U_{n,k}(f)U_{n,l}^*(f)]|$$

$$\le \frac{1}{n} \sum_{k,l \in \mathbb{Z}} |h_k||h_l| \, \| U_{n,k}(f) \| \cdot \| U_{n,l}(f) \|$$

$$\le \frac{2}{n} \left(\sum_{l \in \mathbb{Z}} |h_l|\sqrt{\min(|l|, n)} \right)^2 \tag{S.7}$$

$$\le \frac{2}{n} \left(\sum_{|l| \le m} |h_l|\sqrt{|l|} + \sqrt{n} \sum_{|l| > m} |h_l| \right)^2,$$

where $m < n$. For a fixed $\varepsilon > 0$, we choose m such that $\sum_{|l|>m} |h_l| < \sqrt{\varepsilon}$, and we let n tend towards $+\infty$. It appears that $\lim_{n \to \infty} \| Y_n(f) \|^2 < 2\varepsilon$, and this is true for any positive ε. Therefore, $\lim_{n \to \infty} \| Y_n(f) \| = 0$ and $Y_n(f)$ converges towards 0 in the mean square sense. We note that this convergence is uniform in f. From (S.3), we obtain (S.1) by setting

$$R_n(f) = 2Re[h(e^{2i\pi f})F_{V,n}(f)Y_n^*(f)] + |Y_n(f)|^2. \tag{S.8}$$

It is clear that

$$\mathbb{E}[|R_n(f)|] \le 2Re\left[\mathbb{E}[|h(e^{2i\pi f})F_{V,n}(f)Y_n^*(f)|]\right] + \| Y_n(f) \|^2$$

$$\le 2\sqrt{S_X(f)} \, \| F_{V,n}(f) \| \| Y_n(f) \| + \| Y_n(f) \|^2 \tag{S.9}$$

$$\le (2\sqrt{S_X(f)} \, \| F_{V,n}(f) \| + \| Y_n(f) \|) \, \| Y_n(f) \| .$$

Therefore, in the same way as $\| Y_n(f) \|$, $\mathbb{E}[|R_n(f)|]$ converges uniformly towards 0. $\square$

T. Variance of the Periodogram for a Linear Process

Theorem *For a linear process X whose PSD $S_X(f)$ is strictly positive, $\hat{S}_{X,n}(f)$ converges in distribution towards a random variable, with exponential distribution if $0 < f < 1/2$ and with a $\chi_2(1)$ distribution if $f \in \{0, 1/2\}$. In both cases, the mean of the asymptotic distribution is equal to $S_X(f)$.*

Moreover, if $\mathbb{E}[|V_n|^4] = \nu < \infty$ and $\sum_{k \in \mathbb{Z}} \sqrt{|k|}\|h_k\| < \infty$, we obtain the following asymptotic variances:

$$\text{var}[\hat{S}_{X,n}(f)] \qquad\qquad = 2S_X^2(f) + O(n^{-1/2}) \quad \text{if } f = 0, 1/2,$$

$$\text{var}[\hat{S}_{X,n}(f)] \qquad\qquad = S_X^2(f) + O(n^{-1/2}) \quad \text{if } f \neq 0, 1/2,$$

$$\text{cov}[\hat{S}_{X,n}(f_1), \hat{S}_{X,n}(f_2)] = O(n^{-1/2}) \qquad\qquad \text{if } 0 < f_1 < f_2 < 1/2.$$

$$\tag{T.1}$$

The terms in $O(n^{-1/2})$ decrease towards 0 uniformly in f on any compact set of $]0, 1/2[\times]0, 1/2[-\{(f, f); f \in]0, 1/2[\}$.

Proof We first recall that the convergence of $\mathbb{E}[|R_n(f)|]$ towards 0 implies the convergence in probability of $R_n(f)$ towards 0, since from Markov's inequality,

$$P(|R_n(f)| > \varepsilon) \leq \varepsilon^{-1}\mathbb{E}[|R_n(f)|]. \tag{T.2}$$

We also recall that if $T_n \overset{L}{\to} T$, and $Z_n \overset{P}{\to} c$, where c is a constant, $\forall a \in \mathbb{C}$, $aT_n + Z_n \overset{L}{\to} aT + c$. Consequently, since $\hat{S}_{X,n}(f) = |h(e^{2i\pi f})|^2 \hat{S}_{V,n}(f) + R_n(f)$, and since $\hat{S}_{V,n}(f)$ converges in distribution towards an exponential random variable if $0 < f < 1/2$ and towards a $\chi^2(1)$ distribution if $f = 0, 1/2$, and since $R_n(f)$ converges in probability towards 0, it is clear that $\hat{S}_{X,n}(f)$ converges in distribution towards a random variable of the same kind as the limit of $\hat{S}_{V,n}(f)$ and with mean $|h(e^{2i\pi f})|^2 = S_X(f)$.

To establish the second part of the proof, we begin by showing that under the hypotheses made, $\| R_n(f) \|$ converges uniformly towards 0, with a convergence rate in $O(n^{-1/2})$. Using Cauchy-Schwarz's inequality, we obtain

$$\| R_n(f) \| \le 2 \| h(e^{2i\pi f})F_{V,n}(f)Y_n^*(f) \| + \| \, |Y_n(f)|^2 \, \|$$

$$\le 2\sqrt{S_X(f)} \, \| F_{V,n}(f)Y_n^*(f) \| + (\mathbb{E}[|Y_n(f)|^4])^{1/2}$$

$$\le [2\sqrt{S_X(f)}(\mathbb{E}[|\hat{S}_{V,n}(f)|^2])^{1/4}$$

$$+ (\mathbb{E}[|Y_n(f)|^4])^{1/4}](\mathbb{E}[|Y_n(f)|^4])^{1/4}$$

(T.3)

To show that $\| R_n(f) \| \le O(n^{-1/2})$ uniformly in f, it therefore suffices to show that $\mathbb{E}[|Y_n|^4] \le O(n^{-2})$ uniformly in f. Using the notations of Appendix S,

$$\mathbb{E}[|Y_n|^4] \le \frac{1}{n^2} \sum_{a,b,c,d\in\mathbb{Z}} |h_a h_b^* h_c h_d^* |\mathbb{E}[|U_{n,a}U_{n,b}^* U_{n,c}U_{n,d}^*|]$$

$$\le \frac{1}{n^2} (\sum_{a\in\mathbb{Z}} |h_a|(\mathbb{E}[|U_{n,a}|^4])^{1/4})^4.$$

(T.4)

We now remark that

$$\mathbb{E}[|U_{n,a}|^4] \le 2|a|(\mathbb{E}[|V_n|^4] + 3(2|a|-1) \| V_n \|^4)$$

$$\le 2|a|(\mathbb{E}[|V_n|^4] + 6|a| \| V_n \|^4).$$

(T.5)

Therefore,

$$\mathbb{E}[|Y_n|^4] \le \frac{1}{n^2} (\sum_{a\in\mathbb{Z}} |h_a|[2|a|(\nu + 6|a|)]^{1/4})^4 \le O(n^{-2})$$

(T.6)

and $\| R_n(f) \|$ converges uniformly in f towards 0, with a convergence rate in $O(n^{-1/2})$. Consequently,

$$\mathbb{E}[\hat{S}_{X,n}(f_1)\hat{S}_{X,n}^*(f_2)] = |h(e^{2i\pi f_1})h^*(e^{2i\pi f_2})|^2 \mathbb{E}[\hat{S}_{V,n}(f_1)\hat{S}_{V,n}^*(f_2)]$$

$$+ \mathbb{E}[R_n(f_1)R_n^*(f_2)]$$

$$+ |h(e^{2i\pi f_1})|^2 \mathbb{E}[\hat{S}_{V,n}(f_1)R_n^*(f_2)]$$

$$+ |h(e^{2i\pi f_2})|^2 \mathbb{E}[\hat{S}_{V,n}(f_2)R_n(f_1)]$$

(T.7)

and

$$\left| \text{cov}[\hat{S}_{X,n}(f_1), \hat{S}_{X,n}(f_2)] - |h(e^{2i\pi f_1})h^*(e^{2i\pi f_2})|^2 \text{cov}[\hat{S}_{V,n}(f_1), \hat{S}_{V,n}(f_2)] \right|$$

$$= \left| \mathbb{E}[\hat{S}_{X,n}(f_1)\hat{S}_{X,n}^*(f_2)] - S_X(f_1)S_X(f_2)\mathbb{E}[\hat{S}_{V,n}(f_1)\hat{S}_{V,n}^*(f_2)] \right|$$

$$\leq \left| S_X(f_1)\mathbb{E}[\hat{S}_{V,n}(f_1)R_n^*(f_2)] + S_X(f_2)\mathbb{E}[\hat{S}_{V,n}(f_2)R_n(f_1)] \right|$$

$$\leq S_X(f_1) \parallel \hat{S}_{V,n}(f_1) \parallel \times \parallel R_n(f_2) \parallel$$

$$+ S_X(f_2) \parallel \hat{S}_{V,n}(f_2) \parallel \times \parallel R_n(f_1) \parallel .$$

$$\text{(T.8)}$$

The uniform convergence of $\parallel R_n(f) \parallel$ towards 0 in $O(n^{-1/2})$, and Theorem 12.8, then lead to the convergence results stated, with the uniform decrease of the $O(n^{-1/2})$ terms. $\square$

U. A Strong Law of Large Numbers (I)

Theorem (Kolmogorov) *If a sequence $(Y_n)_{n \in \mathbb{N}^*}$ of random variables is such that the series $\sum_{n \in \mathbb{N}^*} \mathbb{E}[Y_n]$ and $\sum_{n \in \mathbb{N}^*} \| Y_n - \mathbb{E}[Y_n] \|^2$ converge, then the sequence of random variables $S_n = \sum_{k=1,n} Y_n$ converges almost surely towards a random variable, with mean $\sum_{n \in \mathbb{N}^*} \mathbb{E}[Y_n]$ and with variance $\sum_{n \in \mathbb{N}^*} \| Y_n - \mathbb{E}[Y_n] \|^2$.*

Proof We note $Y_n^c = Y_n - \mathbb{E}[Y_n]$, $S_n^c = S_n - \mathbb{E}[S_n]$, and $\alpha_n = \sup_{k \in \mathbb{N}} |S_{n+k}^c - S_n^c|$. We notice that $\mathbb{E}[Y_n]$ converges towards 0. Therefore, the sequence S_n converges almost surely if and only if S_n^c converges almost surely, which can also be expressed by the almost sure convergence of α_n towards 0, which we shall establish.

We note $E_n(\varepsilon) = \{\omega; \alpha_n(\omega) > \varepsilon\}$. The almost sure convergence of α_n towards 0 is also expressed by the fact that $\forall \varepsilon > 0$, $P(\limsup_{n \to \infty} E_n(\varepsilon)) = 0$. We note that

$$P(\limsup_{n \to \infty} E_n(\varepsilon)) = P(\cap_{n=1,\infty}[\cup_{p \geq n} E_p(\varepsilon)]) = \lim_{n \to \infty} P(\cup_{p \geq n} E_p(\varepsilon)).$$
(U.1)

In addition, for $n' > n$,

$$|S_{n'+k}^c - S_{n'}^c| \leq |S_{n'+k}^c - S_n^c| + |S_{n'}^c - S_n^c|,$$
(U.2)

and consequently $\alpha_{n'} \leq 2\alpha_n$. Thus, for $p \geq n$, $(\alpha_p > \varepsilon) \Rightarrow (\alpha_n > (\varepsilon/2))$, and $E_p(\varepsilon) \subset E_n(\varepsilon/2)$. Therefore,

$$\lim_{n \to \infty} P(\cup_{p \geq n} E_p(\varepsilon)) \leq \lim_{n \to \infty} P(E_n(\frac{\varepsilon}{2})).$$
(U.3)

It suffices, therefore, to show that the right-hand term of (U.3) tends towards 0 in order to establish the theorem. We write $T_k = S_{n+k}^c - S_n^c$, and

$$B_k = \{\omega; \forall j < k \ |T_j| < \frac{\varepsilon}{2}, \ |T_k| \geq \frac{\varepsilon}{2}\}.$$
(U.4)

As $E_n(\varepsilon/2) = \cup_{k=0,\infty} B_k$ and as the sets B_k are non-overlapping, $P(E_n(\varepsilon/2)) = \sum_{k=0,\infty} P(B_k)$. But,

$$(\frac{\varepsilon}{2})^2 P(B_k) \leq \int_{B_k} T_k^2(\omega) dP(\omega),$$
(U.5)

and for $p \geq k$,

$$\int_{B_k} T_p^2(\omega)dP(\omega) = \int_{B_k} T_k^2(\omega)dP(\omega)$$

$$+ 2\int_{B_k} T_k(\omega)(T_p(\omega) - T_k(\omega))dP(\omega)$$

$$+ \int_{B_k} (T_p(\omega) - T_k(\omega))^2 dP(\omega) \tag{U.6}$$

$$\geq \int_{B_k} T_k^2(\omega)dP(\omega),$$

since the events $\mathbb{1}_{B_k}T_k$ and $T_p - T_k$ are independent. Then,

$$\int_{B_k} T_k(\omega)(T_p(\omega) - T_k(\omega))dP(\omega) = 0. \tag{U.7}$$

Since the sets B_k are non-overlapping, and the variables Y_k are independent,

$$(\frac{\varepsilon}{2})^2 \sum_{k=0,p} P(B_k) \leq \int_{\cup_{k=0,p} B_k} T_p^2(\omega)dP(\omega)$$

$$\leq \int_{\mathbb{R}} T_p^2(\omega)dP(\omega) \tag{U.8}$$

$$\leq \sum_{k=n+1,n+p} \| Y_k - \mathbb{E}[Y_k] \|^2$$

$$\leq \sum_{k=n+1,\infty} \| Y_k - \mathbb{E}[Y_k] \|^2$$

Therefore, letting p tend towards $+\infty$, we obtain

$$P(E_n(\varepsilon/2)) \leq (\frac{2}{\varepsilon})^2 \sum_{k=n,\infty} \| Y_k - \mathbb{E}[Y_k] \|^2, \tag{U.9}$$

and since the series $\sum_{k=0,\infty} \| Y_k - \mathbb{E}[Y_k] \|^2$ converges, $\lim_{n \to \infty} P(E_n(\varepsilon/2))$ $= 0$, which completes the proof. $\square$

V. A Strong Law of Large Numbers (II)

Theorem (Kolmogorov) *If a sequence* $(Y_n)_{n \in \mathbb{N}^\bullet}$ *of independent random variables is such that*

$$\lim_{n \to \infty} \frac{1}{n} \sum_{k=1,n} \mathbb{E}[Y_k] = a$$

$$\lim_{n \to \infty} \sum_{k=1,n} \frac{\operatorname{var}[Y_k]}{k^2} < \infty, \tag{V.1}$$

then the sequence $\frac{1}{n} \sum_{k=1,n} Y_k$ *converges almost surely towards* a. The proof of the theorem involves the following lemma, called Kronecker's lemma, which is presented here in a weaker but sufficient version for the requirements of the proof.

Lemma V.1 *Let* $(\alpha_n)_{n \in \mathbb{N}}$ *be a real sequence. If the sequence* $\sum_{k=1,n} \alpha_k / k$ *converges, then*

$$\lim_{n \to \infty} \frac{1}{n} \sum_{k=1,n} \alpha_k = 0. \tag{V.2}$$

Proof (of the lemma) We write $\beta_n = \sum_{k=1,n} \alpha_k / k$, and $\gamma_n = \sum_{k=1,n} \alpha_k$.

$$\frac{\gamma_n}{n} = \beta_n - \frac{1}{n} \sum_{k=1,n-1} \beta_k$$

$$= \beta_n - \frac{n-1}{n} \left(\frac{1}{n-1} \sum_{k=1,n-1} \beta_k \right). \tag{V.3}$$

We denote by c the limit of β_n. It is clear that $\lim_{n \to \infty} \gamma_n / n = c - 1 \times c = 0$, which completes the proof of the lemma.

Proof (of the theorem) We write $Z_n = n^{-1}(Y_n - \mathbb{E}[Y_n])$. The sequences $\sum_{k=1,\infty} \mathbb{E}[Z_k] = 0$ and $\sum_{k=1,\infty} \operatorname{var}[Z_k] = \sum_{k=1,\infty} k^{-2} \operatorname{var}[Y_k]$ converge. The version of the strong law of large numbers presented in Appendix U therefore indicates that the series $\sum_{k=1,\infty} Z_k$ converges almost surely towards a certain random variable, that is, the series $\sum_{k=1,\infty} k^{-1}(Y_k - \mathbb{E}[Y_k])$ converges almost surely. Therefore, from Kronecker's theorem, we have almost surely

$$\lim_{n \to \infty} \frac{1}{n} \sum_{k=1,n} (Y_k - \mathbb{E}[Y_k]) = 0, \qquad \text{(V.4)}$$

that is, we have almost surely

$$\lim_{n \to \infty} \frac{1}{n} \sum_{k=1,n} Y_k = a. \square \qquad \text{(V.5)}$$

W. Phase-Amplitude Relationship for Minimum-phase Causal Filters

Theorem W.1 *Let* $h(z) = \sum_{k=0,\infty} h_k z^{-k}$ *be the transfer function of a minimum-phase causal filter, with* $\sum_{k=0,\infty} |h_k| < \infty$. *The modulus and the phase of the frequency response of the filter are linked by the relations*

$$\arg[h(e^{2i\pi f})] = V.P. \int_{\mathcal{I}} \log|h(e^{2i\pi u})| \cot[\pi(u - f)]du$$

$$(W.1)$$

$$\log|h(e^{2i\pi f})| = V.P. \int_{\mathcal{I}} \arg[h(e^{2i\pi u})] \cot[\pi(f - u)]du + h_0,$$

where

$$V.P. \int_{\mathcal{I}} g(u, f)du = \lim_{\varepsilon \to 0+} [\int_{-1/2}^{f-\varepsilon} g(u, f)du + \int_{f+\varepsilon}^{1/2} g(u, f)du]. \qquad (W.2)$$

The transform $g \to V.P. \int_{\mathcal{I}} g(e^{2i\pi u}) \cot[\pi(f - u)]du$ *is called the Hilbert discrete transform of* g.

Proof We consider the discrete Hilbert filter whose impulse response is given by $\text{Hilb}_n = -i.\text{sign}(n)$, for $n \in \mathbb{Z}^*$, and $\text{Hilb}_0 = 0$. The frequency response of this filter is obtained by noticing that the Fourier transform of the distribution $-i \sum_{n \in \mathbb{Z}} \text{Hilb}_n \delta_n$ is the distribution

$$\text{Hilb}: \phi \to V.P. \int_{\mathcal{I}} \phi(u) \cot(-\pi u)du \qquad (W.3)$$

(the verification of this result is left up to the reader). Let $(g_n)_{n\in\mathbb{Z}}$ be the impulse response of a stable causal filter. Since the negative coefficients of the impulse response of the filter are equal to zero, denoting by $g_n = g_{p,n} + g_{i,n}$ its decomposition into an even part and an odd part, it is clear that

$$g_{i,n} = i.\text{Hilb}_n \times g_{p,n}, \text{ and } g_{p,n} = i.\text{Hilb}_n \times g_{i,n} + g_0 \delta_{0,n}. \qquad (W.4)$$

Moreover, the real and imaginary parts of the frequency response $G(f)$ of the filter are the respective Fourier transforms of the sequences $(g_{p,n})_{n\in\mathbb{Z}}$ and $(-i.g_{i,n})_{n\in\mathbb{Z}}$. Therefore,

$$\text{Im}[G(f)] = V.P. \int_{\mathcal{I}} \text{Re}[G(f)] \cot[\pi(u - f)]du,$$

(W.5)

$$\text{Re}[G(f)] = V.P. \int_{\mathcal{I}} \text{Im}[G(f)] \cot[\pi(f - u)]du + g(0).$$

Now, let $h(z)$ be the transfer function of a minimum-phase causal filter. $h(z)$ is holomorphic and is not equal to zero outside the unit disk, otherwise $h'(z) = h(z)(a^* - z^{-1})/(1 - az^{-1})$, where a is a zero of $h(z)$ with $|a| > 1$, would also be a causal factorisation of $|h(e^{2i\pi f})|^2$, with $|h'_0| > |h_0|$, which is impossible (see the proof of Theorem 8.8).

Moreover, as $\sum_{k=0,\infty} |h_k| < \infty$, $h(z)$ is holomorphic and differs from zero on a domain Δ which contains the complex plane except a disk with centre 0 and radius r_0, where $r_0 \leq 1$. We can then show (see the following theorem) that the function $\log[h(z)]$ is also holomorphic for a certain determination of the logarithm. By setting $G(f) = \log[h(e^{2i\pi f})]$, we thus obtain the desired relations by noting that $\text{Im}[G(f)] = \arg[h(e^{2i\pi u})]$, and $\text{Re}[G(f)] = \log|h(e^{2i\pi u})|$, and by using relations (W.5). $\square$

To show that $\log[h(z)]$ is holomorphic on Δ, it suffices to note that $\log[h(z^{-1})]$ is holomorphic on the simply connected domain obtained by transforming Δ by the transform $z \to z^{-1}$, and to conclude by applying the following theorem ([36] p.226):

Theorem W.2 If f is holomorphic in a simply connected domain $\mathcal{D}$, with $f(z) \neq 0$ on $\mathcal{D}$, then there exists a function $\phi(z)$, holomorphic in $\mathcal{D}$ and such that $e^{\phi(z)} = f(z)$ on $\mathcal{D}$.

Proof We define

$$\phi(z) = \int_{z_0}^{z} \frac{f'(u)}{f(u)}du + k,$$

(W.6)

where the integration is done along any curve contained in $\mathcal{D}$, and k is a constant such that $f(z_0) = e^k$. ϕ is a holomorphic function in $\mathcal{D}$ (the proof is straightforward, see, for example, [36] p.209), with derivative

$$\phi'(z) = \frac{f'(z)}{f(z)}.$$

(W.7)

It is then easy to check that the function $e^{-\phi(z)}f(z)$ has a zero derivative on $\mathcal{D}$, and is therefore equal on $\mathcal{D}$ to a constant denoted by M. For $z = z_0$, we obtain

$$M = e^{-k}f(z_0)$$

(W.8)

$$= f(z_0)/f(z_0) = 1.$$

Consequently, $f(z) = e^{\phi(z)}$.

$\square$

X. Convergence of the Metropolis-Hastings Algorithm

Theorem *For the Metropolis-Hastings algorithm, whatever the choice of the instrumental distribution q, the distribution f of interest is a stationary distribution of the sequence $X = (X_k)_{k \in \mathbb{N}}$ generated. We assume, moreover, that f is bounded and strictly positive on any compact set of E, assumed to be connected. Then, if there exists $\varepsilon, \alpha > 0$ such that*

$$|x - y| < \alpha \Rightarrow q(y|x) > \varepsilon, \tag{X.1}$$

the Markov chain X is f-irreducible and aperiodic.

Proof We begin by showing that $P(X_n \in A) = \int_A f(x)dx$.

$$P(X_{n+1} \in A)$$

$$= \int_{E^2} \mathbb{I}_A(x_{n+1})dP(x_{n+1}|X_n = x)f(x)dx \tag{X.2}$$

$$= \int_{E^3} \mathbb{I}_A(x_{n+1})dP(x_{n+1}|X_n = x, Y_n = y)f(x)q(y|x)dxdy.$$

Moreover,

$$dP(x_{n+1}|X_n = x, Y_n = y) = \delta_y(x_{n+1})\rho(x, y) + \delta_x(x_{n+1})(1 - \rho(x, y)). \tag{X.3}$$

Therefore, noting $D = \{(x, y); \rho(x, y) < 1\}$,

$$P(X_{n+1} \in A) = \int_{E^3} \mathbb{I}_A(x_{n+1})\delta_y(x_{n+1})\rho(x, y)f(x)q(y|x)dxdy$$

$$+ \int_{E^3} \mathbb{I}_A(x_{n+1})\delta_x(x_{n+1})(1 - \rho(x, y))f(x)q(y|x)dxdy \tag{X.4}$$

$$P(X_{n+1} \in A) = \int_D \mathbb{I}_A(y) \frac{f(y)q(x|y)}{f(x)q(y|x)} f(x)q(y|x)dxdy$$

$$+ \int_{D^c} \mathbb{I}_A(y)f(x)q(y|x)dxdy$$

$$+ \int_D \mathbb{I}_A(x)(1 - \frac{f(y)q(x|y)}{f(x)q(y|x)})f(x)q(y|x)dxdy \tag{X.5}$$

$$= \int_{E^2} \mathbb{I}_A(x)f(x)q(y|x)dxdy = \int_A f(x)dx.$$

The third equality is obtained by changing (x, y) into (y, x) in the two first integrals of the sum, which, in particular, results in transforming the domain of integration D into D^c.

We show that the chain is f-irreducible. For any Borel set $A \in \mathcal{B}(E)$ such that $\int_A f(x)dx > 0$, and any value x_0 of X_0, there exist an integer m and a sequence of elements x_k of E such that $|x_{k+1} - x_k| \leq \alpha$, and $x_m \in A$. Consequently, $P(X_m \in A | X_0 = x_0) > 0$ for any Borel set A such that $\int_A f(x)dx > 0$, which shows that X is f-irreducible.

Finally, we show that the chain is aperiodic. To show the aperiodicity of X, it suffices to prove that there is $x_0 \in E$, and a neighbourhood V_{x_0} of x_0 such that

$$\forall x \in V_{x_0}, \ P(X_{n+1} \in A | X_n = x) > 0, \tag{X.6}$$

for any set of A such that $\int_A f(x)dx > 0$. For this, we choose any x_0, and we denote by B the ball of E with centre x_0 and radius $\alpha/2$. Then, for any element $x \in B$, and setting $D_x = \{y; \rho(x, y) > 0\}$,

$$P(X_{n+1} \in A | X_n = x)$$

$$\geq \int_{E^2} \mathbb{I}_A(x_{n+1})\rho(x, y)\delta_y(x_{n+1})q(y|x)dy$$

$$\geq \int_{D_x} \mathbb{I}_A(y)\frac{f(y)}{f(x)}q(x|y)dy + \int_{D_x^c} \mathbb{I}_A(y)q(y|x)dy \tag{X.7}$$

$$\geq \int_{D_x \cap B} \mathbb{I}_A(y)\frac{f(y)}{f(x)}q(x|y)dy + \int_{D_x^c \cap B} \mathbb{I}_A(y)q(y|x)dy$$

$$\geq \varepsilon \frac{\inf_{u \in B} f(u)}{\sup_{u \in B} f(u)}\lambda(A \cap B) > 0,$$

where λ here represents Lebesgue's measure. Therefore, X is aperiodic. $\square$

Y. Convergence of the Gibbs Algorithm

Theorem *Whatever the choice of g, the Gibbs algorithm simulates a Markov chain $Y = ([Y_{k,1}, \ldots, Y_{k,p}]^T)_{k \in \mathbb{N}}$ of which g is the stationary distribution. f therefore represents the stationary distribution of the sub-chain $X = (X_k)_{k \in \mathbb{N}}$. Moreover, if there exist $\varepsilon, \alpha > 0$ such that, $\forall i = 1, p$,*

$$|y - y'| < \alpha \Rightarrow g_i(y_i | y_{1:i-1}, y'_{i+1:p}) > \varepsilon, \tag{Y.1}$$

then the Markov chain Y is g-irreducible and aperiodic (X is therefore f-irreducible and aperiodic). Furthermore, if $g_k(y_k | y_{i \neq k}) > 0$ for any value of y, the chain is reversible.

Proof We denote by F the space on which y takes its values, and F_k the space on which the component (scalar or vector) y_k of y takes its values. Moreover, we note $Y_n = (Y_{n,1}, \ldots, Y_{n,p})^T$, and $g^k(y_{n,1:k-1}, y_{n,k+1:p})$ the marginal density defined by

$$g^k(y_{n,1:k-1}, y_{n,k+1:p}) = \int_{F_k} g(y_{n,1:p}) dy_k. \tag{Y.2}$$

We begin by showing that g represents the density of a stationary measure of the sequence generated by the algorithm, that is,

$$P(Y_{n+1} \in A) = \int_A g(y) dy, \tag{Y.3}$$

when $Y_n \sim g$.

$$P(Y_{n+1} \in A) = \int_F \mathbb{I}_A(y_{n+1}) dP(y_{n+1} | Y_n = y) g(y) dy, \tag{Y.4}$$

and

$$dP(y_{n+1} | Y_n = y)$$

$$= g_1(y_{n+1,1} | y_{n,2:p}) g_2(y_{n+1,2} | y_{n+1,1}, y_{n,3:p}) \times \cdots \tag{Y.5}$$

$$\times g_p(y_{n+1,p} | y_{n+1,1:p-1}) dy_{n+1}.$$

Therefore,

$$P(Y_{n+1} \in A)$$

$$= \int_{F \times F} \mathbb{I}_A(y_{n+1}) g_1(y_{n+1,1}|y_{n,2:p}) g_2(y_{n+1,2}|y_{n+1,1}, y_{n,3:p}) \cdots$$

$$\times g_p(y_{n+1,p}|y_{n+1,1:p-1}) g(y_{n,1:p}) dy_{n+1} dy_n$$

$$= \int_{F \times F} \mathbb{I}_A(y_{n+1}) g_1(y_{n+1,1}|y_{n,2:p}) g_2(y_{n+1,2}|y_{n+1,1}, y_{n,3:p}) \cdots \qquad \text{(Y.6)}$$

$$\times g_p(y_{n+1,p}|y_{n+1,1:p-1}) g_1(y_{n,1}|y_{n,2:p})$$

$$\times g^1(y_{n,2:p}) dy_{n+1} dy_n$$

$$= \int_{F \times F_2 \times .. \times F_p} \mathbb{I}_A(y_{n+1}) g_2(y_{n+1,2}|y_{n+1,1}, y_{n,3:p}) \cdots$$

$$\times g_p(y_{n+1,p}|y_{n+1,1:p-1}) g(y_{n+1,1}, y_{n,2:p}) dy_{n+1} dy_{n,2} \dots dy_{n,p}.$$

By iterating the procedure, we finally obtain

$$P(Y_{n+1} \in A) = \int_A g(y_{n+1,1:p}) dy_{n+1}. \qquad \text{(Y.7)}$$

The proof of the g-irreducibility and of the aperiodicity of Y is performed in the same way as in the case of the Metropolis-Hastings algorithm.

We now show that if $g_k(y_k|y_{i \neq k}) > 0$ for any value of y, the chain is reversible. For this, we begin by noting that for fixed $y' \in E^p$

$$g(y_{1:p}) = g_p(y_p|y_{1:p-1}) g^p(y_{1:p-1})$$

$$= \frac{g_p(y_p|y_{1:p-1})}{g_p(y'_p|y_{1:p-1})} g(y_{1:p-1}, y'_p) \qquad \text{(Y.8)}$$

$$= \frac{g_p(y_p|y_{1:p-1})}{g_p(y'_p|y_{1:p-1})} \frac{g_{p-1}(y_{p-1}|y_{1:p-2}, y'_p)}{g_{p-1}(y'_{p-1}|y_{1:p-2}, y'_p)} \times g(y_{1:p-2}, y'_{p-1:p}).$$

By induction, we obtain

$$g(y_{1:p}) = \prod_{k=1,p} \frac{g_k(y_k|y_{1:k-1}, y'_{k+1:p})}{g_k(y'_k|y'_{1:k-1}, y_{k+1:p})} g(y'_{1:p}). \qquad \text{(Y.9)}$$

And by using relation (Y.9),

$$P(\{Y_{n+1} \in A\} \cap \{Y_n \in B\})$$

$$= \int_{F \times F} \mathbb{I}_A(y') \mathbb{I}_B(y) g_1(y_1'|y_{2:p}) \cdots g_p(y_p'|y_{1:p-1}') g(y_{1:p}) dy dy'$$

$$= \int_{F \times F} \mathbb{I}_A(y') \mathbb{I}_B(y) g_1(y_1|y_{2:p}') \cdots g_p(y_p|y_{1:p-1}) g(y_{1:p}') dy dy'$$

$$= P(\{Y_n \in A\} \cap \{Y_{n+1} \in B\}),$$

$$(Y.10)$$

which completes the proof. $\square$

Z. Asymptotic Variance of the LMS Algorithm

Theorem *We denote by T_p the covariance matrix of $\mathbf{X}_n$ and by $U\Lambda U^H$ its eigenvalue decomposition, with $\Lambda = diag(\lambda_0, \ldots, \lambda_p)$. Moreover, we note $\lambda_{\max} = \max\{\lambda_0, \ldots, \lambda_p\}$. When $0 < \mu < 2/\lambda_{\max}$, the covariance of the error $\theta_n - \theta_*$ ($\theta_* = T_p^{-1} r_{XA}$) converges if and only if*

$$\sum_{l=0,p} \frac{\mu\lambda_l}{2 - \mu\lambda_l} < 1. \tag{Z.1}$$

The asymptotic covariance is thus given by

$$\Sigma_\infty = \frac{\mu\sigma_{min}^2}{1 - \sum_{l=0,p} \mu\lambda_l(2 - \mu\lambda_l)^{-1}} \tag{Z.2}$$

$$\times\, U diag((2 - \mu\lambda_0)^{-1}, \cdots, (2 - \mu\lambda_p)^{-1})U^H.$$

Proof We recall that

$$\Sigma_n = (I - \mu T_p)\Sigma_{n-1}(I - \mu T_p)$$
$$+ \mu^2(T_p Tr(\Sigma_{n-1}T_p) + T_p\sigma_{min}^2). \tag{Z.3}$$

If Σ_n has a limit when n tends towards $+\infty$, this limit, denoted by Σ_∞, must satisfy the equation

$$\Sigma_\infty = \Sigma_\infty - \mu(T_p\Sigma_\infty + \Sigma_\infty T_p)$$
$$+ \mu^2(T_p\Sigma_\infty T_p + T_p Tr(\Sigma_\infty T_p) + T_p\sigma_{min}^2), \tag{Z.4}$$

that is,

$$T_p\Sigma_\infty + \Sigma_\infty T_p = \mu(T_p\Sigma_\infty T_p + T_p Tr(\Sigma_\infty T_p) + T_p\sigma_{min}^2). \tag{Z.5}$$

Noting $\tilde{\Sigma}_\infty = U^H\Sigma_\infty U$, where $U\Lambda U^H$ represents the decomposition into eigenvalues of T_p, we obtain

$$\Lambda\tilde{\Sigma}_\infty + \tilde{\Sigma}_\infty\Lambda = \mu(\Lambda\tilde{\Sigma}_\infty\Lambda + \Lambda Tr(\Lambda\tilde{\Sigma}_\infty) + \sigma_{min}^2\Lambda). \tag{Z.6}$$

Consequently,

$$[\tilde{\Sigma}_\infty]_{ab}(\lambda_a + \lambda_b - \mu\lambda_a\lambda_b) = \mu\delta_{a,b}\lambda_a\big(\sum_{l=0,p}\lambda_l[\tilde{\Sigma}_\infty]_{ll} + \sigma_{min}^2\big), \tag{Z.7}$$

where $\delta_{a,b} = 1$ if $a = b$, and 0 otherwise. The mean convergence hypothesis $0 < \mu < 2/\lambda_{\max}$ means that

$$\lambda_a + \lambda_b - \mu\lambda_a\lambda_b > \frac{\lambda_a(\lambda_{\max} - \lambda_b) + \lambda_b(\lambda_{\max} - \lambda_a)}{\lambda_{\max}} \tag{Z.8}$$

$$> 0.$$

Therefore, $[\tilde{\Sigma}_\infty]_{ab} = 0$ for $a \neq b$. For $a = b$,

$$[\tilde{\Sigma}_\infty]_{aa} = \frac{\mu}{2 - \mu\lambda_a}\big(\sum_{l=0,p}\lambda_l[\tilde{\Sigma}_\infty]_{ll} + \sigma_{min}^2\big). \tag{Z.9}$$

From these relations,

$$\sum_{l=0,p}\lambda_l[\tilde{\Sigma}_\infty]_{ll} = \frac{\sigma_{min}^2\big(\sum_{l=0,p}\mu\lambda_l(2 - \mu\lambda_l)^{-1}\big)}{1 - \sum_{l=0,p}\mu\lambda_l(2 - \mu\lambda_l)^{-1}}. \tag{Z.10}$$

Consequently,

$$\Sigma_\infty = \frac{\mu\sigma_{min}^2}{1 - \sum_{l=0,p}\mu\lambda_l(2 - \mu\lambda_l)^{-1}} \tag{Z.11}$$

$$\times U diag((2 - \mu\lambda_0)^{-1}, \cdots, (2 - \mu\lambda_p)^{-1})U^H.$$

It is clear that Σ_∞ represents the asymptotic error covariance of θ only if $\tilde{\Sigma}_\infty > 0$, that is, taking into account the condition $0 < \mu < 2/\lambda_{\max}$, if $1 - \sum_{l=0,p}\mu\lambda_l(2 - \mu\lambda_l)^{-1} > 0$, or in other words, if

$$\sum_{l=0,p}\frac{\mu\lambda_l}{2 - \mu\lambda_l} < 1. \tag{Z.12}$$

We now show the converse of the theorem. We remark that the convergence of Σ_n is ensured if and only if the vector $\tilde{e}_n$ of the diagonal components of the matrix $\tilde{\Sigma}_n$ converges when n tends towards $+\infty$. Relation (Z.3) leads to

$$\tilde{\Sigma}_n = (I - \mu\Lambda)^2\tilde{\Sigma}_{n-1} + \mu^2\Lambda(Tr(\Lambda\tilde{\Sigma}_{n-1}) + \sigma_{min}^2), \tag{Z.13}$$

whence we deduce that the vectors $\tilde{e}_n$ satisfy the recurrence

$$\tilde{e}_n = B\tilde{e}_{n-1} + \mu^2\sigma_{min}^2\lambda, \tag{Z.14}$$

with $\lambda = [\lambda_0, \cdots, \lambda_p]^T$, and $B = diag((1 - \mu\lambda_0)^2, \cdots, (1 - \mu\lambda_p)^2) + \mu^2\lambda\lambda^T$. Consequently,

$$\tilde{e}_n = B^n\tilde{e}_0 + \mu^2\sigma_{min}^2 \sum_{k=0,n-1} B^k\lambda$$

$$= B^n\tilde{e}_0 + \mu^2\sigma_{min}^2(I - B)^{-1}(I - B^n)\lambda. \tag{Z.15}$$

The convergence of $\tilde{e}_n$ will happen if and only if the eigenvalues of B are strictly smaller than 1, the limit being then given by $\tilde{e}_\infty = \mu^2\sigma_{min}^2(I-B)^{-1}\lambda$.

The matrix B is positive, which means that its eigenvalues are positive. To show that the condition (Z.1) is sufficient, it suffices, therefore, to show that the eigenvalues of B are smaller than 1. Let u be a non-zero eigenvector of B and c the corresponding eigenvalue. Since $Bu = cu$,

$$u_k = \frac{\mu^2\lambda_k}{c - (1 - \mu\lambda_k)^2} \sum_{l=0,p} \lambda_l u_l, \quad (k = 0, p), \tag{Z.16}$$

and by adding relations (Z.16) multiplied by factors λ_k, we obtain

$$\sum_{k=o,p} \frac{\mu^2\lambda_k^2}{c - (1 - \mu\lambda_k)^2} = 1. \tag{Z.17}$$

As the function

$$x \rightarrow \sum_{k=o,p} \frac{\mu^2\lambda_k^2}{x - (1 - \mu\lambda_k)^2} \tag{Z.18}$$

is a decreasing and continuous function of the variable x for $x \geq 1$ $(1 - \mu\lambda_k < 1)$ and as, for $x = 1$, this function is equal to $\sum_{l=0,p} \mu\lambda_l(2 - \mu\lambda_l)^{-1}$, which is smaller than 1 by hypothesis, it appears that the eigenvalue c of B is necessarily smaller than 1, which completes the proof. $\square$

References

Probabilities and Processes

1. P.B. Billingsley, (1979). Probability and measure. Wiley Series in Probability and Mathematical Statistics. John Wiley & Sons
2. J.L. Doob, (1953). Stochastic processes. Wiley publications in statistics, John Wiley & Sons
3. P. Doukhan, (1994). Mixing, properties and examples. Lecture Notes in Statistics. Springer-Verlag
4. R.S. Liepster, A.N. Shirayev, (2000). Statistics of random processes. Springer-Verlag
5. B. Oksendal, (1985). Stochastic differential equations. An introduction with Applications. Springer-Verlag
6. A. Papoulis, (1991). Probability, Random Variables, and Stochastic Processes. McGraw-Hill
7. V.S. Pougachev, I.N. Sinitsyn, (1985). Stochastic differential systems, analysis and filtering. John Wiley & Sons
8. L.R. Rabiner, (1989). A tutorial on hidden Markov models and selected applications in speech recognition. Proc. of the IEEE, 77:2:257-286
9. Y.A. Rozanov, (1982). Markov random fields. Springer-Verlag

Stationary Processes

10. P.J. Brockwell, R.A. Davis, (1991). Time series, theory and methods. Springer-Verlag
11. J.A. Cadzow, (1987). Signal enhancement using canonical projection operators. Proceedings of the ICASSP, 673-676
12. P. Delsarte, Y. Genin, Y. kamp, (1978). Orthogonal polynomial matrices on the unit circle. IEEE trans. on circuits and systems, 25:3:149-160
13. S.M. Kay, (1988). Modern spectral estimation, theory and applications. Prentice Hall, Englewood Cliffs
14. T. Kailath, (1977). Linear least-squares estimation. Benchmark papers in electrical engineering and computer science, 17. Dowden Hutchinson & Ross Inc.
15. E.J. Hannan, (1970). Multiple time series. John Wiley & Sons
16. S.L. Marple, (1987). Digital spectral analysis with applications. Signal processing series. Prentice Hall
17. A. Papoulis, (1984). Signal analysis. McGraw-Hill
18. M.B. Priestley, (1981). Spectral analysis and time series. Volume I. Academic Press
19. J.C. Reinsel, (1993). Elements of multivariate time series analysis. Springer-Verlag

20. P. Whittle, (1963). On the fitting of multivariate autoregressions and the approximate canonical factorisation of a spectral density matrix, Biometrika, 50:129-134
21. E. Won, (1971). Stochastic processes in information and dynamic systems. Mc Graw-Hill

Statistics

22. A. Borovkov, (1987). Statistiques Mathématiques. Mir, Moscow
23. S. Degerine, (1992). On local maxima of the likelihood function for Toeplitz matrix estimation. IEEE Trans. on Signal Processing, 40:6:1563-1565
24. S.M. Kay, (1993). Statistical signal processing, estimation theory. Prentice Hall
25. M. Kendall, A. Stuart, (1977). The advanced theory of statistics. C.Griffin
26. J.M. Mendel, (1987). Lessons in digital estimation theory. Prentice Hall
27. D.T. Pham, (1988). Maximum likelihood estimation of autoregressive model by relaxation on the reflection coefficients. IEEE Trans on Acous. Speech, Signal Processing, 36:175-177
28. H.L. Van Trees, (1968). Detection, estimation and modulation theory. John Wiley & Sons

Analysis

29. G. de Barra, (1981). Measure theory and integration. John Wiley & Sons
30. S.K. Berberian, (1976). Introduction to Hilbert space. AMS Chelsea Publishing
31. H. Bremermann (1965). Distributions, complex variables, and Fourier transforms. Addison Wesley
32. J. Dieudonné, (1965). Fondements de l'analyse moderne. Gauthier Villars, Paris
33. S. Lang, (1997). Complex analysis. Addison Wesley
34. D. Mitrovic, D. Zubrinic, (1998). Fundamentals of applied functional analysis: distributions, Sobolev spaces, nonlinear elliptic equations. Addison Wesley Longman
35. T. Myint-U, (1978). Ordinary differential equations. Elsevier North Holland Inc.
36. L.L. Pennisi, (1976). Elements of complex variables. 2nd edition. Holt, Rinehart and Winston, NY
37. W. Rudin, (1970). Real and complex analysis. Mc Graw Hill
38. L. Schwartz, (1976). Analyse Hilbertienne. Editions de l'Ecole Polytechnique
39. M. Willem, (1995). Analyse harmonique réelle. Hermann
40. K. Yosida, (1980). Functional analysis. 6th edition. Springer-Verlag

Fourier Analysis and Trigonometric Moments

41. H. Dym, H.P. Mc Kean, (1972). Fourier series and integrals. Academic Press
42. U. Grenander, G. Szegö, (1958). Toeplitz forms and their applications. Univ. of California Press, Berkley
43. M.G. Krein, A.A. Nudelman, (1977). The Markov moment problems and extremal problems. Translation of the A.M.S.
44. A.Papoulis, (1962). The Fourier integral and its applications. Mc Graw Hill
45. D. Slepian, H.O. Plollack, H.J. Landau, (1961). Prolate spheroidal wave functions. Bell Tech. Journal, 40:43-84

Signal Processing

46. W.A. Gardner (1994). Cyclostationarity in communications and signal processing. IEEE Press
47. S.U. Pillai (1989). Array signal processing. Springer-Verlag
48. L.L. Scharf, C. Demeure, (1991). Statistical signal processing: detection, estimation, and time series analysis. Addison Wesley
49. M. Schwarz, L. Shaw, (1975). Signal processing: discrete spectral analysis, detection, and estimation. Mc Graw Hill
50. S.V Vaseghi, (1996). Advanced signal processing and digital noise reduction. John Wiley & Sons and B.G. Teubner

Wavelets and Time-frequency Analysis

51. L. Cohen, (1995). Time-frequency analysis. Prentice Hall
52. I. Daubechies, (1992). Ten lectures on wavelets. SIAM
53. W. Hardle, G. Kerkyacharia, D. Picard, (1998). Wavelets, approximation, and statistical applications. Springer-Verlag
54. G. Longo, B. Picinbono (Eds), (1989). Time and frequency representation of signals and systems. Springer-Verlag
55. Y. Meyer, (1997). Wavelets. Cambridge University Press

Higher Order Statistics

56. P. Mc Cullagh, (1987). Tensor methods in statistics. Monographs on Statistics and Applied Probability, Chapman and Hall
57. J.L. Lacoume, P.O. Amblard, P. Comon, (1997). Statistiques d'ordres supérieurs pour le traitement de signal, Masson
58. K.S. Lii, M. Rosenblatt, (1982). Deconvolution and estimation of transfer function phase and coefficients for non-Gaussian linear processes. Annals of Statistics, 10:1195-1208
59. J.M. Mendel, (1991). Tutorial on higher-order statistics (spectra) in signal processing and system theory: theoretical results and some applications. Proceedings of the IEEE, 79:3:278-305
60. A.K. Nandi (1999). Blind estimation using higher order statistics. Kluwer Academic Publishers
61. M. Rosenblatt, (1985). Stationary sequences and random Fields. Birkhäuser
62. A.Swami, J.M.Mendel, (1990). ARMA parameter estimation using only output cumulants. IEEE Trans on Acoust. Speech and Sig. Processing, 38:1257-1265
63. J.K.Tugnait, (1991). On the identifiability of ARMA models of non-Gaussian processing via cumulant matching, higher order statistics. J.L.Lacoume Ed.. Elsevier, 117-120

Bayesian Statistics

64. P. Brémaud (1999). Markov chains : Gibbs fields, Monte Carlo simulation and queues. Springer-Verlag.
65. C.K. Carter, R. Kohn, (1996). Markov chain Monte Carlo methods in conditionally Gaussian state space models. Biometrika, 83:589-601
66. G. Celeux, J. Dielbot, (1985). The SEM algorithm, a probabilistic teacher algorithm from the EM algorithm for mixture Problems. Comp. Stat. Quarterly, 2:73-82

67. G. Celeux, J. Dielbot, (1992). A stochastic approximation type EM algorithm for the mixture problem. Stochastics and stochastic reports, 41:119-134
68. A.P. Dempster, N.M. Laird, P.B. Rubin, (1977). Maximum likelihood from incomplete data via the EM algorithm. Journal of the Royal Statistical Society (series B), 39:1-38
69. A. Doucet, (1997). Algorithmes de Monte Carlo pour l'estimation bayesienne de modèles markoviens Cachés. Application au traitement des signaux de Rayonnement. PhD Thesis, n. 97PA112347, Paris 11
70. A. Doucet, N. de Freitas, N. Gordon (Eds) (2001). Sequential Monte Carlo Methods in Practice. Springer-Verlag
71. J. Geweke, (1989). Bayesian inference in econometrics models using Monte Carlo integration. Econometrica, 57:1317-1339
72. N.J. Gordon, D.J. Salmond, A.F.M. Smith (1993). Novel approach to nonlinear/non-Gaussian Bayesian state estimation. IEE proceedings-F 140:107-113.
73. M. Lavielle, B. Delyon, E. Moulines, (1999). On a stochastic approximation version of the EM algorithm. Annals of Stat., vol.8:4:490-503
74. J.S. Liu, W.H. Won, A. Kong, (1994). Covariance structure of the Gibbs sampler with various scans. Journal of the Royal Statistical Society, 57:157-169
75. S.P. Meyn, R.L. Tweedie, (1993). Markov chains and stochastic probability. Springer-Verlag
76. C.P. Robert, (1994). The Bayesian choice. Springer-Verlag
77. C.P. Robert, G. Casella (1999). Monte Carlo Statistical Methods. Springer-Verlag
78. J.O. Ruanaidh, W.J. Fitzgerald, (1996). Numerical Bayesian methods applied to signal processing. Springer-Verlag
79. J.C. Spall, (1998). Bayesian analysis of time series and dynamic models. J.C. Spall ed., Johns Hopkins University, Marcel Dekker Inc.
80. L.Tierney, (1996). Introduction to general state-space Markov chain theory. In Markov Chain Monte Carlo in Practice. Chapman and Hall, 59-74

Optimisation and Adaptive Estimation

81. A. Benveniste, M. Metivier, P. Priouret, (1990). Adaptive algorithms and stochastic approximations. Springer-Verlag
82. P.G. Ciarlet, (1982). Introduction à l'analyse numérique matricielle et à l'optimisation. Masson
83. P.S.R. Diniz, (1997). Adaptive filtering: algorithms and practical implementation. Kluwer
84. M. Duflo, (1997). Random iterative models. Springer-Verlag
85. S. Haykin (1996). Adaptive filter theory. 3rd edition. Prentice Hall
86. J.B. Hiriart-Urruty, C. Lemaréchal, (1993). Convex Analysis and Minimisation Algorithms. Springer-Verlag
87. J. Labat, O. Macchi, C. Laot (1998). Adaptive decision feedback equalisation: can you skip the training period?. IEEE Trans. on Com., 46:7:921-930
88. D.G. Luenberger, (1984). Linear and non linear programming. Addison Wesley
89. J.J. Moder, S.E. Elmaghraby (Eds), (1978). Handbook of operations research. Van Nostrand Reinhold Company.

Index